CAMBRIDGE ZOOLO

THE HOUSE-FLY

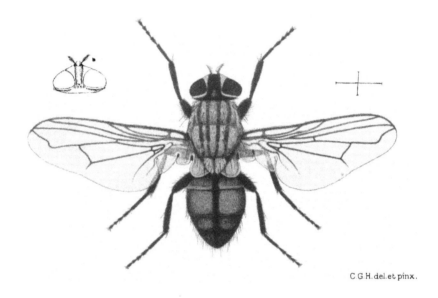

C.G.H.del.et pinx.

Fig. 1. The House-fly, *Musca domestica* L. Female.
Inset, head of male.

THE HOUSE-FLY

MUSCA DOMESTICA LINN.

ITS STRUCTURE, HABITS, DEVELOPMENT, RELATION TO DISEASE AND CONTROL

BY

C. GORDON HEWITT
D.Sc., F.R.S.C.

DOMINION ENTOMOLOGIST OF CANADA; FORMERLY LECTURER IN
ECONOMIC ZOOLOGY IN THE UNIVERSITY OF MANCHESTER

Cambridge:
at the University Press
1914

CAMBRIDGE UNIVERSITY PRESS
Cambridge, New York, Melbourne, Madrid, Cape Town,
Singapore, São Paulo, Delhi, Tokyo, Mexico City

Cambridge University Press
The Edinburgh Building, Cambridge CB2 8RU, UK

Published in the United States of America by Cambridge University Press, New York

www.cambridge.org
Information on this title: www.cambridge.org/9780521232999

© Cambridge University Press 1914

First published 1914
First paperback edition 2011

A catalogue record for this publication is available from the British Library

ISBN 978-0-521-23299-9 Paperback

TO MY WIFE

ELIZABETH BORDEN

PREFACE

THE world-wide interest which has been created during the last few years in the relation which the house-fly bears to the hygienic state of the individual and of the community, as a product of insanitary conditions and as a potential and not infrequent disseminator of certain common and preventable infectious diseases, has rendered the presentation of our knowledge of this insect, its habits and relation to disease most desirable and, indeed, necessary as a means of appreciating its significance from the entomological and medical standpoint and as a basis for further investigation.

In 1907, 1908 and 1909 respectively the three parts of my monograph on the House-fly were published in the *Quarterly Journal of Microscopical Science* under the title: "The Structure, Development and Bionomics of the House-fly, *Musca domestica* Linn." For the convenience of workers the Manchester University Press very kindly republished in volume form two hundred copies of the letterpress and plates of the monograph, which reprints Sir Ray Lankester, the Editor of the *Q. J. M. S.*, permitted me to obtain. With a certain amount of additional matter in the form of appendices this limited edition was issued in 1910 under the title: "The House-fly, *Musca domestica* Linn. A Study of the Structure, Development, Bionomics and Economy." This reprint is now exhausted.

Although the present volume contains the whole of the original matter published in the *Quarterly Journal of Microscopical Science*, the extent of subsequent work by investigators in all the continents has necessitated the preparation of a completely new work. I have naturally endeavoured to review whatever work relating to the house-fly has been carried on, but it is not unlikely, in fact, with the multiplicity of scientific journals, most probable that some contributions to our knowledge may have escaped my notice and any advice in regard to such omissions would be most cordially appreciated.

The volume is not intended as a popular treatise on the subject. Such a function is filled by my small volume "House-flies and how they Spread Disease" in the Cambridge Manuals of Science and Literature, and by Dr L. O. Howard's book "The House-fly : Disease Carrier." It is primarily intended for the use of entomologists, medical men, health officers and others similarly engaged or interested in the subject, and it is hoped that it will be of value to students.

I wish to acknowledge my indebtedness to Mr H. T. Güssow, who took the photographs illustrated in figs. 33, 72, 74, and to Mr C. T. Brues, of Harvard University, for the use of the original photograph, fig. 36. Mr F. W. L. Sladen also has kindly assisted me by taking the photographs, figs. 32, 35, 70, 97. Except where it is otherwise stated, the rest of the illustrations were drawn by me.

C. GORDON HEWITT.

Ottawa, Canada,
July, 1913.

CONTENTS

PART I

THE STRUCTURE AND HABITS OF THE HOUSE-FLY

PART II

THE BREEDING HABITS; LIFE-HISTORY AND STRUCTURE
OF THE LARVA

PART III

THE NATURAL ENEMIES AND PARASITES OF THE
HOUSE-FLY

PART IV

OTHER SPECIES OF FLIES FREQUENTING HOUSES

PART V

THE RELATION OF HOUSE-FLIES TO DISEASE

PART VI

CONTROL MEASURES

LIST OF ILLUSTRATIONS

PART I

THE STRUCTURE AND HABITS OF
THE HOUSE-FLY

CHAPTER I

INTRODUCTION

AMONG the numerous and remarkable advances which have been made in the realm of medical science within the last two decades, none has created so wide a public interest, none has been fraught with consequences affecting so large and wide-spread a proportion of the world's inhabitants, and destined to affect the future welfare and progress of mankind to so great a degree, as the gradual discovery of the *rôle* which insects play in the dissemination of disease. Malaria, which has had a far-reaching effect on the history of the world and on the immigration of the white man into new regions of the earth, and which in India alone imposes a tax of over a million human lives each year, has been shown to be conveyed by the mosquito. Plague, which in all ages has created terrific devastation, sweeping away millions of lives, transforming populous cities into deserted wildernesses, was found to be transmitted by the flea. The "black sickness," or Kala Azar, which has decimated districts and depopulated areas in the tropical countries where it occurs, has been found to be due to a parasitic organism which can be transmitted by the bed-bug. Sleeping sickness, which numbers its victims by the hundred thousand, depends for its distribution upon the tse-tse fly. Lice have been shown to transmit the causative organism of typhus fever. The common stable-fly has been shown to be a possible disseminator of infantile paralysis or poliomyelitis. One by one man's parasitic attendants and blood-sucking visitors have been shown to be potent vehicles of death. Of all revelations perhaps none affects so great a number of people in all countries, both by its significance and effects, as the demonstration of the disease-carrying power of the common house-fly. From the dark ages man has been accustomed to regard his ubiquitous companion in his wanderings over the face of the globe, not only with a marked degree of tolerance, but, if the rhymes of our childhood are to be believed, with some measure of affection. The discovery, therefore, of the fact that the commonest and most widely distributed insect and the animal most closely associated with man was not only begotten and a frequenter of filth, but was also a potent and common carrier of pathogenic and putrefactive organisms, excited an interest in the minds of a larger number of people than many other discoveries of a like character in which both the insect and the disease with which it was associated were more restricted in their distribution and affected a correspondingly small number of people.

So obsessed were people's minds with the idea that the house-fly was of no significance in relation to man's welfare that it was not only deemed unworthy of serious study, but when, as a result of close study, its true character and habits were being revealed, the results of such studies were regarded with considerable surprise and scepticism. The story of the gradual revelation of the disease-carrying powers of the house-fly is recorded in subsequent chapters. The history of the development of our knowledge of the insect itself, its structure, habits and life-history is noteworthy on account of our long continued ignorance concerning these facts.

Naturalists of all ages have briefly referred to the habits and characteristics of the house-fly. Both Réaumur (1738) and De Geer (1758–78) included short accounts of this insect in their classical memoirs, but they contributed little to our knowledge of the structure or development of the fly.

The most comprehensive of the earlier accounts of the house-fly was written by Gleichen (1790)[1]. This most interesting book

---

[1] This is the date of the copy of this rare book which is in my possession; there may have been, however, an earlier edition in 1766.

of thirty-two quarto pages is illustrated by four very striking coloured plates which show the development of the fly and structural details, both external and internal, of the fly and larva (fig. 2). The author gives a somewhat detailed account of the

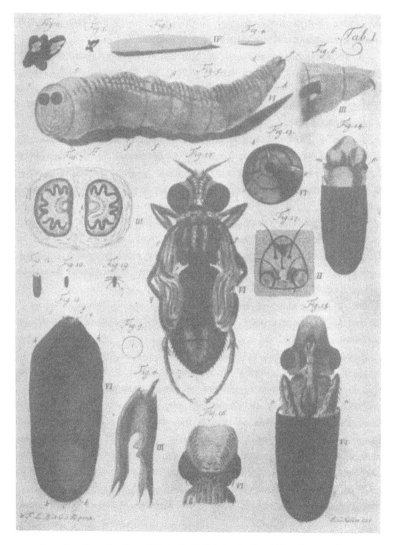

Fig. 2. Reproduction of Plate I of Gleichen's *Geschichte der gemeinen Stubenfliege*, 1790.

habits and development of the fly. It was natural that he should make some errors, such as mistaking the brown testes for the kidneys, in attempting to describe the anatomy at so early a date of our knowledge of morphology and methods of study. Notwithstanding these limitations, Gleichen's was still the most comprehensive and detailed account of the house-fly when I commenced to study the insect.

A short popular account of the house-fly was published, conjointly with that of the earth-worm, by Samuelson and Hicks in 1860, in a book entitled *Humble Creatures*. Though interesting, the account is very superficial and contains much that is inaccurate.

In 1874 Packard wrote a fairly complete account of the developmental history of the house-fly as observed in Massachusetts, U.S.A., and in 1880 Taschenberg gave a good popular account of the insect and its breeding habits in his *Praktische Insektenkunde*.

Howard gave a short account of the life-history of the house-fly in a bulletin on household insects, published in 1896 by the United States Department of Agriculture, and he gave a further account in 1900, in a valuable paper on the insect fauna of human excrement. Newstead, in 1907, gave a preliminary report of his study of the development and breeding-places of the house-fly in the city of Liverpool, and a second report was published in 1909.

Space forbids the enumeration of the countless papers and accounts of the house-fly which have been written during the past two or three years. Some of these contain original information, the majority of them do not. Of recent publications, Howard's popular book on the house-fly (1911) should be mentioned, not only as being a complete account of the insect and its disease-carrying powers, but because it contains many original observations[1]. Special reference should be made to the valuable series of papers published in connection with an inquiry carried out by the Local Government Board on flies as carriers of infection.

---

[1] Since the above lines were written Graham-Smith has given an excellent account of the disease-carrying powers of the non-blood-sucking flies in his *Flies and Disease* (1913); see Bibliography.

This investigation has been the most complete organised study of the fly as a vector of micro-organisms that has yet been made, and in the subsequent pages I have made full use of the contents of the five reports which have appeared up to the time of writing [1].

My own studies were begun in 1905, at which time no complete study of the anatomy or development of the house-fly had been made, and the gradual realisation of the economic status of the fly made such a study not only desirable but necessary in view of our profound ignorance on the subject. A preliminary account of the life-history was published in 1906. In the following year (1907) a detailed account of the anatomy of the fly was published, this being the first part of a monograph on the structure, development and bionomics of *Musca domestica*. The second part of this monograph was published in 1908 and gave an account of the breeding-habits, development and bionomics of the larva. The concluding part of the work was published in 1909, and in it were described the bionomics of the house-fly, its allies and parasites and its relation to human disease.

In addition to the investigations which I have continued since that date, other investigators have added to our knowledge of the bionomics of the house-fly with the result that, although our knowledge of the insect cannot be said to be complete, we have decreased our previous unfortunate ignorance of the commonest insect to a marked degree and have furnished a solid basis for further studies along particular lines, and information necessary to a consideration of the means of control and prevention.

DESCRIPTION OF *MUSCA DOMESTICA*.

Musca domestica was first described in 1758 by Linnaeus in his *Systema Naturae*; his description is as follows:

"Antennis plumatis pilosa nigra, thorace lineis 5 obsoletis abdomine nitidulo tessellato: minor. Habitat in Europe domibus, etiam Americae· Larvae in simo equinae. Pupae parallele cubantes."

It was more fully described by Fabricius in his *Genera Insectorum*.

---

[1] A Further Report, No. 6, has appeared since the above was written.

The house-fly, together with the blow-fly and the blood-sucking flies *Stomoxys* and *Glossina*, belongs to the family Muscidae, which is characterised by having the terminal joint of the antenna, the arista, always combed or plumed, and by the absence of large bristles or macrochaetae on the abdomen. The Muscidae, together with the Anthomyidae and Tachinidae, constitute the group *Muscidae calypteratae*, which are characterised by the possession of squamae, small lobes at the bases of the wings which cover the halters. In the acalyptrate muscids the squamae are absent or rudimentary. These two groups belong to the sub-order *Cyclorrhapha*, one of the two primary divisions of the Diptera. The *Cyclorrhapha* have coarctate pupae, the pupal case being formed by the hardening of the last larval skin, and the flies escaping through a circular orifice formed by the fly pushing off the end of the pupa by means of an inflated sac-like organ—the ptilinum or frontal sac—which is afterwards withdrawn into the head, its presence being marked by a frontal crescentic opening—the lunule. The other sub-order, the *Orthorrhapha*, have obtected pupae.

The most complete specific description of *Musca domestica* has been given by Schiner (1864), of which the following is a free translation:

"Frons of male occupying a fourth part of the breadth of the head. Frontal stripe of female narrow in front, so broad behind that it entirely fills up the width of the frons. The dorsal region of the thorax dusty grey in colour, with four equally broad longitudinal stripes. Scutellum grey, with black sides. The light regions of the abdomen yellowish, transparent, the darkest parts at least at the base of the ventral side yellow. The last segment and a dorsal line blackish brown. Seen from behind and against the light the whole abdomen shimmering yellow, and only on each side of the dorsal line on each segment a dull transverse band. The lower part of the face silky yellow, shot with blackish brown. Median stripe velvety black. Antennae brown. Palpi black. Legs blackish brown. Wings tinged with pale grey, with yellowish base. The female has a broad velvety black, often reddishly shimmering frontal stripe, which is not broader at the anterior end than the bases of the antennae, but becomes so very much broader above that the light dustiness of the sides is entirely obliterated. The abdomen gradually becoming darker. The shimmering areas on the separate segments generally brownish. All the other parts are the same as in the male."

The mature insects measure from 6 to 7 mm. in length and from 13 to 15 mm. across the wings. Frequently dwarfed specimens may be found, normal in every respect but size. This is due, as my breeding experiments demonstrated, to adverse conditions during the larval stage; starvation especially tends to produce undersized individuals.

DISTRIBUTION OF *MUSCA DOMESTICA*.

Musca domestica is undoubtedly the most widely distributed insect to be found; the animal most commonly associated with man, whom it appears to have followed over the entire globe. It extends from the sub-polar regions, where Linnaeus refers to its occurrence in Lapland and Finmark as "rara avis in Lapponia, at in Finmarchia Norwegiae integras domos fere replet," to the tropics, where it occurs in enormous numbers. Referring to its abundance in a house near Para in equatorial Brazil, Austen (1904) says: "At the mid-day meal they swarmed on the table in almost inconceivable numbers," and other travellers in different tropical countries have related similar experiences to me, how they swarm round each piece of food as it is carried to the mouth.

In the civilised and populated regions of the world it occurs commonly, and the British Museum (Natural History) collection and my own contain specimens from the following localities (certain of the localities have, in addition, been obtained from lists of insect faunas):

Asia. Aden; North-West Provinces (India); Calcutta; Madras; Bombay (it probably occurs over the whole of India); Ceylon; Central China; Hong-Kong; Shanghai; Straits Settlements; Japan.

Africa. Port Said; Suez, Egypt; Somaliland; Nyassaland; Uganda; British E. Africa; Rhodesia; Transvaal; Natal; Cape

Colony; Madagascar; Northern and Southern Nigeria; St Helena; Madeira.

America. Distributed over North America; Brazil; Monte Video (Uruguay); Argentine; Valparaiso; West Indies.

Australia and New Zealand.

Europe and the isles of the Mediterranean; it is especially common in Cyprus.

Not only is the world-wide distribution of the house-fly of interest but its local abundance, which will be considered in a subsequent chapter (p. 65), is noteworthy.

CHAPTER II

THE EXTERNAL STRUCTURE OF *MUSCA DOMESTICA*

PREVIOUS to this study the only complete account which has been published on Muscid anatomy was Lowne's comprehensive monograph (1895) on the blow-fly, *Calliphora erythrocephala*, which is an elaboration of his smaller and earlier memoir (1870). Not only are many of Lowne's conclusions untenable, but the value of his work as a comparative study would increase with confirmation.

THE HEAD CAPSULE.

The head capsule of *M. domestica* presents great modifications when compared with the typical insect head. Considerable difficulty is experienced in explaining its structure in the morphological terms employed in the simpler orders of insects. Lowne did not lessen the difficulty in describing the head of the blow-fly by the invention of new terms of little morphological value. The head of the fly is strongly convex in front, the posterior surface being almost flat and slightly conical. For the sake of clearness the composition of the head capsule will be described from behind forwards. The occipital foramen occupies a median slightly ventral position on the posterior surface. It is surrounded by the occipital ring, the inner margin of which projects into the cavity of the head. From the sides of the inner margin of the occipital ring two short chitinous bars bend inwards and approach each other internally, forming a support—the jugum—for the tentorial membrane. On each side of the occipital ring below the jugum a small cavity occurs, into which a corresponding process from the prothorax fits, forming a support for the head.

The occipital ring is surrounded by the four plates, which make up the sides and back of the head capsule. On the ventral side, between the occipital ring and the aperture from which the proboscis depends, a median basal plate, the gulo-mental plate, represents the fused gula and basal portions of the greatly modified second maxillae. The occipital segment is bounded laterally by the genae (Lowne's paracephala) and dorsally by the epicranium. These parts have been divided by systematists into so many regions that a somewhat detailed description will be necessary to make their boundaries clear.

The genae bear the large compound eyes which occupy almost the whole of the antero-lateral region of the head. On the posterior flattened surface of the head the genae are flat, and extend from the gulo-mental plate to the epicranial plate, the sutures of the latter being vertical. On the dorsal side each sends a narrow strip between the inner margin of the eye and the epicranium; this strip surrounds the eye and meets the ventral portion of the gena; it is of a silver to golden metallic lustre. On the ventral side below the eye each gena bounds the proboscis aperture laterally; a number of stout bristles arise from this margin and also from its antero-lateral region, which is often spoken of as the "jowl." In the anterior region, where the genae are in contact with the clypeus, there are two prominent ridges bearing strong setae; these are usually known as the "facialia." Berlese (1909) regards the facialia as representing a portion of the fourth (mandibular) segment. They are certainly distinct from the genae, as may be seen in the head of the newly formed nymph (fig. 43). Strictly speaking they are both facialia and peristomalia, facio-peristomial sclerites in fact.

The epicranium (epicephalon of Lowne) on the posterior surface of the head is flat. On the anterior surface it is convex, and divided into a number of regions. On the top of the head between the eyes it is called the vertex. This contains the three ocelli situated on a slightly raised ocellar triangle, which is surrounded by a second triangle, the vertical triangle. The median region in front of and below the vertex is the frons. In the middle of this there is a black frontal stripe. The frons appears to be composed of two sclerites for which Townsend, in a letter to me, suggests the term "frontalia." In the male the eyes are only narrowly separated by the frontal stripe. In the female the frontal stripe widens out on the vertex. This character provides a ready means of distinguishing the male from the female, as the result of it is that in the male the eyes are close together on the dorsal side, being separated by about one-fifth of the width of the head, whereas in the female the space between the eyes is about one-third the width of the head. The edges of the genae bordering on the frons bear each a row of stout setae—the fronto-orbital bristles.

The antennae are situated below the lower edge of the frons. Each antenna consists of three joints and the arista. The two proximal joints are short and compose the "scape" and arise from a strip representing the antennal segment, situated between the frons and the prefrons. The third joint, the flagellum, is longer, and hangs vertically in front of the clypeus. It is covered with sensory setae, and contains two pits of sensory function (olfactory, I believe). From the upper side the plumose arista arises. This probably represents the terminal three joints of the antenna. The lower edge of the frons represents the anterior margin of the epicranium. It is of interest here to note that, whereas the arista of *M. domestica* is plumose, that is, it bears fine bristles on the upper and lower sides, the arista of the stable-fly, *Stomoxys calcitrans*, bears bristles on the upper side only, and the arista of the lesser house-fly, *Fannia canicularis*, is apparently devoid of bristles, in reality it is minutely pilose.

The rest of the facial region is composed of the clypeus or, as it is usually called, the face, a convenient term, but one which hides its true morphology. The face is depressed, and is covered by the flagella of the antennae. Between the upper and lateral edges of the face and the lower edge of the epicranium a crescentic opening, the lunule, marks the invagination of the ptilinum. The epistomium is a narrow strip below the face bounding the anterior edge of the proboscis aperture.

THE SKELETON OF THE PROBOSCIS.

An account of the proboscis of *M. domestica* was published by Macloskie in 1880. The proboscis of the blow-fly, which is very similar in many respects to that of the house-fly, has been described by Anthony (1874), Kraepelin (1880) and Lowne (*t.c.*). The results of these authors differ in many details. My study of the proboscis of *M. domestica* confirms Kraepelin's results, and as Lowne's is the only complete account of the muscid head, a full description of the anatomy, both internal and external, of the head of *M. domestica* will be given. Recently Graham-Smith (1912) has made a very careful study of the anatomy and function of the oral suckers of the blow-fly, *Calliphora erythrocephala*. His observations are confirmatory of my own study of the oral lobes of *M. domestica* which I made in 1906, but which was not described in great detail in the first part of my monograph (1907).

Lowne regards the greater part of the proboscis as being developed from the first maxillae and not from the labium or fused second pair of maxillae. The latter is the usually accepted view, and one which I support on morphological grounds. On account of the very exceptional nature of his conclusion, he refuted the commonly accepted terms of the various parts of the proboscis and invented new ones, an unfortunate habit to which he was addicted. For the sake of descriptive clearness it will be necessary to refrain from constant reference to Lowne's terms or any discussion as to their merits.

The proboscis consists of two chief parts; a proximal membranous conical portion, the rostrum, and a distal portion which bears the oral lobes and which has been termed the haustellum. The term haustellum has also been used by some authors to designate the distal portion of the proboscis minus the oral lobes.

The *Rostrum* (fig. 3). This proximal membranous portion of the proboscis is attached to the edges of the proboscis aperture, that is to the epistomium, genae and the gulo-mental plate. It has the shape of a truncated cone and bears on the anterior side a pair of palps which bear sensory setae of two sizes.

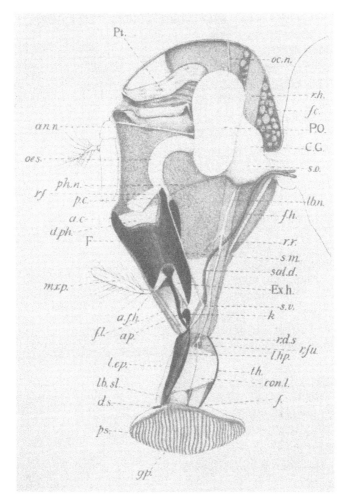

Fig. 3. Interior of the head of *M. domestica*. In this figure the left side of the head capsule and of the proboscis have been removed and the compound eye of the same side, leaving the optic ganglion (periopticon). All the tracheal structures have been omitted.

a.c. Anterior cornu of fulcrum. *a.f.h.* Accessory flexor muscles of haustellum. *ap.* Apodeme of labrum. *an.n.* Antennal nerve. *C.G.* Cephalic ganglion. *con.l.* Dilator muscles of labium-hypopharynx. *d.ph.* Dilator muscles of pharynx. *d.s.* Discal sclerite. *Ex.h.* Extensor muscle of haustellum. *F.* Fulcrum. *f.* Furca. *f.c.* Fat cells. *f.h.* Flexor muscle of haustellum. *f.l.* Flexor muscle of labrum-epipharynx. *g.p.* Gustatory papillae of oral lobes. *k.* Hyoid sclerite of pharynx. *lb.n.* Labial nerve. *lb.sl.* Labial salivary gland. *l.hp.* Labium-hypopharynx. *l.ep.* Labrum-epipharynx. *mxp.* Maxillary palp. *oes.* Oesophagus. *oc.n.* Ocellar nerve. *ph.n.* Pharyngeal nerve. *p.c.* posterior cornu of fulcrum. *P.O.* Periopticon. *ps.* Pseudotrachea. *Pt.* Ptilinum. *r.d.s.* Retractor muscles of discal sclerites. *r.f.* Retractor muscle of fulcrum. *r.fu.* Retractor muscle of furca. *r.h.* Retractor muscle of haustellum. *r.r.* Retractor muscle of rostrum. *s.o.* Suboesophageal ganglion. *sal.d.* Common duct of the lingual salivary glands. *s.v.* Valve of common salivary duct. *s.m.* Muscle controlling valve of salivary duct. *th.* Theca.

The *Haustellum*. This forms the distal portion of the proboscis and is attached to the rostrum. Its distal portion, which
comprises the oral lobes, will be described separately. The posterior side of the proximal portion is formed by a strongly convex
heart-shaped sclerite, the theca (figs. 3 and 29), which morphologically represents a portion of the labium. The lower angle of
the theca is incised by a semicircular sinus. By means of this the
theca rests on a triradiate continuous sclerite, the furca, which
consists of a median, slightly convex rod (fig. 3, *f.*), from the
anterior end of which two arms diverge and form the chief skeletal
structures of the oral lobes. The lower end of the theca rides on
the structure, the bottom of the sinus resting on the median
rod, and the two pointed lateral terminations of the theca rest
on the arms. In this manner these processes, in a state of
repose, keep the arms of the furca closely approximated. The
result of this arrangement will be seen later in studying the
musculature of the proboscis.

The sides of the haustellum are membranous. On its anterior
face, in a groove formed by the overlapping membranous sides, lie
the labrum-epipharynx and labium-hypopharynx. The labrum-
epipharynx (fig. 3, *l.ep.*) is attached at its proximal end to the
membranous rostrum, but is incapable of a labral-like movement
on account of its close connection with the labium-hypopharynx.
Two slightly-curved, hammer-shaped apodemes (fig. 3, *ap.*) are
attached to the proximal end of the labrum-epipharynx. They
assist in folding the proboscis during retraction, as will be shown
later. The labrum-epipharynx is shaped like a blunt arrow-head;
the external surface is somewhat flattened. It is composed of
two pairs of sclerites, an outer pair enclosing an inner pair, which
form the pharyngeal channel. The edges of the inner tube are
connected by a groove with the hypopharyngeal portion of the
labium-hypopharynx, as shown. The labium-hypopharynx (fig. 3,
l.hp.) represents the fusion of the hypopharynx with the greatly
modified and fused second maxillae or labium. It consists of a
sclerite, curved in section, having the chitinous hypopharyngeal
tube (fig. 29, *hp.*) fused to it along the upper half of its length.
The edges of the hypopharyngeal tube engage with those of the
inner pair of sclerites of the labrum-epipharynx, as mentioned

before. Distally, the hypopharyngeal tube becomes free from the labium, as shown in fig. 29, and ends in a point where the lingual salivary duct opens.

Down each side of the labium-hypopharyngeal sclerite a rod-like thickening runs. Distally, these thickened margins (paraphyses of Lowne) articulate with the discal sclerites. The discal sclerites (fig. 3, *ds.*) are united at the posterior end to form, when the oral lobes are expanded, a U-shaped structure, with the limbs constricted in the middle where the ends of the thickened margins of the labium-hypopharynx articulate. They are sunk in deeply between the two oral lobes at the base of the oral pit with the free ends of the U anterior, these being spatulate and curved anteriorly.

The *Oral Lobes*. Normally the two oral lobes or labella are connected by a delicate attachment along the inner anterior edges to form an oral sucking organ, but under pressure this delicate connection is severed and the oral disc presents a heart-shaped instead of the normal appearance. On the upper or outer aboral surfaces the oral lobes bear sensory setae, the larger marginal setae being different in structure from the rest, as will be described later in the account of the internal structure of the oral lobes (p. 61). On the lower and, when the proboscis is withdrawn, the inner oral surface a large number of channels, called the pseudo-tracheae (fig. 3, *ps.*), from their fancied resemblance to the annular tracheae, run from the edges of the oral lobes to the internal margins. These channels are almost circular or oval in section, being incomplete on one side and thereby communicating with the surface of the oral lobe. The channels are kept open by means of small incomplete chitinous rings which give the pseudo-tracheae their annular appearance. Each of these incomplete chitinous rings is bifurcated at one end but single at the other end (fig. 4). The rings are so arranged that the bifid ends alternate with the single ends. The pseudo-tracheal channel communicates with the external surface of the oral lobes through the opening through the bifid extremities of each ring, as is shown in the accompanying figure (fig. 5). From the outer edge of the oral lobe the pseudo-tracheae gradually increase in size as they approach the inner margin of the lobe. The number of pseudo-tracheae traversing each oral lobe is thirty-six, and they are grouped in three sets. One anterior set of twelve pseudo-tracheae run into a single large pseudo-tracheal channel running along the anterior inner margin of the oral lobe, and a posterior set of twenty or twenty-one all run into a common

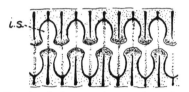

FIG. 4.   Oral aspect of a pseudo-trachea showing interbifid spaces (*i.s.*) between bifid ends of the pseudo-tracheal rings.

FIG. 5.   Chitinous pseudo-tracheal rings.

channel running along the posterior inner margin of the lobe; between these two sets a median set of three or four pseudo-tracheae run direct into the oral aperture.

Graham-Smith (*l.c.*) has made comparative measurements of the pseudo-tracheae of several species of non-biting flies and of the interbifid spaces, as he terms the area enclosed between the bifid extremities of the chitinous rings of the pseudo-tracheae. The average measurements of the various parts are as follows:

	Pseudo-tracheae		Interbifid spaces	
	Diameter at proximal end	Diameter at distal end	Diameter near the proximal ends of the pseudo-tracheae	Diameter near the distal ends of the pseudo-tracheae
Calliphora erythrocephala	·02	·01	·006	·004 mm.
Sarcophaga carnaria	·02	·01	·005	·004
Lucilia caesar	·02	·01	·006	·004
Fannia canicularis	·016	·008	·006	·004
Ophyra anthrax	·016	·008	·006	·004
Musca domestica	·016	·008	·004	·003

The comparatively small size of the interbifid spaces of the house-fly should be noted, as this has some bearing on the feeding habits of the fly.

The *Oral Aperture* lies at the base of the small oral pit, which is a space kept open between the oral lobes by means of the discal sclerites. The median pseudo-tracheae do not extend as far as the discal sclerites, but on entering the oral pit the chitinous rings cease and are replaced by narrow ∧-shaped sclerites for a short distance, while the sides of the oral pit are bordered by a row of teeth, which have been termed the prestomial teeth and which lie at the sides of the openings of the pseudo-tracheal channels. Between the pseudo-tracheae the membranous surface of each oral lobe is thrown, probably in the relaxed state only, into longitudinal sinuous ridges; there appear to be two such ridges between adjacent pseudo-tracheae. Projecting from the bottom of the furrows are several papillae, generally four or five to each interpseudo-tracheal area, of a gustatory nature, the gustatory papillae (fig. 3, *gp.*).

In certain text-books and treatises in which the proboscis of the house-fly is described a misconception of the character and consequent function of the pseudo-tracheae is frequently repeated. The pseudo-tracheae are described as horny "rasp-like" ridges which, by a "rasping" action, remove small particles, of sugar, for example, which the fly can swallow. A careful study of the nature of the pseudo-tracheae and the method of feeding of the house-fly would convince anyone who attempted to verify the above idea of the mistaken interpretation and description.

The *Fulcrum.* This chitinous portion of the pharynx (fig. 3, *F.*) lies on the lower part of the head and in the rostrum. Kraepelin describes it as being shaped like a Spanish stirrup iron. Its structure will be best understood by referring to the figures. It consists of an outer portion, which is U-shaped in section; the basal portion, which is posterior and forms the floor of the pharynx (which Lowne, unfortunately, terms the hypopharynx), is vertical when the proboscis is extended. This basal portion is evenly rounded at both ends, and at the sides of the upper end there is a pair of processes—the posterior cornua (fig. 3, *p.c.*) which serve for the attachment of muscles. The sides of the fulcrum are somewhat triangular in shape; their upper anterior portions are produced to form the anterior cornua (*a.c.*); here the sides bend inwards at right angles, and meet below the epistomium, upon which the fulcrum is hinged.   The fulcrum is therefore quadrilateral in section at the upper proximal end, and trilateral at the lower distal end.   The basal portion (fig. 28, *b.p.*) forms the floor of the pharynx; the roof of the pharynx is formed by another chitinous piece (*r.p.*) with a median thickened raphe.   This roof lies parallel with the basal piece, and is fused with the sides of the fulcrum.   On the membranous wall of the pharynx, between the labium-hypopharynx and the fulcrum, a small chitinous sclerite (fig. 3, *k.*) is developed, which Lowne terms the hyoid sclerite.   It is U-shaped in section, and serves to keep the lumen of the pharynx in this region distended.

THE THORAX (fig. 6).

As in all Diptera the possession of a single pair of wings has resulted in the great development of the mesothorax at the expense of the other thoracic segments, consequently the thorax is chiefly made up of the sclerites composing the mesothorax.   The prothorax and metathorax compose very small portions on the anterior and posterior faces respectively.   Seen from above the thorax is oviform with the blunt end anterior and slightly flattened.   Three transverse sutures on the dorsal side mark the limits of the prescutum, scutum and scutellum of the mesothoracic segment; the mesothoracic scutellum forms the pointed posterior end, and slightly overhangs the anterior end of the abdomen.

The *Prothorax.*   The prothoracic segment has been reduced to such an extent that it is hopeless to attempt to homologise all the separate sclerites with those of a typical thoracic segment.   To obtain a complete view of the prothorax it is necessary to examine it from the anterior end after the removal of the head.   The following sclerites can then be recognised.   The prosternum is a median ventral plate, quadrilateral in shape, having the anterior end rounded and broader than the posterior end.   It does not occupy the whole of the prosternal area, but is bounded by the prosternal membrane. Internally, a ridge runs to the posterior end of the prosternum and bifurcates, each ridge running to the posterior corners, to which two strong processes (the hypotremata of Lowne) are attached.   In front of the prosternum there is a small saddle-shaped sclerite which, on account of its position, may be called the interclavicle (the sella of Lowne). Two lobes at its anterior end are covered with small processes, probably sensory in function. A pair of small sclerites is situated in front of these lobes; these sclerites with the interclavicle no doubt belong to the prosternum. The inter-clavicle is ventral to the cephalothoracic foramen. The jugulares

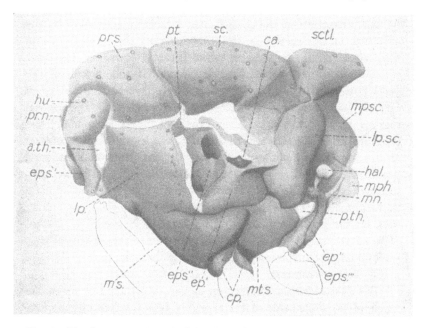

Fig. 6. The thorax seen from the left side. The insertions of the larger setae are shown; for the sake of clearness the sclerites of the wing-base are omitted.

a.th. Anterior thoracic spiracle. *ca.* Costa. *cp.* Intermediate coxal plates. *ep'.*, *ep''.* Epimera of the meso- and meta-thoracic segments. *eps'.*, *eps''.*, *eps'''.* Episterna of the pro-, meso-, and meta-thoracic segments. *hal.* Haltere. *hu.* Humerus. *lp.* Lateral plate of mesosternum. *lp.sc.* Lateral plate of postscutellum. *mph.* Mesophragma. *mpsc.* Median plate of postscutellum. *mn.* Metanotum. *ms.* Mesosternum. *mts.* Metasternum. *p.th.* Posterior thoracic spiracle. *pt.* Parapteron. *pr.n.* Pronotum. *prs.* Prescutum of meso-thorax. *sc.* Scutum. *sctl.* Scutellum.

(*3me jugulaires* of Kunckel d'Herculais) are two prominent pocket-shaped sclerites lying one on each side of the cephalo-thoracic foramen, and having their convex faces external. Lying immediately below each of the jugulares is a small rod-like sclerite—the clavicle. The dorsal region of the prothorax, the pronotum (fig. 6, *pr.n.*), is formed by two sclerites united in the median line, their dorsal sides being curved. From the ventral side of the pronotum a pair of chitinous apodemes project into the thoracic cavity. The lateral regions of the pronotum are in contact with the humeri (*hu.*) and the prothoracic episterna. The humeri are a pair of strongly convexed sclerites situated in the antero-lateral region of the thorax. They are bounded above by the prescutum of the mesothorax, internally and below by the episterna of the prothorax, and externally by the lateral plate of the mesosternum and the anterior thoracic spiracle. Its inner concave surface serves for the attachment of the muscle of the prothoracic coxa. The episterna (*eps'.*) (epitrochlear sclerites of Lowne) are comparatively large sclerites forming the lateral regions of the prothorax. They overhang the attachments of the prothoracic limbs. The internal skeleton of the prothorax consists of the two stout hollow apodemes—the hypotremata mentioned previously. They arise from the postero-lateral edges of the pro-sternum, and run obliquely across the ventral edge of the anterior thoracic spiracle where the hypotreme divides, the posterior branch runs up the posterior margin of the spiracle, between the lateral plate of the mesosternum and the peritreme (the chitinous ring surrounding the spiracle), the anterior branch fuses with the prothoracic episternum.

The *Mesothorax.* The notum of the mesothorax occupies the whole of the dorsal side of the thorax. It is composed of the four sclerites to which Audouin (1824) gave the names of prescutum, scutum, scutellum, and postscutellum. The prescutum (*prs.*) forms the anterior part of the dorsal region of the thorax. Its anterior portion bends down almost vertically to unite with the pronotum. The anterior edge of the prescutum is inflected after the pronotal suture, and is reduced in the median line to a small bifurcating process. The prescutum is bounded laterally by the humerus and a membranous strip—the dorso-pleural membrane. The scutum (*sc.*) is the largest of the mesonotal plates. It occupies the whole of the median dorsal region of the thorax. Anteriorly it is bounded by the prescutum, laterally by the alar membrane and the lateral plate of the postscutellum, and posteriorly by the scutellum. From the lateral region of the scutum a process projects forwards and downwards and articulates with the posterior portion of the wing-base (the metapterygium). The scutellum (*sctl.*) is a triangular pocket-shaped sclerite which overhangs the postscutellum and the base of the abdomen. The posterior surface of the thorax is chiefly composed of the large postscutellum. This is made up of three pieces, a median escutcheon-shaped plate (*mpsc.*) strongly convex to the exterior, and two convex lateral plates (*lp.sc.*). The lateral plates are bounded below by the metasternum and spiracles, and anteriorly by the pleural region of the mesothorax.

The mesosternum is a sclerite of considerable size and forms the keel of the thorax. It consists of a median ventral portion (*ms.*) which is produced laterally to form two large lateral plates (*lp.*). The median portion is bounded in front by the prosternum and the foramina of the anterior coxae, and behind by the median coxal foramina. A short distance behind the anterior end a depression in the mid-ventral line extending to the posterior edge indicates a median inflection forming the entothorax. The lateral regions of the posterior margins of the mesosternum are inflected on each side to form the entopleura. The lateral plates of the mesosternum form the whole of the anterior portion of the pleural region; each is bounded in front by the humerus, spiracle, and prothoracic episternum, above by the dorso-pleural membrane, and behind by the mesopleural membrane. The ventral side of the lateral plate is continuous in front with the median plate of the mesosternum, and behind is united by means of a suture. The remaining portion of the mesopleural region is made up of the episternum, epimeron, and two small sclerites connected with the wing-base—the parapteron and costa. The episternum (*eps''.*) is situated behind the mesopleural membrane and below the alar membrane; below and behind it is bounded by the epimeron. Its surface is marked by two convexities, the ampullae, the upper of the two corresponding to Lowne's great ampulla of the blow-fly. The dorsal side of the episternum is intimately connected with the sclerites[1] of the anterior portion of the wing-base.

The epimeron (*ep'.*) is a triangular sclerite, and is bounded

---

[1] In this account the individual sclerites which compose the wing-base will not be described. Lowne has described them at great length for the blow-fly, and although the wing-base sclerites of *M. domestica* differ slightly in shape from those of *Calliphora*, Lowne's description of the relations hold good for the former insect.

below by the mesosternum and metasternum, behind by the lateral plate of the postscutellum, and above by the episternum and alar membrane. The parapteron (*pt.*) is a sclerite situated at the top of the mesopleural membrane. The greater portion of it is internal, only a small triangular portion can be seen externally. Internally this is continued as a cruriform sclerite to which are attached important muscles controlling the wings. The costa (*ca.*) is a small sclerite situated on the dorsal margin of the epimeron. The internal skeleton of the mesothorax consists of the entothorax, entopleura, mesophragma, and the inflected edges of the episterna and epimera. The entothorax is composed of a median vertical plate subtriangular in shape, on the top of which a median plate produced laterally into wing-like processes rests. On this structure the thoracic nerve-centre lies. The entopleura and the inflected edges of the episterna and epimera all serve for the attachment of wing muscles. The mesophragma (*mph.*) is a convex sclerite fused with the lower edge of the postscutellum. Its posterior edge is incised in the middle and forms the dorsal arch of the thoraco-abdominal foramen.

The *Metathorax*. The largest sclerite of the greatly reduced metathorax is the metasternum (*mts.*). It is a wing-shaped sclerite with the narrow transverse portion situated between the coxal foramina of the median posterior pairs of legs; the expanded lateral portions form the wall of the thorax above the insertion of these legs. The edges of the narrow transverse strip are inflected, and unite the lateral portions of the metasternum. A trough-shaped longitudinal fold—the metafurca—rests on the narrow transverse portion of the metasternum. The posterior end of the metafurca bends downwards and articulates with the posterior coxae on each side. The metafurca serves for the attachment of the thoraco-abdominal muscles. The pleural region of the meta-thorax is a narrow triangular space situated behind the lateral portion of the metasternum and the posterior coxae. It is composed of a narrow triangular episternum and epimeron. The former (*eps'''.*) is bounded in front by the metasternum, the posterior thoracic spiracle and the base of the haltere, below by the posterior coxal foramen, and behind by the epimeron. The epimeron (*ep''.*) is also bounded below by the coxal foramen and behind by the narrow dorsal arch of the metathorax and the first abdominal segment, its apex comes in contact with the base of the haltere. The dorsal region of the metathorax has practically disappeared, all that can be recognised as metanotum is a narrow chitinous strip (*mn.*) on each side between the apex of the meta-pleural area and the dorsal edge of the first abdominal area.

The Wings.

The wings are situated at the sides of the scutum on the alar membrane, to which are attached the sclerites of the wing-base. They are covered with very fine hairs.

In describing the neuration of the wings the nomenclature proposed by Comstock and Needham (1898) for the wings of the whole group of insects will be employed.

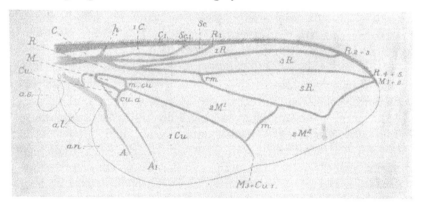

FIG. 7. Wing. The nervures are drawn slightly thicker than they naturally are.

an. Anal lobe. *al.* Alula. *as.* Antisquama. *A.* Anal cell. *A.* 1. Anal nervure. *Cu.* Cubital cell. 1 *Cu.* First cubital cell. *cu.a.* Cubito-anal transverse nervure. *C.* 1. Costa. *C.* Costal cell. 1 *C.* First costal cell. *M.* Medial cell. *m.cu.* Medio-cubital transverse nervure. *m.* Medial transverse nervure. 2 *M*^{1}., 2 *M*^{2}. First and second second medial cells. *M.* 1 + 2. Medial longitudinal nervure. *M.* 3 + *Cu.* Medio-cubital longitudinal nervure. *R.* Radial cell. *R.* 1 to *R.* 4 + 5. Radial longitudinal nervures. *Sc.* Subcostal cell. *Sc.* 1. Subcosta.

The nervures of the wing are ochraceous. The anterior edge of the wing (fig. 7) is formed by a stout nervure, the costa (*C.* 1) which is very setose. The second longitudinal nervure, the sub-costal (*Sc.* 1) joins the costal about half way along its length. A small transverse nervure, the humeral (*h.*) divides the costal cell into costal (*C.*) and first costal (1 *C.*) cells. The next main nervure—the radial—divides into a number of branches (in the typical insect five); some of these have coalesced in the fly. A nervure joining the costal just past the middle is the first radial (*R.* 1) cutting off the sub-costal cell. The next nervure, which joins the costal on the apical curve, represents the fused second and third radial nervures (*R.* 2 + 3). This cuts off the first radial cell (1 *R.*). The last nervure, which joins the costal almost at the apex of the wing, represents the fused fourth and fifth radial nervures (*R.* 4 + 5) and so cuts off the third radial cell (3 *R.*). The fourth main longitudinal nervure is the median, which, in the typical insect, divides into three, but in the fly the nervures have undergone coalescence, as will be shown. The first and second median nervures have coalesced (*M.* 1 + 2), and do not run direct to the margin of the wing, but bend forwards and almost meet *R.* 4 + 5 on the costa. About half way across the wing a transverse nervure, the radio-medial (*rm.*) unites *R.* 4 + 5 and *M.* 1 + 2, and cuts off the fifth radial cell (5 *R.*) from the radial (*R.*). The next longitudinal nervure represents the coalesced third medial and cubital nervures (*M.* 3 + *Cu.* 1). It runs to the posterior margin of the wing about half way along the length of the latter. The nervures *M.* 1 + 2 and *M.* 3 + *Cu.* 1 are united by two nervures; a proximal nervure, the medio-cubital (*m.cu.*) representing part of the original longitudinal vein *M.* 3, cuts off the small triangular medial cell (*M*); the distal transverse nervure *m.* cuts off the first second medial cell (2 *M*^{1}.) from the second second medial cell (2 *M*^{2}.). The last longitudinal nervure, the anal (*A.* 1), is undivided and does not reach the margin of the wing, thus incompletely separating the first cubital (1 *Cu.*) and anal (*A.*) cells. A small nervure, the cubito-anal (*cu.a.*) representing a portion of the original cubital vein (*Cu.* 2) slightly more proximal than the medio-cubital cuts off the small triangular cubital cell (*Cu.*) from the first cubital cell (1 *Cu.*). Running parallel with, and posterior to, the anal longitudinal nervure, there is apparently another nervure. This, however, is not a true nervure but is merely a chitinised furrow giving additional strength to the posterior angle of the wing. The posterior edge of the base of the wing is divided into a number of lobes. These are the anal lobe, and, as Sharp (1895) proposed, the alula, antisquama, and squama. The squama is thicker than the rest, and is attached posteriorly to the wing-root between the mesoscutum and the lateral plates of the postscutellum. It covers the haltere, as in all "calyptrate" Muscidae[1].

The *Halteres*. The halteres or balancers (fig. 6 *hal.*) are generally considered to represent the rudimentary metathoracic wings. They are covered by the squamae, and are situated on the sides of the thorax above the posterior spiracles. Each consists of a conical base on which are a number of chordonotal sense-organs and on this base is mounted a slender rod, at the end of which a small spherical knob is attached. The wall of the distal half of this sphere is thinner than the proximal half, and in preserved specimens is generally indented. Experiments show that the halteres are organs of a static function. They are not balancing organs in the sense that they are equivalent to the balancing pole of a ropewalker. They also have probably an auditory function. They are innervated by the largest pair of nerves in the thorax.

[1] The nomenclature of Comstock and Needham has not yet been adopted by dipterologists in general, but on account of its morphological value, it may in course of time replace the present confused system. It may, therefore, be useful if the nomenclature employed in the foregoing description be compared with those most usually employed.

LONGITUDINAL NERVURES. C_1. Costal. Sc_1. Mediastinal; auxiliary. R_1. Subcostal; 1st longitudinal. $R. 2+3$. Radial; 2nd longitudinal. $R. 4+5$. Cubital; 3rd longitudinal; ulnar (Lowne). $M. 1+2$. Median; 4th longitudinal; discal (Verrall). $M. 3+Cu_1$. Submedian; 5th longitudinal; postical (Verrall). A_1. Anal; 6th longitudinal. Pseudonervure; axillary; 7th longitudinal.

TRANSVERSE NERVURES. *h*. Humeral; 1st transverse; basal cross-vein (Verrall). *rm*. Discal; 2nd transverse; middle cross-vein (Verrall); medial transverse; anterior transverse (Austen). *m.cu.* Anterior basal transverse (Austen); lower cross-vein (Verrall); postical transverse (Lowne). *m*. Posterior transverse (Austen); postical cross-vein (Verrall); discal transverse (Lowne). *cu.a.* Posterior basal transverse (Austen); anal cross-vein (Verrall); anal transverse (Lowne).

CELLS. *C*. Costal. 1 *C*. Second costal. *Sc*. Third costal (Lowne correctly calls this "sub-costal"). 1 *R*. Marginal. 3 *R*. Sub-marginal; cubital (Lowne). 5 *R*. First posterior cell (Austen); sub-apical (Lowne and Verrall). $2 M^2$. Second posterior cell (Austen); apical. 1 *Cu*. Third posterior cell (Austen and Verrall); patagial (Lowne). $2 M^1$. Discal (this term is used also in Lepidoptera, Trichoptera, and Psocoptera, and in each family refers to a different cell!). *R*. Anterior basal cell (Austen); upper or first basal or radical (Verrall); prepatagial (Lowne). *M*. Posterior basal cell (Austen); middle or second basal or radical (Verrall); anterior basal (Lowne). *Cu*. Anal cell (Austen); lower or third basal or radical (Verrall); posterior basal (Lowne).

The Legs.

The three pairs of legs are composed of the typical number of segments. Each consists of coxa, trochanter, femur, tibia, and tarsus. The coxae are the only segments that show any considerable difference in the three pairs of legs. The anterior coxae are comparatively large and boat-shaped, the intermediate coxae are smaller and their separate sclerites more marked; the coxal plates of the intermediate coxae are shown in fig. 6 (*cp.*). The coxal joints of the posterior pair of legs are almost similar to those of the intermediate pair. The anterior femora are shorter and stouter in the middle than those of the intermediate posterior pairs of legs. The anterior tibiae are also shorter than those of the succeeding legs. The anterior tibiae are covered on their inner sides with closely-set, orange coloured setae which serve as a comb by means of which the fly removes particles of dirt adhering to the setae which clothe its body; the first tarsal joints of the posterior legs are also similarly provided. The tarsi consist of five joints, the terminal joints bearing the "feet." These organs, about which so much has been written, consist of a pair of curved lateral claws or "ungues" which subtend a pair of membranous pyriform pads—the pulvilli. The pulvilli are covered on their ventral sides with innumerable, closely-set, secreting hairs by means of which the fly is able to walk in any position on highly polished surfaces. A small sclerite lies between the bases of the pulvilli. The tarsal joints and the other segments of the legs are covered with a large number of setae.

THE ABDOMEN.

The abdomen is oviform with the broad end basal. The total number of segments which compose the abdomen is eight in the male and nine in the female. The visible portion consists of apparently four segments in the male and female, in reality there are five, as the first segment has become very much reduced, and has fused with the second abdominal segment forming the anterior face of the base of the abdomen (see fig. 22). The segments succeeding the fifth are greatly reduced in the male, and in the female they form the tubular ovipositor which, in repose, is tele-scoped within the abdomen. The second, third, fourth and fifth abdominal segments are well developed and each mainly consists of a large tergal plate, which extends laterally to the ventral side. The sternal plates are much reduced, and form a series of narrow plates lying on the ventral membrane along the mid-ventral line. The spiracles are situated on the lateral margins of the tergal plates. The sclerites of the abdomen which are exposed are strongly setose, especially the fourth and fifth dorsal plates, but they do not bear macrochaetae.

The terminal abdominal segments of the male and female are described in detail in the account of their reproduction systems (see pp. 50, 53).

CHAPTER III

THE MUSCULAR SYSTEM.

THE muscular system of the fly is similar to that of *Volucella*, described by Kunckel d'Herculais (1881), and of the blow-fly, described by Lowne and Hammond, and consequently it will be but briefly described. The muscles may be divided into the following groups: 1. Cephalic; 2. Thoracic; 3. Segmental; 4. Those controlling the thoracic appendages; and 5. Special muscles.

1. The cephalic muscles will be considered in the detailed description of the head (see p. 58).

2. The thoracic muscles are enormously developed and almost fill the thoracic cavity. They are arranged in two series. The dorsales (figs. 17, 18) are six pairs of muscle-bands on each side of the median line, attached posteriorly to the postscutellum and mesophragma, and anteriorly to the prescutum and anterior region of the scutum. The sternodorsales (*st.do.*) are vertical and external to the dorsales and are arranged in three bundles on each side. The first two pairs have their upper ends attached to the prescutum and scutum, and their lower ends inserted on the mesosternum; the third pair is attached dorsally to the scutum and ventrally to the lateral plate of the postscutellum above the spiracle. As Hammond has shown in the blow-fly (1881), all these muscles are mesothoracic. The dorsales by contraction loosen the alar membrane and so depress the wing, the sterno-dorsales have the opposite effect.

3. The segmental muscles. These muscles, which are so prominent in the larva, have almost disappeared in the imago. They are represented by the cervical muscles, certain small thoracic muscles, the thoraco-abdominal muscles, and the segmentally-arranged abdominal muscles, together with the muscles controlling the ovipositor and male gonapophyses.

4. The muscles controlling the thoracic appendages, the wings, legs and halteres. There is an elaborate series of muscles controlling the roots of the wing, but in order to avoid too much detail they will not be described here. The flexor muscles of the anterior coxae have their origin on the inner surfaces of the humeri, a fact supporting the prothoracic nature of these sclerites; the flexors of the middle pair of legs have their origin on the sides of the posterior region of the prescutum. The internal muscles of the leg are similar to those of the blow-fly and *Volucella*.

5. Special muscles. These are the muscles controlling the spiracular valves, the penis, and other small muscles.

THE NERVOUS SYSTEM.

The central nervous system (fig. 8) consists of: (1) the brain or supraoesophageal ganglia, which are closely united with the suboesophageal ganglia, the whole forming a compact mass which I propose to call the cephalic ganglion (fig. 8, *C.G.*), perforated by a small foramen for the passage of the narrow oesophagus, and (2) the thoracic compound ganglion, which is composed of the fused thoracic ganglia with the abdominal ganglia. The two compound nerve centres are united by a single median ventral cord running from the suboesophageal ganglia to the anterior end of the thoracic nerve-centre.

The cephalic ganglion consists of the supraoesophageal ganglia and the suboesophageal ganglia so closely united that the commissural character of the circumoesophageal connectives is completely lost. Externally on the dorsal side of the brain three longitudinal fissures can be seen, a median fissure and two lateral fissures marking the origin of the optic lobes.

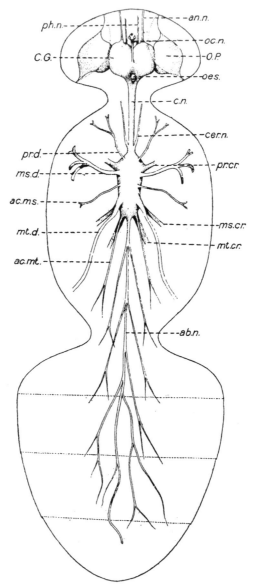

FIG. 8. Nervous system. The very fine nerve which runs along the dorsal side of the oesophagus to the proventricular ganglion (*Pv.g.*, fig. 20) has been purposely omitted.

ab.n. Abdominal nerve. *ac.ms.* Accessory mesothoracic dorsal nerve. *ac.mt.* Accessory metathoracic dorsal nerve. *cer.n.* Cervical nerves. *c.n.* Cephalothoracic nerve cord. *O.P.* Optic peduncle. *pr.cr.*, *ms.cr.*, *mt.cr.* Pro-, meso-, and metathoracic crural nerves. *pr.d.*, *ms.d.*, *mt.d.* Pro-, meso-, and meta-thoracic dorsal nerves.

The *Supraoesophageal ganglia.* The characters of the ganglia composing the brain are hidden by the sheath of cortical cells which fill up the spaces between the ganglia; the characters of these can be ascertained by a study of the serial sections. The median mass, the procerebrum, is formed by the fusion of the procerebral lobes. These are united before and behind, and enclose a central ganglionic mass—the central body. Behind the cerebrum two pairs of fungiform bodies arise. On the anterior face of the procerebrum the antennal or olfactory lobes which represent the deutocerebrum are situated laterally. Each sends a nerve (figs. 3, 8, *an.n.*) to the antenna. Above these and on the dorsal side is a pair of lobes, the frontal lobes which are contiguous with each other in the median line; these belong morphologically to the tritocerebrum. Posterior to these in the median dorsal line of the cerebrum a single median nerve, the ocellar nerve (figs. 3, 8, *oc.n.*), arises; this runs vertically to the ocelli. A pair of lobes, which correspond to Lowne's thalami of the blow-fly, are situated external to and between the frontal and antennal lobes. The peduncles of the optic lobes have their origins from the sides of the procerebrum. Each optic peduncle (fig. 8, *O.P.*) contains three ganglionic masses which Hickson (1885) has termed from the brain peripherally the opticon, epi-opticon and periopticon (fig. 3, *P.O.*) respectively.

The *Suboesophageal ganglia* (fig. 3, *s.o.*). The commissures uniting the supraoesophageal ganglia to the oesophageal mass cannot be recognised as such, owing to the extreme state of cephalisation of the cephalic ganglia. They are represented by the regions of the oesophageal foramen, and from the anterior side of each of them arises a pharyngeal nerve (fig. 3, *ph.n.*). From the ventral side of the suboesophageal ganglia a pair of nerves, the labial nerves (fig. 3, *lb.n.*), arise and run down the proboscis, innervating the muscles of that organ; on reaching the oral lobes they bifurcate and branch freely, supplying the numerous sense organs in those structures. The cortical cells (Leydig's *Punktsubstanz*), which fill up the spaces between the ganglia and form an investing sheath round the whole ganglionic mass, are of two kinds. The smaller cells are rounded, their nuclei are large in proportion to the protoplasm, and their protoplasmic fibres anastomose with each other. Among these smaller cortical cells, and also occasionally in the ganglionic substance, large ganglionic cells occur, their protoplasm taking the stain very readily. Unipolar, bipolar, and tripolar ganglion cells are found.

The *Eyes.* Each eye contains about 4000 facets. They are similar in all respects to the eyes of the blow-fly, which have been fully described by Hickson (*loc. cit.*), whose results my study confirms; consequently a description of their structure will not be given. It should be noted that, in spite of the fact that Hickson corrected many mistaken views held by Lowne in his memoir (1884), these are repeated in his later monograph of the blow-fly.

The cephalo-thoracic nerve cord (fig. 8, *c.n.*) unites the cephalic and thoracic ganglia. Near its junction with the thoracic ganglion a pair of cervical nerves (*cer.n.*) arise, innervating the muscles of the neck.

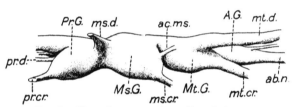

FIG. 9.   Thoracic compound ganglia.   Left aspect.
Lettering as in figs. 8 and 10.

The *Thoracic ganglion* (figs. 8, 9, 10) is pyriform, with the broad end anterior, and rests on the entothoracic skeleton of the mesothorax. As in the cephalic ganglion, the component ganglia are ensheathed in a cortical layer, which is of the same nature as that of the cephalic ganglion. The nerves of the three pairs of legs (*pr.cr., ms.cr., mt.cr.*) arise from three large ganglia, which are the prothoracic (*Pr.G.*), mesothoracic (*Ms.G*) and metathoracic (*Mt.G.*) ganglia. These are united by a median longitudinal band of nerve tissue, which runs dorsal to them, and behind the metathoracic ganglia swells out into a ganglionic mass (*A.G.*) which represents the abdominal ganglia. In this median dorsal band there is a median dorsal fissure stretching posteriorly from above the middle of the mesothoracic ganglia. The dorsal regions of the mesothoracic and metathoracic ganglia show ganglionic swellings. From the antero-dorsal sides of the prothoracic ganglia a pair of prothoracic dorsal nerves (*pr.d.*) arises and supplies the muscles of that region, including those of the anterior thoracic spiracle. The nerves supplying the mesothoracic legs (*ms.cr.*) arise from the postero-ventral sides of the mesothoracic ganglia. Between the mesothoracic ganglia there is a median ganglionic

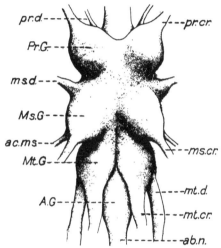

Fig. 10. Thoracic compound ganglion after the removal of the cortex. Seen from the ventral side. This and fig. 9 were drawn from models reconstructed from sections.

Pr.G., *Ms.G.*, *Mt.G.* Pro-, meso-, and meta-thoracic ganglia. *A.G.* Abdominal ganglion. Other lettering as in fig. 8.

mass, situated slightly dorsal, from the middle region of which the nerve fibres of the large pair of dorsal mesothoracic nerves (*ms.d.*) arises; Lowne, in the blow-fly, calls these prothoracic. The roots of these nerves are broad dorso-ventrally. These nerves innervate the sterno-dorsales muscles of the middle region. In this median mesothoracic nerve centre, posterior to the origin of the dorsal mesothoracic nerves, the fibres of a pair of nerves, the accessory dorsal mesothoracic nerves (*ac.ms.*), have their origin; externally these appear to arise dorsal to the roots of the mesothoracic crural nerves. The dorsal metathoracic nerves (*mt.d.*), which innervate the halteres and are the largest pair of thoracic nerves, have their origin from the median dorsal band in front of the metathoracic ganglia, so that they appear to be almost mesothoracic in origin. The metathorax crural nerves (*mt.cr.*) arise from the posterior ventral sides of the metathoracic ganglia. Posterior to these a pair of slender nerves, the accessory dorsal metathoracic nerves, have their origin, and innervate the muscles at the posterior end of the thorax.

The dorsal band becomes much thinner posterior to the abdominal ganglion, and runs into the abdomen as a median abdominal nerve (*ab.n.*). In the thorax two pairs of abdominal nerves arise. In the abdomen the abdominal nerves arise alternately and irregularly from the median abdominal nerve. The median abdominal nerve finally terminates in the genitalia.

THE ALIMENTARY SYSTEM (figs. 11 and 12).

The alimentary canal of the house-fly is shorter than that of the blow-fly, and also than that of *Glossina* described by Minchin (1905), and slightly longer than the alimentary tract

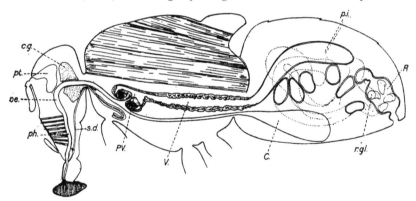

FIG. 11.   Longitudinal section of the Alimentary Canal of *M. domestica*.

ph. Pharyngeal suction pump. *oe.* Oesophagus. *pt.* Ptilinum. *c.g.* Supra-oesophageal ganglion. *s.d.* Lingual salivary duct. *PV.* Proventriculus. *V.* Ventriculus. *C.* Crop. *p.i.* Proximal intestine. *R.* Rectum. *r.gl.* Rectal gland.

of *Stomoxys* described by Tulloch (1906). It serves as a good example of the Muscid digestive canal. It is of a suctorial character, and consists of pharynx, oesophagus, crop, proventri-

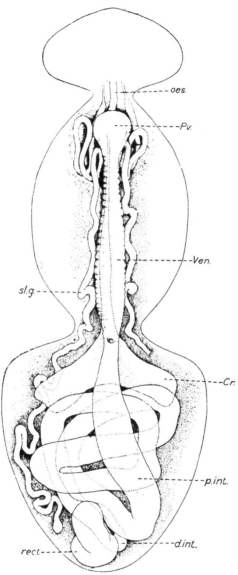

FIG. 12. The alimentary canal as it is seen on dissection from the dorsal side. The malphigian tubes have been omitted, and also the distal portion of the lingual salivary gland (*sl.g.*) of the right side. The duct of the crop (*Cr.*) is shown by the dotted line beneath the proventriculus (*Pv.*) and ventriculus (*Ven.*).

p.int. Proximal intestine. *d.int.* Distal intestine. *rect.* Rectum.

culus, ventriculus or chyle stomach, proximal and distal intestine and rectum.

The *Pharynx* has already been described, and will be further referred to in the detailed description of the head (pp. 56 *et seq.*). At the proximal end of the fulcrum, where the oesophagus arises, there is usually a small mass of cells, which Kraepelin has described as glandular, but which I believe to be simply fat-cells.

The *Oesophagus* (*oes.*) commences at the proximal end of the pharynx, and describes a curve before passing through the oesophageal foramen in the cephalic ganglion, where it narrows slightly. It then passes through the cervical region into the thorax in the anterior region of which it opens into the pro-ventriculus (*Pv.*) continuous with, and in the same line as the oesophagus, the duct leading to the crop (fig. 13, *d.cr.*) passes along the thorax dorsal to the thoracic nerve-centre, and entering the abdomen it leads into the crop, which lies on the ventral side of the abdomen. The oesophagus has a muscular wall, enclosing a layer of flat epithelial cells, and is lined by a cuticular intima, which is thrown into several folds at the anterior end.

The *Crop* (*Cr.*) is a large bilobed sac, capable of considerable distension, and when filled with the liquid food, it loses its bilobed shape and occupies a large portion of the antero-ventral region of the abdomen. Its walls exhibit muscular (unstriped) fibres; the flat epithelial cells have a very thin cuticle. The function of the crop will be more fully described later when an account of the method of feeding is given. Graham-Smith has shown (1910) that the capacity of the crop of *M. domestica* varies between ·003 and ·002 c.c.

The *Proventriculus* (*Pv.*) is circular and flattened dorso-ventrally. Its structure will be understood by reference to fig. 13. In the middle of the ventral side it opens into the oesophagus, and on the dorsal side the outer wall is continued as the wall of the ventriculus (*Ven.*). The interior is almost filled up by a thick circular plug (*Pv.p.*), the cells of which have a fibrillar structure, and is pierced through the centre by the oesophagus. The neck of the plug is surrounded by a ring of elongate cells, external to which the wall of the proventriculus begins, and, enclosing the plug at the sides and above, it merges into the wall of the ventriculus. I do not agree with Lowne, who regards the proventriculus as "a gizzard and nothing more," but its structure suggests a pumping function and also that of a valve. This interpretation of a combined pump and valve operated at will by the fly is supported by the fly's method of feeding and regurgitating its food, which habits will be described later (p. 80).

On the dorsal side of the oesophagus, at its junction with the proventriculus, a small ganglion, the proventricular ganglion

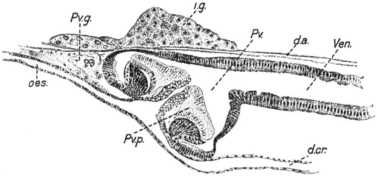

Fig. 13. Section through the proventriculus and the anterior end of the ventriculus, to show the structure of the proventricular plug (*Pv.p.*) and the ducts of the oesophagus (*oes.*) and crop (*d.cr.*). (Camera lucida drawing.)

(*Pv.g.*) lies, communicating by means of a fine nerve with the cephalic ganglion; this forms a part of the sympathetic or visceral nervous system.

The *Ventriculus*, or *Chyle Stomach* (*Ven.*), represents the anterior region of the mesenteron, the posterior region of the latter being formed by the proximal intestine. It is narrow in front, and widest in the posterior region of the thorax, where it again narrows in passing through the thoraco-abdominal foramen into the abdomen to become the proximal intestine. Except in the anterior and posterior regions, where columnar cells compose the digestive epithelium, the walls of the ventriculus are thrown into a number of transverse folds, which are again subdivided longitudinally, the result being the formation of small crypts or sacculi, which are lined by large cells. These sacculi correspond to the digestive coeca of other insects.

The *Proximal Intestine (p.int.)* is the longest region of the gut. It varies in length considerably. In the normal-sized condition its course is as follows: Beginning at the anterior end of the abdomen, it runs dorsally beneath the heart to the posterior region, where it curves downwards, turns to the left, and runs forward for a short distance, curving to the right, where it doubles back transversely to the left. Here it doubles sharply back to the right, from whence it runs forward for a little way

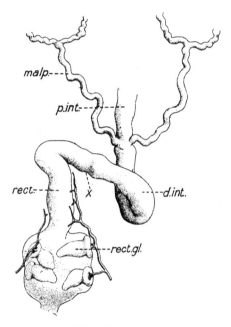

Fig. 14. The posterior region of the alimentary canal, to show the rectal glands (*rect.gl.*) with their tracheal supply, the origin of the malpighian tubes (*malp.*), and the position of the rectal valve indicated at X.

and crosses over to the left. Curving, it runs posteriorly to become the distal intestine. Its walls are lined by an epithelium of large columnar cells.

The *Distal Intestine (d. int.)*. The junction of this with the proximal intestine is marked by the entrance of the ducts of the malpighian tubes. It runs posteriorly and curves dorsally and forwards to become the rectum, from which it is separated by a cone-shaped valve, the rectal valve, the position of which is marked externally (fig. 14, *X*.). The epithelium of the distal intestine consists of small cubical cells, which project into the lumen, and are covered by a fairly thick chitinous intima. The epithelial wall of the distal intestine is thrown into usually about six longitudinal folds.

The *Rectum* (*rect.*) is composed of three parts, an anterior region, an intermediate region which is swollen to form the rectal cavity, and a shorter region posterior to this which opens externally by the anus. The anterior region is lined by cubical cells, whose internal faces project into the lumen of the rectum, and give the chitinous intima a tuberculated structure. The intermediate region, which forms the rectal cavity, contains the four *rectal glands* (*rect.gl.*). Its walls are lined by a thin cuticle supported by a flattened epithelium. The posterior portion of the rectum is short, and has thick muscular walls. The cuticular intima is continuous with that of the external skeleton.

Salivary glands.

There are two sets of salivary glands—a pair of labial and a pair of lingual glands. The structure of the labial glands will be described in the account of the anatomy of the head (p. 63).

The *Lingual Glands* (fig. 12, *sl.g.*), though considerably longer than the total length of the body, are of the simplest tubular type. They are of uniform width throughout their whole length, except the slightly swollen blind termination. These blind ends lie one on each side of the ventral and posterior region of the abdomen, generally embedded in the fat body. They take a sinuous course forwards through the abdomen into the thorax, where they run alongside the ventriculus. At the sides of the proventriculus they are thrown into several folds, which appear to be quite constant in character. They pass forwards at the sides of the oesophagus, and on entering the cervical region the ducts lose their glandular character and assume a spiral thickening; before leaving the cervical region the two ducts unite below the oesophagus, and the single median duct enters the head ventral to the cephalothoracic nerve cord and runs direct to the proximal end of the hypopharynx, at the end of

which it opens.   A short distance before entering the hypopharynx the salivary duct (fig. 3, *sal.d.*) is provided with a small
valve controlled by a pair of fine muscles (*s.m.*) which serve to

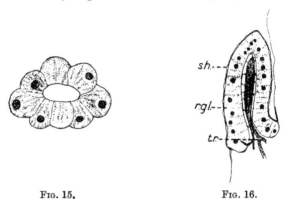

Fig. 15.   Fig. 16.

Fig. 15.   Transverse section of the lingual salivary gland, showing the fibrillar
character of the gland cells.   × 220.   (Camera lucida drawing.)

Fig. 16.   Vertical section of one of the rectal glands, to show its structure.   × 56.
(Camera lucida drawing.)

sh. Perforate chitinous sheath.   *r.gl.* Gland cell.   *tr.* Trachea.

regulate the flow of the salivary secretion.   The glands are composed of glandular cells (fig. 15), which are convex externally and
have a fibrillar appearance in section.   No vacuoles have been
found in the cells.

The malpighian tubes.

A pair of malpighian tubes (fig. 14, *malp.*) arise at the point
of junction of the proximal and distal intestines, that is, where
the mesenteron joins the proctodaeum.   Each malpighian tube
is shortly divided at an angle of 180° into two malpighian tubules.
The malpighian tubules are very long and convoluted, and intimately bound up with the diffuse fat body, so that it is a matter
of considerable difficulty to dissect them out entire.   They have
a moniliform appearance and are of uniform width throughout;
never more than two cells can be seen in section.   They are
generally yellowish in colour.   As in most insects, they are undoubtedly of an excretory nature, as the contents of the cells

and tubules show. Lowne's view that, in the blow-fly, they are of the nature of a hepato-pancreas is untenable on both morphological and physiological grounds.

The rectal glands.

The four rectal glands (*rect.gl.*) are arranged in two pairs, two each side of the rectal cavity. Each rectal gland (fig. 16) has a conical or pyriform apex with a swollen circular base. It is composed of a single layer of large columnar cells (*r.gl.*); the papilla is hollow and its cavity is in communication with the general body cavity. It is covered externally by a perforate chitinous sheath (*sh.*), which is continuous with the intima of the rectum. A number of tracheae (*tr.*) enter the cavity of each gland, and fine tracheae may be found penetrating the wall. The cavity of the gland is filled with a loose tissue of branching cells. As the gland is capable of pulsation, there is no doubt a constant interchange of blood between the cavity of the gland and the body cavity (which is a haemocoel). By this means waste products may be extracted from the blood by the large gland cells and excreted into the rectum through the pores on the external sheath of the gland. The rich supply of tracheae probably assists the cells in the process of excretion, as we find the tracheae very numerous, and intimately connected with the malpighian tubules.

THE RESPIRATORY SYSTEM.

The respiratory or tracheal system is developed to a very great extent in the fly and occupies more space than any other anatomical structure. Only by dissection of the freshly-killed insect can one obtain a true conception of its development and importance. It consists of tracheal sacs of varying size, having extremely thin walls and tracheae which may arise from the sacs, or, in the case of the abdominal tracheae, independently from the spiracles.

The *Anterior Thoracic Spiracles* (figs. 6, 17, *a.th.*). Each is a large vertical opening behind the humeral sclerite and above the anterior legs. It is surrounded by a chitinous ring, the peritreme, and the opening is guarded by a number of dendritic processes which prevent the entrance of dust and other foreign bodies. It leads into a shallow chamber or vestibule which communicates with the rest of the spiracular system through a valvular aperture.

The anterior thoracic spiracles supply the whole of the head, the anterior and median regions of the thorax, the three pairs of legs, and by means of the abdominal air sacs a large part of the viscera.

Internal to the valve the tracheal system divides. The tracheal sacs springing from the posterior side are as follows: Ventrally a rather narrow tracheal duct leads into a sac—the anterior ventral thoracic sac (fig. 17, *a.v.s.*)—situated at the side of the thoracic ganglion which it supplies. Above the origin of this another tracheal duct leads to a vertical sac supplying the anterior sterno-dorsales muscles. Dorsally the ducts of two sacs take their origin; the smaller and more dorsal is a flat sac closely apposed to the anterior ends of the dorsales muscles (*do.*) which it supplies; the more ventral of the two is one of the two most important branches of the anterior thoracic spiracle (the other being the branch supplying the head). In the thorax it takes the form of an elongated sac lying below the dorsales muscles, and by the side of the alimentary canal. From the dorsal side of this the longitudinal thoracic sac (*l.tr.s.*) a number of branches arise which supply the lower dorsales muscles. It is constricted about the middle of its length and anterior to the constriction a branch is given off which supplies the ventral portion of the median sterno-dorsales muscles. In the posterior region of the thorax another ventral branch is given off from which branches arise, one supplying the ventral portions of the posterior sterno-dorsales muscles, the other opening into the posterior ventral thoracic sac (*p.v.s.*) which supplies the intermediate and posterior legs.

The longitudinal thoracic sac then narrows, and passes through the thoraco-abdominal opening into the abdomen. In the abdomen it immediately dilates to form one of the large abdominal air-sacs (*ab.s.*). The pair of abdominal air-sacs in some cases occupies about half the total space of the abdomen. When the

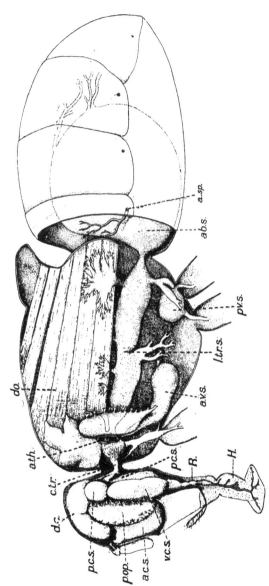

FIG. 17. The tracheal sacs supplied by the anterior thoracic spiracle (*a.th.*). In this figure the tracheal sacs supplied by the posterior thoracic spiracle and the sterno-dorsales muscles of the left side have been removed. The left side of the head and proboscis have also been removed. The first abdominal segment has been removed to show the large abdominal air sacs (*ab.s.*) and an abdominal trachea which is supplied by the second abdominal spiracle (*a.sp.*).

a.c.s. Anterior cephalic sac.   *a.v.s.* Anterior ventral thoracic sac.   *c.tr.* Cervical tracheal duct.   *d.c.* Dorsal cephalic sac.   *do.* Dorsales muscles.   *H.* Haustellum.   *l.tr.s.* Longitudinal tracheal sac.   *p.c.s.* Posterior cephalic tracheal sacs.   *p.v.s.* Posterior ventral thoracic sac.   *p.op.* Periopticon.   *R.* Rostrum.   *v.c.s.* Ventro-lateral cephalic sac.

fat-body is not greatly developed they occupy almost the whole
of the basal portion of the abdomen. They give off internally
a large number of tracheae which ramify among the viscera
and provide a large portion of the contents of the abdomen
with air.

From the anterior side of the anterior thoracic spiracle a

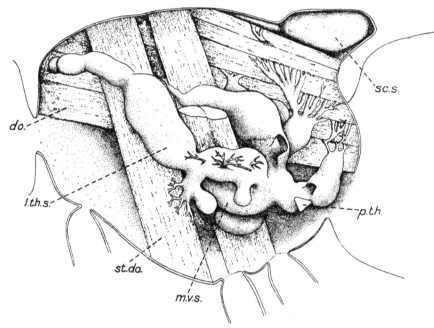

Fig. 18. The tracheal sacs supplied by the posterior thoracic spiracle. In this
figure the left side of the thorax has been removed, together with the wing
muscles and the posterior sterno-dorsales. It must be imagined that this
figure is superimposed on fig. 17.

do. Dorsales.   *l.th.s.* Lateral thoracic sac.   *m.v.s.* Median ventral sac.
    p.th. Posterior thoracic spiracle.   *sc.s.* Scutellar sac.   *st.do.* Sterno-
    dorsales.

flattened sac arises. On its ventral side this gives off a branch
which supplies the muscles of the neck and the anterior leg. The
sac then narrows into a rather thick-walled cervical tracheal duct
(*c.tr.*), which passes through the neck alongside the cephalo-
thoracic nerve-cord and enters the head.

Tracheal Sacs of the Head. The tracheal sacs of the head
occupy the greater portion of the head capsule. They entirely

fill up all the space which would otherwise be haemocoel. These tracheal sacs are supplied by the tracheal ducts which, on entering the head capsule, curve dorsally behind the cephalic ganglion. Before curving upwards each gives off a large ventral duct (fig. 19, *tn.d.*) which spreads out beneath the cephalic ganglion, forming a structure of a tentorial nature upon which the ganglion rests. The dorsal cephalic ducts unite behind the cephalic ganglion above the oesophagus. From the point of junction three ducts arise, two lateral ducts and a median dorsal duct. The median dorsal duct (*m.d.*) opens into a large bilobed dorso-cephalic sac lying on top of the ganglion, and occupying the dorsal region of the head capsule. It gives off branching tracheal twigs supplying the antero-dorsal portion of the optic ganglion (periopticon). Each of the lateral ducts (*l.d.*) sup-

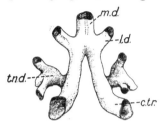

Fig. 19.   Posterior view of the tracheal ducts which supply the cephalic sacs and tracheae. *c.tr.* Cervical tracheae which fuse above the oesophagus on the posterior side of the cephalic ganglion.   *l.d.* Lateral duct. *m.d.* Median dorsal duct.   *tn.d.* Tentorial tracheal ducts which spread out beneath the cephalic ganglion.

plies the posterior cephalic sacs. It first communicates with a sac (fig. 17, *p.c.s.*) lying behind the dorsal portion of the optic ganglion to which it gives off a large number of tracheal twigs. This sac opens into an elongate vertical sac which occupies the ventro-posterior region of the head capsule. The remaining tracheal sacs of the head are supplied by the tentorial tracheal ducts (fig. 19, *tn.d.*), which spread out beneath the cerebrum in a fan-shaped manner and are bilaterally distributed. Each half, in addition to giving off internally tracheal twigs to the optic ganglia, communicates with two tracheal sacs. An internal duct leads into a large spherical sac, the anterior cephalic sac (fig. 17, *a.c.s.*), situated in the anterior region of the head dorsal to the fulcrum. From the dorsal side of this sac a branch is given off which supplies the antenna of its side; the ventral side is continued down the fulcrum as a narrow tracheal sac. The lateral portion of the tentorial tracheal duct opens into the ventro-lateral cephalic sac (*v.s.c.*) situated posterior to the optic ganglion. The lower end of this sac gradually narrows as it enters the rostrum which it traverses, giving off half-way along its length a trachea which supplies the palp of that side. On reaching the haustellum it takes the form of a trachea proper, having annular thickenings. Shortly after entering the haustellum it gives off two branches to the muscles of this region. The main trachea is continued into the oral lobe of its side, where it divides into anterior and posterior branches, and these again divide into numerous small tracheae running to the edges of the oral lobes. Lowne, in his description of the tracheal system of the blow-fly, describes and figures the tracheal supply of the proboscis as being of the nature of tracheal sacs and capable of distension; he also describes a trefoil-shaped tracheal sac at the base of the oral lobes giving off very regular branches, the dilation of which, he claims, causes the inflation and tension of the oral lobes. The mechanism of the proboscis will be discussed later (p. 62), but it may be noticed here that in *M. domestica* there is no trace of a trefoil-shaped sac at the base of the oral lobes, and that *all the tracheal structures of this the haustellum region are definite annular tracheae, and therefore incapable of distension.*

The *Posterior Thoracic Spiracle* (figs. 6, 18, *p.th.*) is triangular in shape and is guarded by dendritic processes. The tracheal sacs of this system (fig. 18) have not the extended range of those supplied by the anterior thoracic spiracle, but are confined to the thorax, chiefly in the median and posterior regions which are not aerated to any great extent by those of the other system. They supply chiefly the large muscles of the thorax. Laterally a series of sacs (*l.th.s.*) extends anterodorsally in an oblique direction, external to the sterno-dorsales muscles, to the humeral region. From the first of these sacs a large number of tracheal twigs arise and supply the muscles of the wing and the anterior sterno-dorsales muscles. Ventral to this sac a large sac (*m.v.s.*) penetrates internally between the anterior and median sterno-dorsales muscles and supplies the lower dorsales muscles.

From the dorsal side of the distributing sac a number of sacs arise, some of which penetrate between the sterno-dorsales muscles and supply the upper dorsales muscles. A more posterior set supplies the posterior regions of the dorsales muscles, ramifying between them in a very extensive manner, some ultimately terminating in the tracheal sacs beneath the scutum and the scutellar sac (*sc.s.*).

The *Abdominal Spiracles* differ in number in the two sexes. In the male there are seven pairs of abdominal spiracles; in the female I have only been able to find five pairs. In both sexes each of the large tergal plates which cover the abdomen has near its lateral margin a small circular spiracle. The first abdominal segment which has fused with the second pair has a pair of small spiracles (see fig. 22) slightly anterior to those of the second (apparent first) abdominal segment. In addition to these the male possesses two pairs of spiracles in the membrane at the lateral extremities of the rudimentary sixth and seventh abdominal segments (see fig. 25). In the female I have been unable to find any additional spiracles. Each of the abdominal spiracles is provided with a vestibule and atrium which are separated by a valve controlled by a minute chitinous lever. All the spiracles of the abdomen communicate with tracheae which ramify among the viscera and fat-body. There are no tracheal sacs in connection with these spiracles.

THE VASCULAR SYSTEM AND BODY CAVITY.

By the great development of the tracheal sacs in the head, the muscles in the thorax, and the fat-body and air-sacs in the abdomen, the haemocoelic space in the fly is greatly reduced. The blood is colourless, and is crowded with corpuscles, mostly containing substances of a fatty nature.

The *Fat-body* varies greatly in the extent of its development. In some cases it may almost fill the body-cavity, pushing the intestine back into a postero-dorsal position: this is generally the case in flies before hibernating; in other cases it may be only moderately developed. The fat-body receives a very rich tracheal supply, and stores the products of digestion which are conveyed to it by the blood with which it is bathed. It consists chiefly of very large cells, both uninucleate and multinucleate; the fat-cells of the head are not so large.

The *Dorsal Vessel* or *Heart* lies in the pericardial chamber, immediately beneath the dorsal surface. It extends from the posterior end to the anterior end of the abdomen, and four large chambers, corresponding to the four visible segments, and a small anterior chamber can be recognised; the last represents the chamber of the first abdominal segment. The chambers are not separated by septa, but each has a pair of dorso-lateral ostia situated at its posterior end where the alar muscles of the pericardium arise. The walls of the heart are composed of large cells. The pericardium contains fat-cells and tracheae, and its floor is composed of large cells of a special nature. The alar muscles run laterally in the floor of the pericardium to the sides of the dorsal plates where they are inserted. The anterior end of the heart is continued as a narrow tube (fig. 13, *d.a.*) along the dorsal side of the ventriculus, where it terminates in a mass of cells (*l.g.*) which are usually considered to be of a lymphatic nature.

THE REPRODUCTIVE SYSTEM.

The two sexes are slightly different in size, the females being larger than the males; the sexual dimorphism of the width of the frontal region of the head has already been noticed. There does not appear to be any great disparity in the numerical proportions of the sexes; near breeding-places there is naturally a preponderance of females, but in houses the sexes are approximately equal in number. In this respect they differ from the lesser house-fly *Fannia canicularis*.

The female reproductive organs.

The generative organs of the female consist of ovaries, spermathecae or vesiculae seminales, accessory glands and their ducts.

The *Ovaries*, when containing mature ova, occupy the greater part of the abdominal cavity (fig. 20, *ov.*). They lie ventral to the gut, occupying the whole of the ventral and lateral regions, the gut resting on the V-shaped hollow between them. Each ovary contains about seventy ovarioles, in each of which ova in various stages of development can be seen. The two short thin-walled oviducts (*ov.d.*) unite on the ventral side of the abdomen to form the common oviduct (*c.o.d*). The walls of the common oviduct are muscular, and when the ovipositor is in a state of rest, retracted into the abdominal cavity, the oviduct curves forwards and dorsally to enter the ovipositor (*ovp.*) ventral to the rectum (*rect.*). Here it swells slightly to form a sacculus

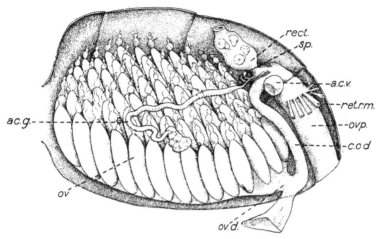

Fig. 20. Female reproductive organs *in situ*; the left ovary and the viscera have been removed. The ovipositor (*ovp.*) is shown retracted, in which state the common oviduct (*c.o.d.*) is doubled back.

ac.g. Accessory gland.　　*a.c.v.* Accessory copulatory vesicle.　　*ov.* Ovary composed of about seventy ovarioles, and containing ova in various stages of development.　　*ov.d.* Oviduct.　　*retr.m.* Retractor muscles of the ovipositor. *sp.* Spermathecae or vesiculae seminales.

(fig. 21, *sac.*), which leads into the muscular vagina (*vag.*). The vagina opens into the ventral side of the ovipositor immediately behind the sub-anal plate.

The *Spermathecae* (*sp.*) or *Vesiculae Seminales* are three in number, two on the left side and a single one on the right. Each consists of a small black, oviform, chitinous capsule, the lower half of which is surrounded by a follicular investment continuous with the cellular wall of the duct, the whole having the appearance of an acorn with a long stalk. The ducts of the spermathecae are lined by a thin chitinous intima continuous with the chitinous capsule, and they open at the posterior end of the sacculus on the dorsal side.

There is a single pair of *accessory glands* (*ac.g.*) which are fairly long, and on nearing the vagina they become narrower to form a slender duct which opens on the dorsal side of the vagina immediately behind the ducts of the spermathecae. The accessory glands are closely united with the fat-body. They probably secrete the adhesive fluid which covers the eggs when they are

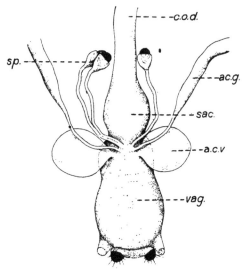

Fig. 21.   Terminal region of the female reproductive organs, showing the accessory glands, etc.
sac. Sacculus.   *vag.* The muscular vagina which evaginates during copulation; a pair of retractor muscles are shown.   Other lettering as in fig. 20.

laid, and causes them to adhere to each other and to the material upon which they are deposited. Behind the accessory glands there is a pair of thin-walled transparent vesicles (*tasche dell' ovidutto* of Berlese), which I propose to name the *accessory copulatory vesicles* (*a.c.v.*) on account of the part they take in ensuring firm coitus with the male when copulating, during which process they expand to a much greater extent.

The ovipositor (fig. 22).

The terminal abdominal segments of the female are much reduced in size to form a tubular ovipositor, the chitinous sclerites being reduced to form slender chitinous rods. When extended it equals the abdomen in length. It is composed of segments vi, vii, viii, and ix, each being separated from the adjacent segments by an extensible intersegmental membrane, which is covered with fine spines. When the ovipositor is retracted (fig. 20, *ovp.*) it lies

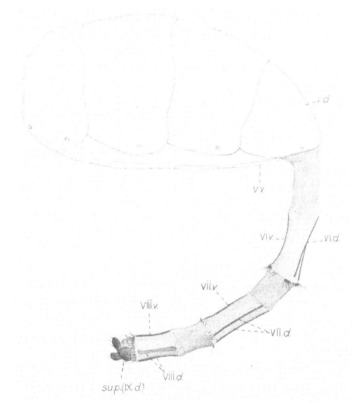

Fig. 22. Abdomen of female showing the extended ovipositor.
v, *d.* to ix, *d.* Fifth to ninth dorsal arches or plates of the abdomen.   v, *v.* to viii, *v.* Fifth to eighth ventral plates or arches.   *su.p.* The suranal plate (ninth dorsal arch).
The anus is situated between the two lateral terminal tubercles.

in the interior of the posterior end of the abdomen, the segments being telescoped the one within the other, so that only the terminal tubercles are visible from the exterior. The dorsal arch of the sixth abdominal segment is reduced to a **Λ**-shaped sclerite (vi, *d*) lying on the dorsal side of the segment. The ventral arch of this

4—2

segment is reduced to a slender chitinous rod (vi, *v.*) in the mid-ventral line. The dorsal arch of the seventh segment is represented by two slightly-curved sclerites (vii, *d.*) with their concave faces opposite ; the ventral arch (vii, *v.*) is similar to that of the sixth segment. At the junction of the posterior ends of the sixth and seventh segments with the inter-segmental membranes succeeding them there are several setose tubercles arranged more or less in pairs, but they vary in development in different individuals. The dorsal arch of the eighth segment consists of two parallel and slender sclerites (viii, *d.*), not so narrow as those of the two preceding segments. A pair of slender sclerites (viii, *v.*) also represents the ventral arch. The terminal anal segment, which I consider represents the reduced ninth segment, has a dorsal chitinous sclerite, the sub-anal plate (*su.p.*) which is triangular in shape, and a ventral sub-anal plate of the same shape. The female genital aperture is situated at the anterior end of the latter plate, between the eighth and anal (ninth) segments. A pair of terminal setose tubercles is situated laterally at the apex of the anal segment.

The male reproductive organs.

The male reproductive organs (fig. 23) are situated ventral to the alimentary canal, and lie within the fifth abdominal segment. They consist of a pair of testes, vasa deferentia, ejaculatory duct and sac, and the terminal penis. There are no accessory genital glands in the male.

The *Testes* (*te.*) are a pair of brown pyriform bodies, with their long axes placed transversely, and their pointed ends facing. In young males they have a bright red appearance. They are covered with a follicular investment of cells, which varies in thickness apparently according to age. The thin brown chitinous capsules contain the developing spermatozoa. The pointed end of each testis is continued as a fine vas deferens (*v.d.*) which meets that of the other testis in the median line, where they open into the common ejaculatory duct (*d.e.*). This runs forwards for a short distance, and then bends to the left ventrally, and, after several convolutions on the left ventral side of the abdomen, the duct narrows considerably, forming a narrow ejaculatory duct. This crosses over the dorsal side of the rectum to the right side, where it runs forwards for a short distance and then curves back in the median ventral line, opening into a pyriform ejaculatory sac (*e.s.*). The walls of this ejaculatory sac are muscular, longitudinal muscles, giving the walls a striated appearance. It contains a phylliform,

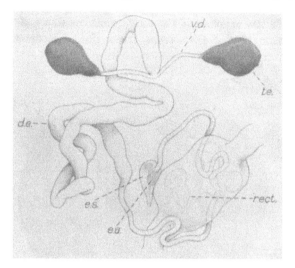

FIG. 23. The male reproductive organs. They have been slightly spread out, and the rectum (*rect.*) has been turned over to the right side.

d.e. Ejaculatory duct.　　　*e.a.* Ejaculatory apodeme.　　　*e.s.* Ejaculatory sac.
te. Testis.　　　*v.d.* Vas deferens.

chitinous sclerite—the ejaculatory apodeme (*e.a.*) which has a short handle at the broad end. This sclerite is no doubt of great assistance in propelling the seminal fluid along the ejaculatory duct during copulation. A short distance behind the ejaculatory sac the duct opens into the penis.

The male gonapophyses.

The extremity of the abdomen in the male (fig. 24) has undergone considerable modification in the formation of the external genitalia. The visible portion of the abdomen, as seen from above, consists of the first five abdominal segments; the remaining three segments are slightly withdrawn into the fifth segment, and on looking at the abdomen from the posterior end, only the terminal segment, the eighth, surrounding the anus, can be seen.   The sixth and seventh segments have been greatly reduced.   The sternal portion of the fifth segment consists of a cordiform sclerite (v, *v.*), the apex of which is directed forwards, and each of the lateral margins is produced to form a short process, swollen at the tip— these lateral processes form the primary forceps (*p.f.*), and lie at each side of the aperture of the male genital atrium (*g.a.*), of which the posterior edge of the sclerite forms the lower or anterior lip.

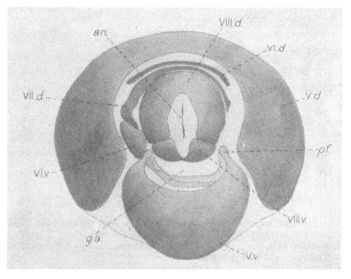

Fig. 24.   The posterior end of the abdomen of the male seen from behind, showing the pronounced sinistral asymmetry.

v, *d.* to viii, *d.* Fifth to eighth dorsal plates or arches.   v, *v.* to viii, *v.* Fifth to eighth ventral plates or arches.   *an.* Anus.   *g.a.* Aperture of genital atrium.   *p.f.* Primary forceps.

The dorsal plates of the sixth and seventh segments lie on the membrane which is tucked underneath the posterior edge of the fifth abdominal segment.   The dorsal plate of the sixth segment (vi, *d.*) is a narrow transverse sclerite; its lateral edges, which do not extend down the sides, are slightly produced anteriorly.   The ventral plate of the sixth segment (vi, *v.*) is asymmetrical, and, with the dorsal plate of the seventh segment, produces a pronounced asymmetry of the posterior end of the male abdomen. It consists of a spatulate plate on the left side, the anterior or ventral side of which is produced into a narrow bar extending across the ventral side of the aperture of the genital atrium, its distal extremity bifurcating. The dorsal plate of the seventh segment (vii, *d.*) is asymmetrical. It consists of a narrow sclerite, which, on the dorsal side, is similar to the sixth dorsal plate, but the left side (see fig. 25) extends down the side, and broadens out into a somewhat triangular-shaped area; the anterior edge of this is incised, and receives the seventh spiracle (vii, *a.sp.*); the ventral edge is internal to the spatulate portion of the sixth ventral plate. The ventral arch of the seventh sclerite has been completely withdrawn into the abdomen, and consists of a pair of curved sclerites

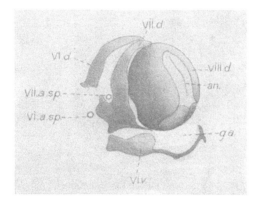

Fig. 25. Lateral view of the terminal segments of the abdomen of the male after their removal from the fifth segment.

vi, *a.sp.* and vii, *a.sp.* Sixth and seventh abdominal spiracles. Lettering as in fig. 24.

(fig. 26, vii, *v.*) somewhat rhomboidal in shape, lying dorsal to the fifth ventral arch and ventral to the penis (*P.*); they form the secondary forceps. Their lateral edges, which are thickened, articulate with the alar processes of the body of the penis (*c.pe.*), and with the dorsal arch of the eighth abdominal segment (viii, *d.*). Their inner edges are curved, and almost meet in the mid-ventral line. The dorsal arch of the eighth and last abdominal segment (viii, *d.*) forms the apex of the abdomen. It consists of a strongly convex sclerite, deeply incised on the ventral side; in this incision the vertical slit-like anus (fig. 25, *an.*) lies. The ventral portion of the segment is completed by a pair of convex sclerites (fig. 26, viii, *v.*)

which are united in the mid-ventral line, forming the ventral border of the anal membrane and the dorsal side of the entrance to the genital atrium.

All the sclerites of the posterior segments except the sixth and seventh are setose.

Berlese (1902) in his account of the copulation of the house-fly describes the genitalia. From his account of the male genitalia he appears to have missed the narrow dorsal arch of the sixth segment, or what is very probable, he may have mistaken it for the fifth dorsal arch, as he terms the seventh dorsal arch the sixth, and

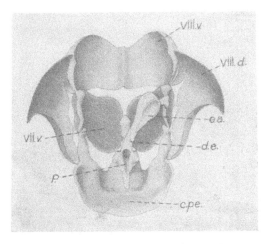

Fig. 26. Dorsal view of the penis and the ventral half of the terminal abdominal segments. The median portion of the eighth dorsal arch has been removed, leaving the lateral portions attached to the body of the penis (*c.pe.*) and the ventral arch of the seventh segment (vii, *v*.). Lettering as in fig. 24.

describes what I have called the ventral arch of the seventh as the dorsal arch of that segment. This mistake in nomenclature has probably arisen from the fact that he considered the visible portion of the abdomen as consisting of four segments instead of five, in which case the narrow dorsal arch of the sixth segment would naturally be taken for that of the fifth[1].

The *Penis* (figs. 26, 27) lies internally on the ventral side of

[1] Berlese describes a sinistral asymmetry of the posterior segments, but his figures show a dextral asymmetry, a mistake probably in the reproduction of his figures which has escaped the author's notice.

the abdomen, dorsal to the ventral arches of the fifth and seventh segments. It is composed of several sclerites. A median sclerite (*c.pe.*), the anterior and ventral edge of which is roughly semi-circular in outline, forms the body of the penis. This is produced laterally to form two alar processes ; at the bases of these processes the lateral extremities of the dorsal arch of the eighth segment articulate with the body of the penis ; the extremities of the pro-cesses are attached to the lateral extremities of the ventral sclerites of the seventh segment, the secondary forceps. The penis proper consists of a hollow cylindrical tube, the theca, which receives the

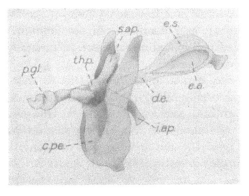

Fig. 27. Penis seen from the right side after it has been removed from within the terminal abdominal segments.
i.ap. Inferior apophysis.   *th.p.* Theca of penis.   *p.gl.* Glans.   *s.ap.* Superior apophysis.   Other lettering as in fig. 26, etc.

ejaculatory duct. The theca articulates with the body of the penis by means of a pair of small chitinous nodules ("*cornetti*" of Berlese); posterior to the attachment the theca is constricted slightly. Below the aperture of the entrance of the ejaculatory duct, the theca is produced into a ventrally directed curved process, the inferior apophysis (*i.ap.*); above the aperture a short cylindrical process, the superior apophysis (*s.ap.*) arises. The anterior end of the theca is continued as a slightly inflated hyaline structure, the glans (*p.gl.*), at the curved extremity of which the ejaculatory duct opens.

CHAPTER IV

THE INTERNAL STRUCTURE OF THE HEAD AND PROBOSCIS
OF *MUSCA DOMESTICA*

THE exo-skeleton and tracheal system of the head and proboscis have already been described. An account will now be given
of the internal structure and musculature of the head and pharynx
and also of the oral lobes.

The posterior region of the head (fig. 3) not occupied by
tracheal sacs is usually filled up with small multinucleate fat-cells
(*f.c.*), which are also occasionally found on the proboscis. The
frontal sac or *ptilinum* (*Pt.*) fills up the anterior portion of the
head not occupied by the air-sacs. Its crescentic opening, the
lunule, has already been described. It is attached to the wall of
the cephalic capsule by muscles which vary considerably in the
extent of their development. In recently emerged flies the muscle-
supply of the ptilinum is considerable, as they have served to
retract the sac after it has been inflated to assist the exclusion of
the imago, but in older specimens it becomes less. The walls
of the ptilinum are muscular and lined by a chitinous intima
covered with small broad spines.

THE MUSCULATURE OF THE PROBOSCIS.

The chief muscles controlling the movements of the pharynx
and proboscis are as follows:

The *Dilators of the Pharynx* (Figs. 3, 28, *d.ph.*). This pair of
muscles occupies the interior of the fulcrum. Each muscle is
attached to the antero-lateral regions of the fulcrum and inserted
into the dorsal plate of the pharynx (*r.p.*). These muscles are the
chief agents in pumping the liquid food into the oesophagus, and
in drawing it up through the pharyngeal tube.

The *Retractors of the Fulcrum* (*r.f.*). These muscles are attached to the internal anterior edges of the genae, and are inserted into the posterior cornu (*p.c.*) of the fulcrum. Their contraction causes the rotation of the fulcrum on the epistome as a hinge in the retraction of the proboscis.

The *Retractors of the Haustellum* (*r.h.*). These muscles have their origin on the dorso-lateral regions of the occiput. They are long and narrow, and running on each side of the common salivary duct are inserted into the dorsal margin of the theca.

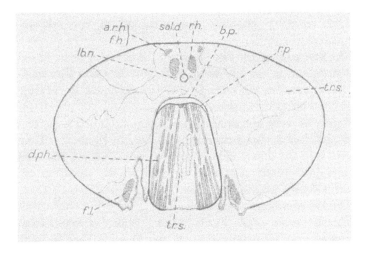

FIG. 28.   Transverse section through the lower portion of the head-capsule, showing the muscles and tracheal sacs in this region and the fulcrum in section. (Camera lucida drawing.)

b.p. Floor of pharynx.   *r.p.* Roof of pharynx.   *tr.s.* Tracheal sac.   Other lettering as in fig. 3.

The *Retractors of the Rostrum* (*r.r.*). This pair of muscles has its origin at the sides of the occipital foramen, and is inserted into the posterior side of the membranous rostrum about half way down its length. In the retraction of the proboscis these muscles draw in the rostrum.

The last two pairs of muscles acting together assist in the retraction of the whole proboscis.

The *Flexors of the Haustellum* (*f.h.*) have their origin close to that of the retractors of the rostrum at the sides of the occipital foramen. They are inserted into the base of the labral apodeme (*ap.*) and serve to flex the haustellum on to the anterior face of the rostrum.

The *Extensors of the Haustellum* (*ex.h.*). Each of these muscles arises from the distal cornu of the fulcrum, and is inserted into the head of the labral apodeme.

The *Accessory Flexors of the Haustellum* (*a.f.h.*) are attached to the lower (distal) anterior margin of the fulcrum, and inserted with the extensors into the head of the labral apodeme.

The *Flexors of the Labrum-epipharynx* (*f.l.*). These muscles have their origin on the anterior and upper edge of the fulcrum, and are inserted into the proximal end of the labrum-epipharynx.

The first pair of the last three sets of muscles serve to extend the haustellum in the extension of the proboscis, and the remaining two pairs assist in the retraction of the proboscis by flexing the haustellum on to the rostrum.

A pair of very fine muscles (*s.m.*) have their origin at the base of and internal to the posterior cornua of the fulcrum. They are inserted into the dorsal side of a small valve (*s.v.*) on the common salivary duct which regulates the flow of the secretion of the lingual salivary glands.

The muscles of the haustellum are:

The *Retractors of the Furca* (*r.fu.*). A pair of muscles having their origin on the upper part of the theca. Each is inserted along the upper proximal half of the lateral process of the furca. When the muscles contract the lateral processes of the furca, which, in a state of repose, are brought together by the elasticity of the ventral cornua of the theca, are diverged, and thus cause the divergence and opening of the oral lobes.

The *Retractors of the Discal Sclerites* (*r.d.s.*). These muscles have their origin on the lateral edges of the upper part of the theca, and are inserted upon the sides of the discal sclerites. They work together with the retractors of the furca, their contraction causing the divergence of the discal sclerites, and the consequent opening of the oral pit.

The *Dilators of the Labium-hypopharynx* (*di.l.*). These fan-shaped muscles arise in the middle region of the theca on either side the median line, and diverging are inserted in the lateral edges of the labium-hypopharyngeal sclerite. By their contraction they will widen the channel of the labium-hypopharynx.

The *Dilators of the Labrum-epipharynx* (fig. 29, *di.l.*). These form a series of short muscles attached to the anterior and posterior walls of the labrum-epipharynx. The size of the pharyngeal channel will be regulated by these muscles.

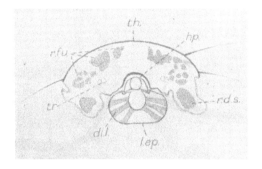

FIG. 29. Transverse section through the lower half of the haustellum, where the hypopharynx (*hp.*) has become free from the labium. (Camera lucida drawing.)

di.l. Dilator muscles of the labrum-epipharynx.   *tr.* Trachea.   Other lettering as in fig. 3.

THE ORAL LOBES.

The external structure of the oral lobes has already been described. Their internal structure and histology will now be given.

The setigerous cuticle and the pseudo-tracheae lie on a hypodermis of cubical cells (fig. 30, *hy.*). Beneath the hypodermis of the aboral surface is another layer of cells containing a large amount of dark pigment. Each of the large marginal sensory bristles (*g.s.*) of the aboral surface has a fine channel running down the whole length of the seta. This channel communicates with the cavity of a pyriform mass of nerve-end cells (*s.p.*), consisting of five or six cells. These masses of cells occupy a large part of the interior of the oral lobes. As these gustatory bristles are exposed and directed ventrally when the proboscis is retracted, they may assist the fly in testing the nature of its food before extending its proboscis. On the oral side of the oral lobes the nipple-like gustatory papillae (figs. 3, 30, *g.p.*) have already been described. The aperture at the end of the papilla leads into a fine duct, which ends in a pyriform sensory bulb (*s.g.p.*). The tracheae (*tr.*) can be seen running through the cells, some of which contain several nuclei, and from their appearance are probably derived from the fat-body. No tracheal sacs could be found either in the oral lobes

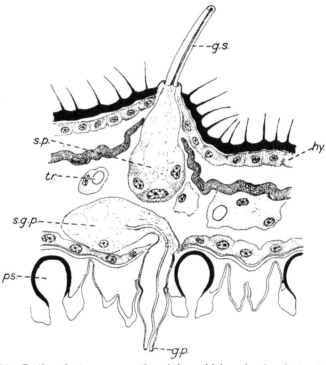

Fig. 30.   Portion of a transverse section of the oral lobes, showing the two types of gustatory sense organ, etc.

g.s. Gustatory seta.   *g.p.* Gustatory papilla.   *hy.* Hypodermis under which lies a pigmented layer.   *p.s.* Pseudo-trachea in section.   *s.g.p.* Sensory bulb of gustatory papilla.   *s.p.* Sensory bulb of gustatory seta.   *tr.* Trachea.

or at their bases, but the annular tracheae are continuous with those of the proboscis. The haemocoel of the oral lobes is well developed. This supports the view set forth by Kraepelin, and with which I agree, that the inflation of the oral lobes is due to the blood. I consider that the extension of the proboscis is due to the inflation of the tracheal sacs of the head. The proboscis having been protruded the oral lobes are then diverged by the contraction of the retractor muscles of the furca and discal sclerites, and distended by the inrush of blood which keeps them turgid.

The *Labial Salivary Glands* (figs. 3, 31, *lb.sl.*). These salivary glands lie in the haustellum at the base of the oral lobes. The glands, which are spherical in shape, are composed of a large number of gland cells somewhat triangular in shape. Each gland cell

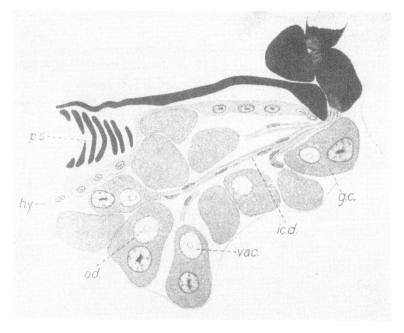

Fig. 31. Transverse section of labial salivary gland, to show the structure of the gland cells (*g.c.*). (Camera lucida drawing.)

hy. Hypodermis.   *ic.d.* Intracellular duct.   *p.s.* Pseudo-trachea.   *od.* Opening of intracellular duct into the permanent vacuole (*vac.*) of the gland cell.

is 40μ in size, and possesses a large nucleus (12μ), and internal to this a permanent circular vacuole (*vac.*) which is 16μ in size, and is lined by a thin chitinous intima. The duct of each gland cell opens into the side of the vacuole (*od.*). The ducts (*ic.d.*) are intracellular, and run from the centre of the gland, some of them uniting, to form a number of fine ducts on the ventral sides of the discal sclerites, which unite and open into the oral pits by a median pair of pores.   Kraepelin, in his description of the proboscis of the blow-fly, described the labial glands and their ducts (but not their histology) of that insect, his descriptions being similar to the condition I find in *M. domestica*.   Lowne, however, states that in the blow-fly he traced the ducts of the gland cells through the oral lobes to the apertures of the gustatory papillae, which he regarded therefore as the apertures of the labial salivary glands.   Graham-Smith (1911) figures what he calls the "salivary gland of the oral disc" but does not refer to its structure or relations in his paper on the proboscis of the blow-fly.

The appearance of the labial salivary glands of *M. domestica* calls to mind the maxillary glands of the ant *Myrmica levinodis* which Janet has described and figured.   The secretion of the labial salivary glands serves, I believe, to keep the surface of the oral lobes moist.

CHAPTER V

THE HABITS AND BIONOMICS OF THE HOUSE-FLY

LOCAL DISTRIBUTION AND COMPARATIVE ABUNDANCE.

THE local distribution of flies is almost entirely governed by two factors: the presence of breeding places and of food, the former being undoubtedly the most important. This fact is supported not only by my own observations extending over a number of years but by those of such observers as Niven and Hamer who, in particular, have studied the question of the factors governing the local distribution of flies.

Compared with the other species of flies which inhabit or occasionally visit houses, the house-fly, *Musca domestica*, is by far the most abundant. Nevertheless, even in houses a slight variation in the relative abundance of *M. domestica* and *Fannia canicularis* may be found, for whereas the former will be more numerous in warm places such as the kitchen and dining room where food is present, a large proportion of *Fannia canicularis* will occasionally be found in the other rooms of houses. In country houses the proportions sometimes vary by the intrusion of *Stomoxys calcitrans*[1]. In 1905 in a certain northern country cottage, out of several hundred flies captured, *S. calcitrans* constituted about 50 per cent. of the total, the rest being chiefly *F. canicularis* together with a few *Anthomyia radicum*, whose larvae breed on horse manure with those of *M. domestica*.

[1] Jennings and King (1913), in discussing the occurrence of *S. calcitrans* in dwellings in South Carolina, U.S.A., state: "Strong preference was shown for the living rooms and in more than half of the houses studied these were the only rooms infested."

The following records taken from a "fly census" that was made in 1907 may be taken as illustrative of the proportional abundance of the different species in different situations; although the numbers of these records are small the proportions are more obvious.

Place	M. do-mestica	F. cani-cularis	Other species
Restaurant, Manchester	1869	14	2 (M. stabulans, C. erythrocephala)
Kitchen, detached suburban house (six records), Lancashire	581	265	14
Kitchen, detached suburban house in Manchester	682	7	14
Stable, suburban house	22	153	14 (12 S. calcitrans)
Bedroom, suburban house	1	33	4 (M. stabulans)

Out of a total of 3856 flies caught in different situations, such as restaurants, kitchens, stables, bedrooms and hotels, 87·5 per cent. were *M. domestica*, 11·5 per cent. *F. canicularis*, and the rest were other species such as *S. calcitrans, Muscina stabulans, C. erythrocephala*, and *Anthomyia radicum*. These figures are comparatively small, but are representative of the average occurrence, as I have observed, of the different species.

In a collection of flies caught in rooms where food supplies were exposed in different cities of the United States, Howard (1900) found that out of a total of 23,087 flies, 22,808 or 98·8 per cent. were *Musca domestica* and of the remaining 1·2 per cent. *F. canicularis* was the commonest species. Hamer (1908) found that more than nine-tenths of the flies caught in the kitchens and "living rooms" of houses in the neighbourhood of depots for horse-refuse, manure, etc., were *M. domestica*. In a further report Hamer gives more details as to the different species that were found. In one lot of 35,000 flies caught on four fly papers exposed in similar positions, 17 per cent. were *F. canicularis*, less than 1 per cent. were *C. erythrocephala* and considerably less than 1 per cent. were *Muscina stabulans*; whereas of nearly 6000 flies caught in another situation in four fly balloons 24 per cent. were *F. canicularis*, 15 per cent. were *C. erythrocephala* and nearly 2 per cent. were *M. stabulans*. In his report for 1909 he gives an excellent diagram illustrating the seasonal prevalence of the six principal genera of flies caught in houses.

Niven found that out of 8553 flies caught in six different

localities in Manchester 8196 were *M. domestica,* 293 were *F. canicularis* and 64 were other species.

In Birmingham Robertson (1909) found the different species of flies in the following proportions in a collection of 24,562 flies: *M. domestica* 22,360, 91 per cent.; *F. canicularis* 1154, 47 per cent.; the blow-fly *C. erythrocephala* 840, 3·4 per cent.; and other species, 218 or 9 per cent.

From the foregoing figures it may be taken that as a general rule *Musca domestica* constitutes more than 90 per cent. of the total fly population of a house. The proportions of the other species vary, the variation depending largely on the situation of the house, whether it is urban, suburban or rural and also upon the sanitary conditions prevailing in the neighbourhood.

Seasonal Prevalence of Flies.

As a general rule house-flies are most abundant during the hottest months of the year. In Europe and North America, north of Mexico, they are most numerous during the months of July, August and September. In South Africa they are abundant from October to February. The fly season in Australia extends from October to March. In more tropical countries they are prevalent during a more extended period.

Flight or Distance travelled by House-Flies.

The distance that house-flies are able to travel either by their own exertions or by the aid of the wind is manifestly an important question in view of its connection with the spread of infection by these insects and the location of their breeding places. We are not concerned here with the ability of flies to travel in electric street cars, trains or steamboats; the fact that they are able to be transported by these means is a matter of common observation. The question I am about to discuss is: how far do flies travel under natural conditions?

Normally they do not fly great distances. I have previously compared them to domestic pigeons which hover about a house and the immediate neighbourhood. On sunny days they may be found in large numbers out-of-doors but when it becomes cloudy, or should it rain, they retire indoors.

In August 1906, Dr M. B. Arnold (see Niven, 1907) carried out some exact experiments at the Monsall Fever Hospital, Manchester, on the distance travelled by flies. Three hundred flies were captured alive and marked with a spot of white enamel on the back of the thorax. These marked flies were then liberated in fine weather. Out of the 300 five were recovered in fly traps at distances varying from 30 to 190 yards from the place of liberation and all the recoveries were within five days. The maximum distance of 190 yards was determined by the limitations of the hospital grounds and did not indicate the possible limit of flight.

In the summer of 1907 when visiting the Channel Islands I found *M. domestica* from $1\frac{1}{2}$ to 2 miles from any house or any likely breeding place so far as I was able to discover. The fact that the house-fly is able to fly at a considerable height above ground is indicated by the fact that I have frequently found them flying at an altitude of 80 feet above the ground. Flight at so great a height conjoined with a steady wind would enable them to cover a considerable distance.

Howard (1911) records an experiment of J. S. Hine who caught 350 flies and marked them with gold enamel before liberation. Flies so marked were observed about dwellings from 20 to 40 rods (600 to 1200 yards) from the point of liberation up to the third day. Hine states: " It appears most likely that the distance flies may travel to reach dwellings is controlled by circumstances. Almost any reasonable distance may be covered by a fly under compulsion to reach food or shelter. When these are close at hand the insect is not compelled to go far and, consequently, does not do so." The experiments which I am about to mention, however, show that flies will travel a considerable distance even where houses occur. The same author also states that Prof. S. A. Forbes had experiments carried out in which it was shown that marked flies spread naturally for at least a quarter of a mile.

An interesting and valuable series of experiments on the range of flight of flies under rural conditions was carried out by Copeman, Howlett and Merriman (1911), an opportunity being afforded by an unusual plague of house-flies in the neighbourhood of a small village, Postwick in Norfolk, where the observations were made. The work was much hampered by the meteorological conditions, which were for the most part unfavourable, the temperature remaining low while rain showers were frequent and high winds often prevailed. The flies were caught in a net and were marked by being placed in a paper bag containing finely powdered coloured chalk of which that of a yellow colour was found to give the best results. Describing the results of the experiments the authors state : " Nevertheless, on each occasion on which several hundred marked flies were liberated, a certain number were subsequently recovered, within forty-eight hours or less, from human habitations in Postwick, at points of the compass which were apparently dependent, for the most part, on the direction of the prevailing wind, and at distances ranging from 300 yards to 1700 yards from the refuse deposit. On one particular occasion on which chalk of a bright canary yellow colour was employed for marking the flies, the day being fine and sunny with a gentle north-west breeze, the results were specially interesting. Of these flies several were observed and captured along the stretch of river bank between a point opposite the refuse heap and Postwick Hall, within half an hour of their liberation. Two other marked flies were caught in a large open-fronted house on the lawn of Postwick Hall, a distance of 800 yards in a direct line from the point at which they were set free—one thirty-five minutes and another forty-five minutes after being liberated. Within the next four days, more than forty of these yellow-coloured flies were caught, on hanging fly papers, in the kitchen and outbuildings at Postwick Hall, while isolated specimens were also trapped in various parts of the village, at greater distances and at different points of the compass, from the refuse deposit of the marked flies caught on " tanglefoot " papers and a certain proportion probably were overlooked, owing to the fact that the colour of the chalk soon gets considerably obscured by the glutinous material becoming spread over the bodies of the flies in their struggles to free themselves."

In view of the results of the foregoing experiments on the range of flight of house-flies under rural conditions, it seemed to me very desirable that experiments should be carried out under city conditions, where so many factors are present which may affect

TABLE GIVING DETAILS OF THREE EXPERIMENTS ON THE RANGE OF FLIGHT OF FLIES AT POSTWICK

(from Copeman, Howlett and Merriman)

Date	Wind	Weather	Time of liberation	Marking colours	Approximate number of flies liberated	Date of recovery of flies	No. of flies recovered	Where observed	Direction of flight	Distance travelled in yards
Aug. 20	S. to S.W. moderate	Dull	12 to 2 p.m.	Blue	2000	Aug. 21	1	Observed over river (S.M.C.)	S.	400
			3 to 4 p.m.	Blue & Orange	500	,, 22	1	,, 4 p.m.	E.	1000
					500	,, 21	1	,, in church, 11.30 a.m.	E.N.E.	1085
						,, 21	1	Caught in Rectory Cottage, 5 p.m.	E.	980
,, 21	S.W. to W. strong	Bright	4 p.m.	Orange	1500	,, 22	1	Caught in Rectory Cottage on fly-paper, 1 p.m.	E.	980
						,, 22	1	Caught in Summer house, Postwick Hall, 3 p.m.	S.S.E.	800
						,, 23	2	Caught at Postwick Hall, 12 noon	S.S.E.	850
						,, 23	1	,, Bungalow adjoining Heath Farm	N.N.E.	1408
,, 22	N.W. moderate Breeze	Bright	3 p.m.	Yellow	2000	,, 22	4	On river bank opposite refuse deposit	S.E.	300/700
						,, 22	2	Caught in Summer house, Postwick Hall, 3.35 p.m. and 3.45 p.m.	S.S.E.	800
						,, 23	10	Caught at Postwick Hall	S.S.E.	850
						,, 23	1	,, Appleton's Cottage	E.	1050
						,, 23	1	,, The Lodge	E.N.E.	1027
						,, 23	1	,, Bungalow, Heath Farm	N.E.	1173
						,, 24	2	,, Post Office	E.	997
						,, 24	1	,, Lodge Farm	E.N.E.	1027
						,, 24	6	,, Postwick Hall	E.S.E.	850
						,, 24	1	,, Rectory Cottage	E.	980
						,, 25	1	,, Heath Farm	N.N.E.	1408
						,, 25 and 26	24	,, Postwick Hall (house and out-buildings)	S.S.E.	850

the flies' ability and desire to travel. The range of flight of flies in cities has a direct bearing on the possibilities of the carriage of infection and location of breeding places and nuisances. Accordingly in the summer of 1911 I arranged a series of experiments in the city of Ottawa which were carried out, under my direction, by Mr G. E. Sanders, a field officer of the Division of Entomology, to whom belongs the credit of devising the excellent method of marking the flies.

The point of liberation of the flies was on Porter's Island, a small island about 1000 feet long, lying in the Rideau River, which runs through a part of the city and is a tributary of the Ottawa River which it joins a short distance further along its course. The surrounding district forms a portion of the north-eastern part of the city and consists chiefly, especially on the northern side of the river which is known as New Edinburgh, of working-class dwelling houses. On this island which is connected with the bank by a small bridge, small-pox cases were isolated in a small wooden house used as a hospital, or in tents. The land rises gently from the river on the southern side.

The flies used in the experiments were obtained and marked in the following manner: Stable refuse, in which the larvae were found in large numbers, was placed in breeding boxes, provided with a circular aperture at the top and balloon fly-traps were placed over these apertures. The flies, on emerging from the pupae, entered these cages and were thus obtained in a healthy and uninjured condition. They were marked by spraying them while in the wire cages with a solution of rosalic acid (rosaurin or methyl-aurin $C_{20} H_{16} O_3$) in 10 per cent. alcohol, applied by means of a fine spray. This method of marking insects was devised by Mr Sanders, who first used it in experiments with ants. It is simple, harmless and reliable as a means of detection. The presence of a marked fly on a sticky fly-paper is indicated by its producing a scarlet coloration when the paper is dipped into water made slightly alkaline. In these experiments the flies were reared and marked in the Division of Entomology, which is situated about three miles from Porter's Island, to which they were carried in the cages and liberated on arrival. "Tanglefoot" fly-papers were placed in as many as possible of the houses in the neighbouring district on both sides of the river. The papers were placed chiefly in the kitchens of houses and were collected one or two days after being distributed. They were usually collected in that portion of the district towards which the wind had been blowing from the direction of the island, as it was found that the wind was the chief factor in determining the direction of the distribution from day to day.

EXPERIMENTS.

Flies liberated on Porter's Island.

Date	Number of marked flies liberated	Number of marked flies recovered	Place of recovery	Distance in straight line from point of liberation
August 29, 1911	8000	—	——	
August 30, 1911	4000	1	647 St Patrick St.	180 yards
September 1, 1911	1000	1	35 Cobourg St.	
		1	106 Cobourg St.	600 yards
		2 marked flies observed	{37 Cobourg St. St Patrick St.	
September 3, 1911	100	1	38 Beechwood Av.	520 yards
September 6, 1911	500	—	——	
September 7, 1911	—	43	647 St Patrick St.	
		12	612 St Patrick St.	
		4	681 St Patrick St.	
		21	619 St Patrick St.	
		3	565 St Patrick St.	
		2	559 St Patrick St.	
		2	553 St Patrick St.	
		5	35 Cobourg St.	
		8	19 Cobourg St.	
		1	106 Cobourg St.	
		1	608 Rideau St.	700 yards
		1	55 Augusta St. (Butcher's shop)	
		3	4a Anglesea Sq. (Grocer's shop)	
		6	355 Mackay St.	
		12	337 Mackay St.	
		16	305 Mackay St. (Grocer's shop)	
		7	316 Crichton St.	
		16	197 Crichton St.	
		2	143 Crichton St.	
		2	134 Beechwood Av.	
		1	Stanley Av.	600 yards

As will be seen from the dates given, these experiments extend over a short time only, having been terminated by the advent of a period of cold weather which checked the flies' activity. Nevertheless, they are of value as indicating the possibilities in the way of the range of flight of flies under normal city conditions. There is no doubt that given the necessary conditions with regard to wind and elevation above the ground the range would be considerably greater than was actually found in these experiments. The

Map of portion of City of Ottawa showing range of flight of marked flies.
X Point where marked flies were liberated. Shaded portions **A, B, C,** etc. indicate breeding places of flies. Numbers in circles indicate number of marked flies recovered at the various points.

greatest range of flight obtained in these experiments, namely 700 yards, represents an actual flight of considerably greater distance than is represented by a straight line from the place of liberation to the point of capture.

The chief breeding places of house-flies on a large scale are shown on the accompanying map as shaded areas. The western extremity of Porter's Island was used as a garbage dump (C) until June 1911, this being about 170 yards from the isolation tents and hospital. Between the end of Water Street and Rideau River garbage was being dumped (A), a large proportion of which consisted of stable refuse (horse manure). At this place, which is about 530 yards from the hospital, flies were found breeding in considerable numbers. At the foot of St Andrews Street, adjoining the river and about 270 yards from the hospital, about 100 tons of horse manure and compost had been dumped (B). There were in addition numerous breeding places apart from an unusually large number of unprotected heaps of horse manure in stable yards. Consequently flies were extremely abundant on the island and throughout the district. The presence of countless numbers of flies in the district which were bred in these natural breeding places made the recovery of 172 marked flies out of a total of 13,500 which were liberated all the more remarkable.

The results of these experiments indicate in a significant manner the distances which, under city conditions, flies are able to travel from their breeding places or from a source of infection[1].

FEEDING HABITS AND THE INFLUENCE OF FOOD.

The fact that the fly ingests micro-organisms on account of its indiscriminate feeding habits and that bacteria remain in a viable condition in the alimentary canal for a greater length of time and in larger numbers than externally renders a consideration of the

---

[1] Since the above was written the results of further experiments on the range of flight of house-flies have been published.

Nuttall, Merriman and Hindle (1913) carried out a series of experiments in Cambridge. They liberated upwards of 25,000 flies marked with coloured chalk dust. The maximum flight in thickly-housed localities of the flies recovered was about a quarter of a mile; in one case a single fly was recovered at a distance of 770 yards, part of which distance was across fen-land. It is considered by the authors that the chief factors favouring the dispersal are fine weather and warm temperature. The nature of the locality is a considerable factor.

Zetek (in a paper presented at the meeting of the Ent. Soc. of America, Atlanta, Ga., Dec. 31st, 1913) describes experiments carried out in the Panama Canal Zone. He showed that marked flies had flown from the breeding place (cow manure) to dwelling places situated 2500 feet away and at a lower level of 150 feet.

Hodge (1913) in explanation of the discovery of flies 1¼, 5 and 6 miles respectively out from the shore of Lake Erie (U.S.A.) suggests that they were carried by the wind.

feeding habits of the fly essential to a thorough comprehension of the possibilities of bacterial dissemination by flies. This problem will be fully discussed later when it will be shown that the transportation of bacterial and other organisms in the alimentary tract is of greater import than the transportation of such organisms externally, on the legs, proboscis, etc., of the fly.

Most people are acquainted in a general way with the feeding habits of the house-fly. Its visits to the dining table are a matter of common knowledge; the general ignorance in regard to its visits to repositories of filth and excrementous matter has been largely responsible for the indifference of the majority of people to the hygienic aspect of the house-fly.

The anatomy of the proboscis and of the alimentary tract, a knowledge of which is essential to understanding the feeding habits of the fly, have already been described. Although I made extended observations on the feeding habits of the house-fly from the beginning of my study of this insect, these observations were not recorded in my earlier papers and to Graham-Smith belongs the credit of first recording any extensive and accurate observations on this subject. In his papers (1910, 1911 a and 1911 b) he has given an excellent account of the general behaviour of flies during feeding and of the digestion of the food and his observations fully confirm those which I had made; he also makes a large number of new observations in connection with his bacteriological studies and his study of the structure and function of the oral sucker of the blow-fly is invaluable.

The proboscis of the house-fly is adapted to a sucking function and the absorption of liquid or liquefied food. Except under certain circumstances it does not and cannot take in solid particles of food. Nicoll (1911) in his experiments on feeding flies on the eggs of tapeworms found that flies were "apparently unable to ingest particles of larger size than ·045 mm." A careful examination of the structure of the oral lobes or sucker makes plain the reason for this. As I have shown in describing the structure of the oral lobes, and as Graham-Smith in his study of the oral sucker of the blow-fly also pointed out, communication between the surface of the oral lobes and the pseudo-tracheal channels is effected by way of the spaces between the bifid extremities of the chitinous rings which keep open the pseudo-tracheal channels. In the house-fly these interbifid spaces, as Graham-Smith aptly termed them, measure from ·004 to ·003 mm. in diameter, indicating how impossible it would be for particles of food to be absorbed with the food in the normal manner and pass along the pseudo-tracheal channels to the mouth. The only possible manner in which solid particles such as the eggs of tapeworms could gain access to the pharynx would be by direct entrance into the mouth. In experimenting with Indian ink I found that through the sucking action of the oral lobes the solid particles were heaped up in a slight ridge in the channel between the lobes and remained between the lobes while the fly continued feeding. Graham-Smith observed the same process in feeding the blow-fly with pollen grains. It is not difficult to understand how, when feeding on solid food which is being liquefied or on liquid food containing minute particles, such particles which were not too large might be sucked up into the oral pit and into the mouth. This undoubtedly happens in the case of the absorption of the tapeworm eggs. Ordinarily, however, the entrance of food into the oral pit is prevented by the pre-stomial teeth and the close apposition of the inner edges of the cushion-like oral lobes.

In order to feed on dry substances such as sugar, dried specks of milk or sputum, etc., the fly has first to liquefy the substance. This is accomplished by the secretion of what I termed the lingual salivary glands. This is poured down the labium-hypopharynx into the oral pit and on to the surface of the solid substance, both directly and through the pseudo-tracheal channels. This process can be observed by inverting a film of dried up sugar solution on a slide over a glass capsule containing a fly, the process being watched through a Zeiss binocular microscope. The solid matter is soon liquefied by the rapid sucker-like movements of the oral lobes moistened by the salivary secretion. Frequently the liquefaction is aided or brought about by means of the regurgitated food when the fly has fed previously. Graham-Smith found carmine stains in the proboscis marks of flies feeding on semi-fluid material as long as 22 hours after they had been fed on carmine coloured food, indicating that some of the previous meal had been apparently regurgitated, thereby assisting the salivary secretion.

In flies which have never previously fed or have not fed for some time, I believe the solid food is liquefied solely by the action of the salivary secretion; this would specially apply to flies which have recently emerged from the pupal case or from hibernation. In feeding on dried films of food material such as sugar, milk, etc., the fly generally sucks clean the surface to which the oral lobes are applied, leaving a heart-shaped area and the whole of a small area is rapidly cleaned up in this manner (see fig. 32). In some cases only the impressions of the pseudo-tracheal channels are left (fig. 33). In the case of fluids the resulting fluid is sucked up into the oral pit through the pseudo-tracheal channels and by the action of the powerful pharyngeal pump, the action of which may be seen when the feeding fly is observed from the front[1], it is sucked up into the oesophagus through which it runs into the crop. If the food is coloured with nigrosin or carmine the filling of the crop can be readily observed. The slightly concave, ventral surface of the abdomen of the hungry fly rapidly becomes convex

FIG. 32. Proboscis marks of *M. domestica* allowed to feed upon film of sweetened Indian ink. Enlarged.

in the anterior region where the crop lies. The filling of the crop takes place very rapidly in the case of liquid food; sometimes it is gorged in less than a minute, as Graham-Smith also observed. If feeding continues the food begins to flow directly into the ventriculus, and the abdomen, as a whole, becomes rather distended. Such gorges, however, frequently prove fatal, as I find that there was greater fatality among abnormally gorged flies than among those which fed, as one might say, rationally. As a rule flies do not appear to cease feeding until they are gorged. Their meals are frequently disturbed and here, I believe, we see the natural function of the crop as a food reservoir. Food can be taken into

---

[1] I have found that the action of the pharyngeal pump can be studied most advantageously in *Stomoxys* when the same is allowed to feed on the back of the hand and watched through the Zeiss binocular microscope.

the crop rapidly. The fly then rests in a quiet place   The absorbed food now begins to flow through the proventriculus, which acts as a pump and valve, into the ventriculus[1]. I have compared this to the feeding of the ruminating animal, as the crop of the fly calls to mind the storage stomach of the ruminant.

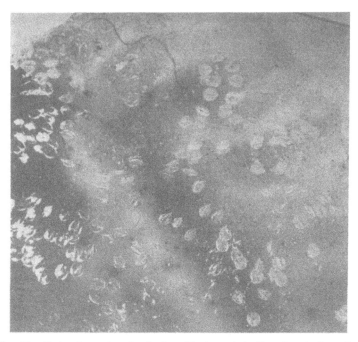

Fig. 33.   Proboscis marks of a fly (not *M. domestica*) allowed to feed on film of Indian ink (sweetened); showing the alternating light and dark lines, the former being where the film has been removed by the pseudo-tracheae. Greatly enlarged. (Photographed by H. T. Güssow.)

Graham-Smith found that the rate at which food passes into the intestine from the crop appears to vary, depending to some extent on the nature of the food and the temperature. If flies were kept in an incubator at 57° C. and fed on carmine gelatine much of the food may reach the rectal valve within one hour of feeding. He gives the following tables which indicate the rate

[1] Wheeler (1910) mentions the fact that in ants the proventriculus not only passes the liquid food back from the crop to the stomach but also fills the crop in the first place.

at which the food passes from the crop and the period during
which coloured food may remain in the crop.

TABLE SHOWING THE RATE AT WHICH FOOD PASSES THROUGH
THE CROP INTO THE INTESTINE.

(from Graham-Smith.)

Time after feeding	Number of flies dissected	Results
3 minutes	1	Crop full of red fluid, but none found in ventriculus or intestine
6 minutes	1	Crop full of red fluid, but none found in ventriculus or intestine
10 minutes	1	Crop full of red fluid and some fluid beginning to pass into ventriculus
15 minutes	1	Crop full of red fluid and upper third of intestine red
20 minutes	1	Crop full of red fluid and upper third of intestine red
2 hours	3	Crop full of red fluid and upper third of intestine red
	4	Crop full of red fluid and upper half of intestine red
	1	Crop full of red fluid and upper three-quarters of intestine red

TABLE SHOWING THE PERIOD DURING WHICH COLOURED
FOOD MAY REMAIN IN THE CROP.

(from Graham-Smith.)

Time after feeding	Number of flies dissected	Results
24 hours	2	Crop red and distended. Intestine red throughout
48 hours	3	Crop red and distended. Intestine red throughout
3 days	3	Crop red and distended. Intestine red throughout
4 days	2	Crop pink and distended. Intestine red throughout

In most cases on dissection the crop was found to be nearly
empty on the third day after feeding, although the intestine still
contained large quantities of red material. The above results indicate that the crop is not completely emptied for many hours and
in some cases for days, even though no further food is given. This
fact has been confirmed in my own observations. Flies which have

been allowed to feed naturally on coloured jam have been found with jam still in their crops on the second day after feeding.

After feeding the fly usually retires to a quiet spot. It invariably cleans its head and proboscis, as is also the case with *Stomoxys*. Very frequently it regurgitates its food from the crop in the form of large drops of fluid which may equal in diameter the depth from front to back of the head (fig. 34). This regurgitation of "vomit" drops conveys, as Graham-Smith, who has also observed and recorded the fact states, the impression that the flies "have distended their crops to an uncomfortable degree and that some of the food is regurgitated to relieve the distension." While this is conceivable, it is not unlikely that the regurgitation of the food may be primarily concerned in the digestion, as I have observed it to take place when there had been no unusual distension of the

Fig. 34.   *M. domestica* in the act of regurgitating food.   × 4½.

crop.   One can understand that the regurgitation would enable the food to become mixed with an additional amount of salivary fluid which would further facilitate the digestion on the absorption of the drop.   The drops are slowly extruded and then withdrawn. One fly which I had under observation alternately and regularly regurgitated and absorbed a drop of fluid eight times, each regurgitation and absorption lasting one and a half minutes.   In some cases these "vomit" drops are deposited upon the surface on which the fly is resting and they may be easily recognised as light-coloured opaque spots (see fig. 35).   Their colour will, of course, vary according to the nature of the flies' last meal.   Graham-Smith found that flies fed on coloured syrup often regurgitated coloured fluid 24 or more hours later, though fed in the interval on uncoloured syrup.   In his infection experiments he observed that "when infected food has been given, the infecting organisms are usually found in great numbers in these 'spots' and moreover, as will be shown later, fluid regurgitated from the crop is used to dissolve or moisten sugar and other similar dry food materials."

Fig. 35. Portion of window pane from fly-infested cow-shed, showing dark-coloured faecal spots and more numerous light-coloured vomit spots. Natural size.

The importance of this habit, from the point of view of bacterial contamination, cannot be emphasized too strongly.

The rate of digestion depends chiefly on the temperature and the nature of the food. At ordinary room temperatures the coloured faeces produced by feeding flies on coloured syrup are not deposited until several hours after the meal. Graham-Smith in his experiments found they were not deposited within two hours after feeding. He made a number of interesting observations on the rate of defaecation and the effect of different kinds of food on the same. The yellowish to dark brown faecal spots or fly "specks" are well known. A series of ten flies which had been given one feed of milk produced in a period of 22 hours an average of 12·5 vomit spots and 3·9 faecal spots. A series which were allowed to feed on milk whenever they wished produced an average of 17·9 vomit spots and 4·5 faecal spots in 22 hours. The average number of "spots" both faecal and vomit, produced by the two series in 22 hours was 16·4 and 20·4 respectively. In a further experiment the average number of spots, faecal and vomit, which were deposited per fly in 24 hours was 30·7. Three series of flies were fed on syrup, milk and sputum for several days. Those fed on syrup produced an average of 4·7 deposits (faecal and vomit) per fly per day, those fed on milk 8·3 and those fed on sputum 27·0. In the latter case Graham-Smith points out that the faeces were much more voluminous and liquid than usual and in fact the flies seemed to suffer from diarrhoea.

The rate of deposition and number of the faecal vomit spots deposited is highly significant in connection with the question of the bacterial contamination of food and especial attention is drawn to the bearing of the abundance and character of the spots deposited by flies feeding on sputum on account of its bearing on the dissemination of the tubercle bacillus.

Mention will be made later of the influence of food on the development of the larvae; the experiments which were carried out showed that the larvae develop more rapidly in certain kinds of food, such as horse manure, than in others. It has yet to be discovered what are the chemical constituents which favour the more rapid development. It was found that insufficient food in the larval state retarded development and produced flies which were subnormal in size. Bogdanow (1908), in an interesting experiment, fed *M. domestica* through ten generations on unaccustomed food such as meat and tenacetum in different proportions, and he found that the resulting flies did not show any change.

.The Influence of Temperature.

The influence of temperature on the development of the larva and on the life-history will shortly be discussed. Temperature also directly affects the adult insect; house-flies are most active at a high summer temperature and their metabolism is correspondingly active. Cold produces inactivity and torpor. They are able, however, to withstand a comparatively low temperature. Bachmetjew (1906) was able to submit *M. domestica* to as low a temperature as − 10° C., and vitality was retained, as they recovered when brought into ordinary room temperature. Donhoff (1872) performed a number of experiments previous to this with interesting results. He submitted *M. domestica* for five hours to a temperature of − 1·5° C., and they continued to move. Exposed for eight hours to a temperature of first − 3° C. and then − 2° C. they moved their legs. On being submitted for twelve hours to a temperature first of − 3·7° C. and then − 6·3° C. they appeared to be dead, but on being warmed they recovered. When exposed for three hours to a temperature of − 10° C. which was then raised to − 6° C., they died. These experiments show that *M. domestica* is able to withstand a comparatively low degree of temperature.

The Influence of Light and Colour Preference.

The adult house-fly seeks the light, that is, it is positively heliotropic, and as Felt showed, is unwilling to enter very dark places to deposit its eggs, although in placing the eggs it endeavours to place them as far away from the light as possible.

A series of experiments were carried out by Galli-Valerio (1910) with a view to ascertaining whether flies had preferences for certain colours. Flies were placed in a cubical glass-sided box measuring 35 cm. each way. On the sides of the box pieces of coloured paper of equal size were pasted. The flies resting on the different colours were then counted, the cage being turned in different positions to avoid error. The colour preferences which he found were as follows, the preference for clear light colours being strongly indicated :

Clear green	18	Azure	... ...	13
Rose	17	Clear red...	...	10
Clear yellow	14	Dark grey	...	9

White... ...	9	Pale rose	3
Dark red ...	8	Very clear green	2
Pale grey ...	5	Black	1
Dark yellow	5	Blue	1
Dark green...	5	Pale violet ...	1
Red	4	Dark brown ...	1
Orange ...	3	Lemon yellow ...	1
Clear brown	3		

While these results are interesting further experiments would be required before any general conclusions could be drawn[1].

HIBERNATION.

The disappearance of flies towards the end of October and during November is a well known fact and the question is frequently asked, what becomes of them? Observations on this question were made from the beginning of my study of this insect.

Three causes contribute to the disappearance of the flies, namely, retreat into hibernating quarters or into permanently heated places, natural death and death from the parasitic fungus *Empusa muscae*. The last cause of disappearance is fully considered later and it accounts for a large proportion of the summer-bred flies. The natural death of flies may be compared, I think, to the like phenomenon that occurs in the case of the hive-bee, *Apis mellifica*, where many of the workers die at the end of the season by reason of the fact that they are simply worn out, their function having been fulfilled. The flies which die naturally have probably lived for many weeks or months during the summer and autumn, and in the case of the females have deposited many batches of eggs; their life work, therefore, is complete. Those flies which hibernate are, I believe, the most recently emerged, and therefore the youngest and most vigorous. On dissection it is found that the abdomens of these hibernating individuals are packed with fat cells, the fat-body having developed enormously. The alimentary canal shrinks correspondingly and occupies a very small space; this is rendered possible by the fact that the fly does

---

[1] Hindle (1913) has more recently carried out a series of experiments in which strips of coloured cardboard were exposed to catch flies which settled on them. The experiments appear to demonstrate that the flies did not display any marked colour preference.

not take food during this period. In some females it was found that the ovaries were very well developed, while in others they were small, and mature spermatozoa were found in the males. Like most animals in hibernating, *M. domestica* becomes negatively heliotropic and creeps away into a dark place. In houses they have been found in various kinds of crevices such as occur between the woodwork and the walls. They have been found behind pictures, books and curtains. A favourite place for hibernation is between wall paper which is slightly loose and the wall. A certain number hibernate in stables, where, owing to the warmth, they do not become so inactive, and they emerge earlier at the latter end of spring. During the winter the hibernating flies are sustained by means of the contents of the fat-body, which is found to be extremely small in hibernating flies if dissected when they first emerge in May and June. The abdominal cavity is at first considerably decreased in size, but the fly begins to feed and soon the alimentary tract regains its normal size, and together, with the development of the reproductive organs, causes the abdomen to regain its normal appearance. The emergence from hibernation appears to be controlled by temperature, as one may frequently find odd flies emerging from their winter quarters on exceptionally warm days in the early months of the year. A few flies may occasionally be found active throughout the winter. I have found active flies frequently during the months of December to February in such warm places as kitchens, restaurants and stables during which time they are able to breed. Jepson (1909) caught flies in the bakehouse of one of the Cambridge colleges in February and used these flies in his breeding experiments. He records the reported occurrence of flies in the college sculleries throughout the winter months. These active and periodically active flies, together with the wholly dormant flies are the progenitors of the summer millions[1].

[1] Skinner (1913) asserts on the circumstantial evidence of a single recently emerged fly, found entering a window, unsupported by experimental evidence, that house-flies pass the winter in the pupal stage. So far, I have been unsuccessful in carrying pupae of *M. domestica* over the winter months. Skinner's fly was probably one bred in a warm place during the winter.

Copeman (1913) calls attention to the desirability of securing further evidence as to where and under what circumstances surviving flies pass the winter.

REGENERATION OF LOST PARTS.

If the wings or legs of *M. domestica* are broken off they do not appear to be able to regenerate the missing portions, as in the case of some insects, notably certain Orthoptera. Kammerer (1908), however, experimenting with *M. domestica* and *C. vomitoria*, has found that if the wing is extirpated from the recently pupated fly it is occasionally regenerated. The new wing is at first homogeneous, and contains no veins, but these appear subsequently.

LONGEVITY.

The difficulty of experimentally determining the length of the natural life of a house-fly can be appreciated only by those who have attempted it. Those who have not done so will hardly realise it. Few insects are more difficult to deal with under experimental conditions. One would imagine that these insects flying about everywhere as they do, could be easily kept in a large roomy cage if given the necessary food and water. In my experience, however, this has not been the case; I have never succeeded in keeping flies alive in captivity for a longer period than seven weeks. In the winter it is apparently more easy to keep them alive in captivity. Jepson (1909 *b*) kept alive for eleven and a half weeks flies which had been reared in confinement in February. Flies which had been caught in the kitchens during the same month, and were, therefore, probably flies of the previous autumn, were kept in captivity for ten weeks. Jepson states that in the summer he was unable to keep flies in captivity for more than three weeks. Griffith (1908) succeeded in keeping a male fly alive for sixteen weeks. The evidence which is available clearly indicates, I think, that the late autumn bred flies, if they escape death from *Empusa* and other causes, live through the winter to produce eggs in the following spring. I am inclined to regard the winter breeding of *M. domestica* as unnatural, as it is induced by conditions which are, strictly speaking, artificial. The summer flies are probably shorter lived owing to their extremely active lives which endure probably for about two months or so.

PART II

THE BREEDING HABITS; THE LIFE-HISTORY AND STRUCTURE OF THE LARVA

CHAPTER VI

THE BREEDING HABITS OF *MUSCA DOMESTICA*

THE meagre nature of the information concerning the life-history and the breeding habits of the house-fly which was available at the time my investigations were begun has been indicated already in the introductory chapter, in which the history of our knowledge of this insect was traced. Gleichen and Taschenberg in Europe, Packard and Howard in the United States, had been the chief contributors to our knowledge of the breeding habits.

Carl de Geer (1776) was one of the first to describe the breeding habits. He stated that the house-fly developed in warm and humid dung, but did not give the time occupied by the different developmental stages. He refers to the enormous quantities of flies occurring from July to August. His statement concerning their development is especially interesting, as he appears to be the first investigator who called attention to what I consider to be one of the most important factors in the development of the fly, namely, the process of fermentation occurring in the substance in which development is taking place. He says (p. 76): "Les larves de cette espèce vivent donc dans le fumier, mais uniquement dans celui qui est bien chaud et humide, ou pour mieux dire *qui se trouve en parfaite fermentation*" (the italics are mine).

Gleichen (*t.c.*) found that the eggs hatched from twelve to twenty-four hours after deposition. He reared the larvae in decaying grain where, no doubt, fermentation was taking place; also in small portions of meat, slices of melon, and in old broth. His observations are extremely interesting, and, excluding mistakes which were due to the lack of modern apparatus, his account is still a valuable contribution to our knowledge of the subject. Bouché (1834) describes the larvae as living in horse-manure and fowl-dung, especially when warm. He does not give the time occupied by the earlier developmental stages, but states that the pupal stage lasts from 8—14 days.

Packard (*l.c.*), working at Salem, Massachusetts, U.S.A., found that the larvae emerge from the eggs twenty-four hours after deposition; the times taken by the three larval stages—for he found that there were two larval ecdyses—were: first, about twenty-four hours; the second stage, he thought, was from twenty-four to thirty-six hours; and the third was probably three or four days; the entire larval life being from five to seven days. The pupal stage was from five to seven days, so that in August, when the experiments were carried on, the time from the deposition of the egg to the exclusion of the imago was ten to fourteen days.

Taschenberg (*t.c.*) incorporates the work of Gleichen and Bouché, and he does not appear to add materially to the facts already mentioned. He states that the female flies deposit their eggs in damp and rotting food-stuffs, bad meat, broth, slices of melon, dead animals, cesspools, and manure-heaps. He further says that they have also been observed laying their eggs in spittoons and open snuff-boxes. With reference to the last statement, I find that the larvae will feed on expectorated matter mixed with a solid substance, such as earth, if they are kept warm, though they cannot feed on salivary sections merely. It is interesting to note in connection with the statement as to flies depositing their eggs in snuff-boxes that Forbes, as recorded by Howard (1911), reared *M. domestica* in 1889 from larvae found in a box of snuff at Kensington, Ill., U.S.A.

Howard (1896—1906) first studied the breeding habits of the fly in 1895 in Washington, U.S.A., and he described them in

1896, and more fully subsequently. He found that they could be rarely induced to lay their eggs in anything but horse-manure and cow-dung, and that they preferred the former. The periods of development he found were as follows: from the deposition of the egg to the hatching of the larva about eight hours; the first larval stage one day; second larval stage one day; third larval stage—that is, from the second ecdysis to pupation—three days, and the flies emerge five days after the pupation of the larvae, thus making the whole period of development about ten days. The same author, in a valuable study of the insect fauna of human excrement (1900), describes experiments in which he was successful in rearing *M. domestica* from human excrement both in the form of loose faeces and in latrines.

My own studies were commenced in 1905, and a short pre-liminary account of some of the results were published in the following year (1906). The complete account of my investigations on the breeding habits and development of the fly was not published until 1908.

Newstead (1907) found that horse-manure, spent hops, and ashpits containing fermenting materials and old bedding, or straw and paper, paper mixed with human excreta or old rags, manure from rabbit hutches, all constituted permanent breeding places. He also found that the following served as temporary breeding places: collections of straw mixed with other vegetable matter and feathers lying in a fermenting condition in open spaces in poultry yards, accumulations of manure on wharves, bedding in poultry pens. He was unable apparently to confirm the observations made by Howard and myself as to the breeding of *M. domestica* in human excreta.

In my experiments, which were carried out under both arti-ficial and natural conditions, the larvae of *M. domestica* were successfully reared in, and the flies bred from, the following sub-stances: horse-manure, cow-dung, fowl-dung, human excrement, both as isolated faeces and in ashes containing or contaminated with excrement, obtained from ashpits attached to privy middens, and such as is sometimes tipped on to public tips. I found that horse-manure is preferred by the female flies for oviposition to all other substances, and that it is in this material that the great majority of larvae are reared in nature; manure-heaps in stable yards sometimes swarm with the larvae of *M. domestica*. It was also found that the larvae will feed on paper and textile fabrics, such as woollen, cotton garments, and sacking which are fouled with excremental products if they are kept moist and at a suitable temperature. They were also reared on decaying vegetables thrown away as kitchen refuse, and on such fruits as bananas, apricots, cherries, and peaches, which were mixed, when in a rotting condition, with earth to make a more solid mass. Although they can be reared in such food-stuffs as bread soaked in milk and boiled egg, when these are kept at a temperature of about 25° C., I was unable to rear them to maturity in cheese, although they fed on the substances for a few days and then gradually died; my failure may have been due to the nature of the cheese which was used, only one kind being tried. In addition to rearing the larvae on isolated human faeces, such as are frequently found in insanitary courtyards and similar places, they were found in privy middens, and also on a public tip among the warm ashes and clinker where the contents of some privy middens had also evidently been emptied; I bred the flies out from this material.

In Canada I have further found that *M. domestica* can be reared in germinating wheat, no doubt owing to the heat engendered by the fermentation which takes place.

Jepson (1909) reared *M. domestica* during the winter months on moist bread in which the process of fermentation had begun.

Nash (1909) records that in 1904 he found the spaces round moveable excreta boxes in privies swarming with fly larvae. He refers to an interesting observation of Austen's, communicated by the latter in 1908. Austen found the larvae of *M. domestica* in rubber which was suspended in a drying-room at a temperature of 100° F. They were apparently full-grown, and the circumstances indicated that they could not have been more than three days in developing from the egg stage, which indicated a rapid growth at this exceedingly high temperature. Nash records the breeding of the house-fly in stored house-refuse, and he reared them on bread, pear, potato, banana peelings, boiled rice and old paper.

Theobald has reported the breeding of flies on refuse tips. In India Surgeon-Major Smith has also found that horse-manure is the commonest breeding place of *M. domestica*, especially around the military camps. He has also reared it from cow-dung.

Orton (1910), in some investigations quoted by Howard (1911), found house-fly larvae in a mixture of horse and cow-manure underneath a farm barn, and it is interesting to note that the larvae and puparia were more abundant in that portion of the pile where the horse-manure was either pure or predominated. Pupae were also found in piles of pig-manure mixed with straw bedding exposed to the air and rain. An ounce of this material taken from a point a few inches below the surface showed 868 pupae. Flies were also found breeding in spent hops and brewery waste (malt); an ounce of the latter contained 1018 maggots. In an experiment one pound of material constituting these breeding places was taken and kept in screen-covered glass jars in the laboratory for ten days. The following was the result:

Stable-manure	... ...	no adult flies issued
Farm barn, horse end	... ...	77 „ „ „
„ mixed ...	... ...	19 „ „ „
„ cow end	... ...	1 „ fly „
Piggery manure pile	... ...	361 „ flies „
Spent hops ...	... ...	129 „ „ „
Barley malt	... ...	539 „ „ „

As Howard points out, these results were no doubt in the main correct, although the identification of the flies was by no means thorough, as Orton admits.

During the summers of 1908 and 1909 Prof. S. A. Forbes, State Entomologist of Illinois, U.S.A., had observations carried out in Urbana, Ill., by Mr A. A. Girault, and in Chicago by Mr J. J. Davis, on the breeding habits of the house-fly with a view to discovering what substances, other than the usual breeding place, horse-manure, served as breeding places. The results of these observations have been published by Howard (1911) and are as follows:

Date	Breeding material	No. of House-flies bred
Sept. 1—3	Rotten water-melon and musk-melon	14
Aug. 18 and } Sept. 8—11	Rotten carrots and cucumbers ...	23
Sept. 7	Rotten cabbage stump	1
Sept. 7	Banana peelings	1
Aug. 30	Rotten potato peelings	12
Sept. 25	Cooked peas	1
Oct. 1	Ashes mixed with vegetable waste ...	1
Sept. 7—14	Rotten bread or cake	8
Aug. 22	Kitchen slops and offal	193
Sept. 10—26	Mixed sawdust and rotting vegetables	41
Aug. 30—Sept. 4	Old garbage, city dump	15
Aug. 14 and 28	Rotten meat, slaughter-houses ...	40
Aug. 30—Sept. 11	Carrion in street	267
Sept. 7	Seepage from garbage pile	1
Aug. 17—20	Hog's hair, slaughter-house waste ...	9
Aug. 23—28	Sawdust sweepings, stock yards slaughter-house	110
Aug. 23	Sawdust sweepings, meat market ...	4
Aug. 16—28	Animal refuse, stock yards	39
Aug. 14	Contents of paunches of slaughtered cattle	168
Sept. 2—11	Rotten chicken feathers	258
Aug. 16	Chicken manure, stock-car dump ...	3
Aug. 31—Sept. 7	Cow-dung, stable, Urbana, Ill. ...	997
Sept. 7—10	Cow-dung, outdoor yard	22
Sept. 6	Cow-dung, pasture	1
Aug. 24—Sept. 16	Human excrement	196

The above observations are of great interest and importance as indicating the breeding habits of the flies when they have a choice of substances and under natural conditions, and are confirmatory of my own experiments under laboratory conditions where flies were confined in cages with the various substances with a view to ascertaining not only whether they would deposit their eggs on the substances in question, but also whether the larvae could feed on them and the effect of the different substances upon the rate of development, of which mention is made later. I also confirmed most of the laboratory results by breeding flies from material exposed or collected in the open. As my experiments were mainly carried out in England, many having been since confirmed in Canada, the observations of Forbes, recorded above, made in Illinois under different conditions as regards climate, etc., are of additional interest. They indicate that the breeding habits are practically the same whatever the geographical position or climatic conditions may be.

Milliken (1911) found the eggs and larvae of *M. domestica* in alfalfa or lucerne ensilage in Kansas, U.S.A. The fermentation taking place no doubt attracted the adult flies.

Pratt (1912) bred fifty specimens of *M. domestica* from cow-manure at Dallas, Texas, U.S.A. He states that "this is one of the most common species in stables. Fresh manure attracts it in great numbers."

Paine (1912) made a study of the species of flies breeding in garbage in the city of Boston, Mass., U.S.A. Larvae were collected from the contents of garbage-pails as they were emptied into the scavenger's waggon. In some cases they were so abundant that the interior of the receptacle appeared as a wriggling mass. The larvae were allowed to complete their development under laboratory conditions, being fed on moist bread ; they pupated in a small quantity of earth. The following table gives the results of these rearings :

Number and species of Muscids reared from city garbage, Boston, Massachusetts, U.S.A.

Number of Lot	Date of collection	Number of specimens					Percentage of *Musca domestica*
		Musca domestica	*Phormia regina*	*Lucilia sericata*	*Calliphora erythrocephala*	*Muscina stabulans*	
I	Sept. 5	80		40			66·6
II	,,	9		1			90
III	,,	1		7			12·5
IV	,,	8	5	60			10·9
V	,,	4		16		1	19
VI	Sept. 9			3			0
VII	,,	26		1			96·3
VIII	,,	7					100
IX	Aug. 10	1		100	27		·78
X	,,		139	110	2		0
Totals		136	145	338	29	1	
Per cent.		22·4	22·3	50·5	4·4	·1	22·4

It will be seen that, out of 649 flies of various species which were bred from larvae found in garbage, 136 or 22·4 per cent. were *Musca domestica.*

SUMMARY OF SUBSTANCES IN WHICH *MUSCA DOMESTICA* BREEDS.

All the foregoing observations on the breeding places and substances in which flies breed may be conveniently summarised as follows:

Excrementous

Horse-manure	Cow-manure
Human excrement	Pig-manure
Fowl excrement	Rabbit-manure

Vegetable

Spent hops	Barley malt
Decaying grain	Excreta-soiled straw
Cooked peas	Bread or cake
Rotten water melon	„ and milk
„ musk or other melons	Rotten apricots
„ cucumber	„ bananas
„ carrots	„ cherries
„ cabbage	„ plums
„ potato and peelings	„ peaches
Boiled rice	

Animal matter

Rotten meat	Rotten fowl feathers
Carrion	Old broth
Cattle paunch contents	Boiled egg

Miscellaneous

Kitchen refuse	Snuff
Fermenting substances in ashpits	Expectoration with earth
Sawdust and excrementous refuse	Excreta-soiled paper, rags
Garbage pile drainage	Ensilage
Cesspool	Rubber

All observations which have been made indicate that the chief breeding place of the house-fly is horse-manure or stable-refuse, and that, in addition, it is able to breed in other forms of excrement and in rotting or decaying animal and vegetable substances, especially when they are in a fermenting condition.

LOCATION OF BREEDING PLACE.

The female flies, as a rule, prefer to deposit their eggs on substances or heaps of breeding material exposed to the light. General observations on the occurrence of the larvae have shown that fewer larvae are found in proportion as the situation of the substance in which they are living is further away from the light. Felt (1910) has demonstrated experimentally that the absence of light has a very marked influence on the ovipositing impulse of the fly; the darker the situation was, the fewer were the number of flies which were to be found there. The fact that flies are disinclined to penetrate a dark place to oviposit has some practical significance in their control.

BREEDING SEASON.

All my experiments and observations on the breeding habits of the house-fly indicated the important fact that, if suitable larval food were present and the temperature of the surrounding air were sufficiently high to permit the prolonged activity of the flies, the female house-flies will deposit their eggs and the larvae will develop at any time of the year. Under ordinary circumstances, however, the condition regarding temperature is not satisfied in temperate climates during the whole year round. Consequently, we find that in the temperate region the breeding season is confined to the period June to October; exceptional circumstances may extend this period a few weeks earlier or later. During these months, as the breeding and activity of the house-fly is directly dependent upon the temperature, the greatest breeding activity takes place during the hottest months, July, August and September.

During the winter months, under natural conditions, the outside temperature will not permit breeding to take place, but under such conditions as are found, for example, in warm stables and cowsheds, restaurants and kitchens, female flies may be often found either ovipositing or able to do so. On dissecting such flies I have found mature ova in the ovaries and living spermatozoa in the spermathecae. I have obtained eggs from flies caught in restaurants in December. Griffith (1908) has succeeded in rearing *M. domestica* from eggs in November, December and January under artificial conditions as regards temperature. Jepson (1909) caught flies in January in a Cambridge bakehouse and, under artificial conditions, namely, in a greenhouse in the laboratory with a temperature varying from 65° F. in the morning to 75° F. in the evening, he was able to rear the flies successfully, the whole life-history lasting about three weeks. Gleichen (*t.c.*) mentions the fact that he obtained eggs in January.

Hermes (1911) states that at Berkeley in California he has seen house-flies emerging from their breeding places during every month of the winter season. In early March a veritable plague of flies was encountered while on a trip through the Imperial Valley in California. In sub-tropical and tropical climates the breeding season is continuous throughout the year.

ABUNDANCE OF FLIES IN RELATION TO BREEDING PLACES.

To anyone who studies the breeding places and habits of the house-fly one of the most striking facts is the enormous number of flies which are able to develop in a certain quantity of breeding material such as horse-manure or excrement. Some interesting observations have been made by observers in different countries on this point.

Faichne (1909) records experiments carried out in India in which he reared about 4000 flies from one-sixth of a cubic foot of ground from a latrine and as many as 500 from a single dropping of human excreta.

Herms (1909) in California took samples from a pile of manure after an exposure of four days. The larvae in these samples were counted with the following results: the first sample, 4 lbs., contained 6873 larvae; second sample, 4 lbs., contained 1142 larvae; third sample, 4 lbs., contained 1585 larvae; fourth sample, 3 lbs., contained 682 larvae; the total quantity examined, comprising 15 lbs, contained, therefore, 10,282 larvae. All these larvae were nearly or quite full-grown, and the average number per pound of manure was 685 larvae. The weight of the entire pile of manure was estimated at not less than 1000 lbs., of which the observer estimated that two-thirds was infested. The pile would contain, therefore, about 450,000 larvae in round numbers.

Howard (1911) states that 160 larvae and 146 puparia were counted in a quarter of a pound of rather well-infested horse-manure taken from a manure pile in Washington, D.C., U.S.A., on August 9th. He rightly points out, however, that this number cannot be taken as an average, since no larvae are found in perhaps the greater part of horse-manure piles; nor does it show the limit of what can be found since, on the same date, about 200 puparia were found in less than one cubic inch of manure taken from a spot two inches below the surface of the pile where the larvae had evidently congregated in considerable numbers prior to pupating[1].

The practical importance of the foregoing observations, while requiring no further emphasis, will be considered in a future chapter.

---

[1] In a series of experiments carried on during the summer of 1913, an account of which will be published in the *Journal of Economic Entomology*, Vol. 7, 1914, I was able to obtain considerable evidence of the uneven distribution of the larvae of *M. domestica* in the manure pile. My observations and temperature records showed that the distribution of the larvae was *peripheral* and that the internal heat of a well piled manure heap, in which the manure had been piled at one time, was too great to permit the larvae to penetrate to any depth.

CHAPTER VII

THE LIFE-HISTORY OF THE HOUSE-FLY

PROPORTION OF SEXES.

THE proportion of males to females is approximately equal of flies which are caught in and about the house. There is usually a very slight preponderance of females during the summer. Hamer (1909) found that in the early summer the males were slightly in the majority. In the neighbourhood of the breeding places, however, the females greatly outnumber the males, often exceeding the proportion of nine to one, as was observed by Herms (1911). The breeding experiments also confirmed the observation that the sexes are approximately equal in number, the females being slightly in excess.

COPULATION.

The copulation of *M. domestica* appears to have been first described by Réaumur (1738). Berlese (1902) has given a careful and illustrated account of the operation which I have confirmed by observation and dissection and by means of serial sections, as the process is one of some complexity. I have never observed copulation to take place in the air. The sexes will often "come to earth" in an agitated state but I have never seen them, under such conditions, actually *in coitu*; it is usually a male which has seized a female in the air. The male may perform a few tentative operations before copulation takes place, and these have been mistaken for the actual act. The male alights on the back of the female by what appears to be a carefully calculated leap from a short distance, and this act would appear to indicate on the part of the fly a faculty of being able to judge distance. It then caresses the head of the female, bending down at the same time the distal or apical portion of the abdomen. The male fly, however, is peculiarly passive during the operation, its influence apparently being only tactual; it is only when the female exserts her ovipositor and inserts it into the genital atrium of the male that copulation can successfully take place. When the ovipositor has been inserted into the genital atrium of the male, the accessory copulatory vesicles (see fig. 21) of the female become turgid and retain the terminal segment in this position, in which the female genital aperture is situated opposite to the male genital aperture at the end of the penis, the latter depending from the roof of the genital atrium. Reference to figs. 26 and 27 will make this description clearer. The attachment of the penis to the female genital aperture is made still firmer by the dorsal sclerites of the eighth abdominal segment of the female and the ventral sclerites of the seventh abdominal segment, the so-called secondary forceps of the male, acting respectively above and below the penis. The ventral sclerite of the fifth abdominal segment of the male, called the primary forceps, assists the accessory copulatory vesicles of the female in preventing the withdrawal of the ovipositor before the spermatozoa have been injected into the female genital aperture, by way of which they enter the spermathecae. The whole act may be over in a few moments or the sexes may remain *in coitu* for several minutes. While they are *in coitu* the flies are usually at rest, the male grasping the sides of the female by means of the fore and middle pairs of legs, while the tibiae and tarsi of the hind pair of legs are folded crosswise and forward across the ventral side of the abdomen of the female.

OVIPOSITION.

The females begin to deposit their eggs a few days after copulation. I found that oviposition may take place as early as the fourth day. Taschenberg (*t.c.*) states that the female lays on the eighth day after copulation.

When about to deposit its eggs the fly alights on the substance which it selects as a suitable nidus and, if possible, crawls down a crevice out of sight. By extending its ovipositor it is able to deposit its eggs still further away from the light and

7—2

where they will be more protected from drying up. On the other hand, one may frequently find that eggs have been deposited in large numbers on the surface of the breeding substance.

During a single day a fly, if undisturbed, will deposit the whole batch of eggs which are mature in the ovaries. From actual count of eggs in the ovaries of dissected flies and of batches which have been deposited, I found the number varied from 100 to 150, the usual number being about 120. This number is

FIG. 36.   Mass of eggs of *M. domestica*.
(Photo by H. T. Brues.)

confirmed by other observers. During its lifetime as many as four such batches of eggs may be laid, as shown by actual observation, by the same fly, and a careful study of the ovaries of a large number of female flies has led me to believe that five or even six batches of eggs may be deposited.

The eggs are deposited in clumps, sometimes the whole batch of eggs in a single clump. They are usually placed vertically on their posterior ends and closely packed together (fig. 36).

EGG.

The egg of *M. domestica* (fig. 37) measures 1 mm. in length, sometimes slightly less. It is cylindrically oval; one end, the posterior, is broader than the other, towards which end the egg tapers off slightly. The outer covering or chorion is pearly white in colour, the polished surface being very finely sculptured with minute hexagonal markings. Along the dorsal side of the egg are two distinct curved rib-like thickenings having their concave faces opposite. In the hatching of the eggs which I have observed, the process was as follows: A minute split appeared at the anterior end of the dorsal side to the outside of one

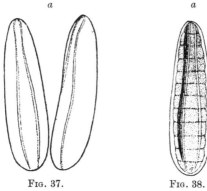

FIG. 37.   FIG. 38.

FIG. 37.   Eggs of *M. domestica*, × 40, dorsal and dorso-lateral views.   *a*. Anterior end.

FIG. 38.   Egg immediately before emergence of the larva which can be seen through the dorsal split of the chorion through which it emerges.

of the ribs; this split was continued posteriorly (fig. 38) and the larva crawled out, the walls of the chorion collapsing after its emergence. The time of hatching varies according to the temperature. With a temperature of 25° C.—35° C. the larvae hatched out from eight to twelve hours after the deposition of the eggs; at a temperature of 15° C.—20° C. emergence takes place after about twenty-four hours, and, if kept as low as 10° C., two or three days elapse before the larvae emerge.

Beclard (1858) showed that the eggs develop more quickly under blue and violet glass than under red, yellow, green or white.

LARVA.

There are three larval stages or instars, the larva moulting twice during the course of its development.

The first larval stage or first instar.

The newly-hatched larva (fig. 39) measures 2 mm. in length. It contains the same number of segments as the mature larva and at the anterior end of the ventral surface of each of the posterior eight segments there is a spiny area (*sp.*). The posterior end is obliquely truncate, and bears centrally the only openings of the two longitudinal tracheal trunks, each trunk opening to the exterior by a pair of small oblique slit-like apertures situated on a small prominence (*p.sp.*). There are no anterior spiracular processes in the first larval stage. The oral

FIG. 39.   Larva shortly after hatching (1st instar).

m.s. Mandibular sclerite.   *p.sp.* Posterior spiracle raised on short tubercle.
sp. Spiniferous pad.

lobes are relatively large, and on the internal ventral surface of each there is a small T-shaped sclerite, the labial sclerite (fig. 57, *t.s.*). These sclerites lie lateral to the falciform mandibular sclerite (*m.s.*). The cephalo-pharyngeal skeleton[1] of the first larval instar is slender and, in addition to the sclerites already mentioned, consists of a pair of pharyngeal sclerites or plates (*l.p.*) deeply incised posteriorly, forming well-pronounced dorsal and ventral processes. The lateral plates are connected antero-dorsally by a curved dorso-pharyngeal sclerite (*d.p.s.*). The anterior edges of the pharyngeal plates are produced ventrally into a pair of slender processes (*h.s.*), the anterior portions of these processes, which represent the hypostomal sclerite, are

---

[1] The cephalo-pharyngeal skeleton of the mature larva is described on page 134.

involute and articulate with the mandibular sclerite. The alimentary canal of the first larval instar is relatively shorter than that of the adult, and consequently it is not so convoluted; the salivary glands are proportionately large.

The first larval instar may undergo ecdysis as early as twenty hours after hatching, but it is usually twenty-four to thirty-six hours before the ecdysis takes place. Under unfavourable conditions with regard to the factors governing the development, the first larval instar sometimes lasts three or four days. Ecdysis begins anteriorly, and the larva not only loses its skin, but also the cephalo-pharyngeal sclerites which are attached to the stomodaeal portion of the ecdysed chitinous integument; the chitinous lining of the proctodaeal portion of the alimentary tract is also shed.

The second larval stage or second instar.

This stage is provided with a pair of anterior fan-shaped spiracular processes similar to those of the mature larva. The posterior spiracular orifices are shown in fig. 60. They are slit-like apertures rather similar to those of the first instar, but larger in size. The cephalo-pharyngeal skeleton is thickened and less slender in form than that of the first instar. It resembles the cephalo-pharyngeal skeleton of the mature larva except that the posterior sinuses of the pharyngeal sclerites are much deeper, thus making the dorsal and ventral posterior processes more slender than in the mature larva. The second larval instar may undergo ecdysis in twenty-four hours at a temperature of 25° C.—35° C., but under cooler conditions or with a deficiency of moisture the period is prolonged and may take several days.

The third larval stage or third instar. (Fig. 40.)

This is the last larval stage, during which growth is very rapid. The anatomy of the mature larva is fully described later.

Larvae incubated at a temperature of 35° C. complete this larval stage and pupate in three to four days; on the other hand, under less favourable developmental conditions, it sometimes extended over a period of eight or nine days.

Incubated larvae ceased feeding at the end of the second day of this stage. They gradually assume a creamy colour, which

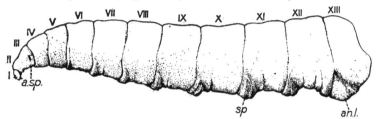

Fig. 40.   Mature larva of *M. domestica*.

a.sp. Anterior spiracular process.   *an.l.* Anal lobe.   *sp.* Spiniferous pad.
I—XIII. Body segments.

colour is due to the large development of the fat-body and to the histolytic changes which are taking place internally; larvae

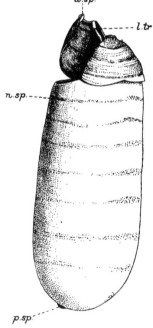

Fig. 41. Pupal case or puparium of *M. domestica* from which the imago has emerged, thus lifting off the anterior end or "cap" of the pupa; ventro-lateral aspect.

a.sp. Remains of the anterior spiracular process of larva.   *l.tr.* Remains of the larval lateral tracheal trunk.   *n.sp.* Temporary spiracular process of nymph. *p.sp.* Remains of the posterior spiracles of larva.

dissected at this stage contain a very large amount of adipose tissue cells. Between the third and fourth day the larva contracts to form the pupa or puparium.

The larvae of the house-fly are strongly negatively heliotropic and move rapidly away from the light, as Loeb (1890) also proved in the case of the larvae of the blow-fly.

THE PUPA OR PUPARIUM.

The process of pupation may be completed in so short a time as six hours. The larva contracts, the anterior end especially being drawn in, with the result that a cylindrical pupal case or puparium is formed (fig. 41), the posterior region being very slightly larger in diameter than the anterior; the anterior and posterior extremities are evenly rounded. The average length of the pupa is 6·3 mm. Owing to the withdrawal of the anterior segments the anterior spiracular processes (*a.sp.*) are now situated at the anterior end, and the posterior spiracles (*p.sp.*) form two flat button-like prominences on the posterior end. The pupa changes from the creamy-yellow colour of the larva to a rich dark brown in a few hours. As the last larval skin has formed the pupal case, it being a coarctate pupa, in addition to the persistence of the spiracular processes the other larval features such as spiny locomotory pads can be seen.

During the first twelve hours or so of pupation the larva loses its tracheal system, which appears to be withdrawn anteriorly and posteriorly, the latter moiety being the larger; the discarded larval tracheal system lies compressed against the interior of the pupal case (*l.tr.*). Communication with the external air is formed for the nymphal[1] developing tracheal system by means of a pair of temporary *pupal spiracles*, which appear as minute spine-like lateral projections between the fifth and sixth segments of the pupal case (*n.sp.*). Each of these communicates with a knob-like spiracular process (fig. 42, *n.sp.*), attached to the future prothoracic spiracle of the fly. The proctodaeal and stomodaeal portions of the alimentary tract are also withdrawn, and with the

---

[1] The word "nymph" is used here to designate that stage in the development which begins with the appearance of the form of the future fly, and ends when the exclusion of the imago takes place.

latter the cephalo-pharyngeal skeleton, which lies on its side on the ventral side of the anterior end of the pupal case.

The histogenesis of the nymph is extremely rapid and at the end of about thirty to forty-eight hours, in rapidly developing specimens, the nymph has reached the stage of development shown

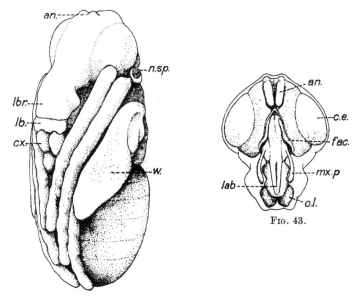

Fig. 42. "Nymph" of *M. domestica* dissected out of pupal case about 30 hours after pupation.

an. Swellings of nymphal sheath marking bases of antennae.   *cx.* Coxa of leg.
    lb. Labial portion of proboscis sheath.   *lbr.* Labral portion of same.
    n.sp. Spiracular process of nymph.   *w.* Wing in nymphal alar sheath.

Fig. 43. Head of "nymph" (about 48 hours after pupation).   Enclosed in nymphal sheath.   To show the development of the imaginal proboscis.

an. Antenna.   *c.e.* Compound eye.   *fac.* Facialia.   *lab.* Labrum.
    mx.p. Maxillary palp.   *o.l.* Oral lobe.

in fig. 42, in which most of the parts of the future fly can be distinguished although they are ensheathed in a protecting nymphal membrane.   Perez[1], in his description of the metamorphosis of the blow-fly, does not divide the nymphal period into two successive phases of histolysis and histogenesis, as he finds a progressive substitution takes place and the beginning of histogenesis actually

[1] *Arch. Zool. Exper.* IV, pp. 1—274, 1910.

precedes that of histolysis. The parts that disappear during this breaking-down process are the parts most specialised for larval life; the parts that are built up entirely *de novo* from embryonic histoblasts are the parts most specialised for adult life. The less specialised parts, which are more plastic, are reorganised *in situ*. The importance of phagocytosis, which some observers have doubted, is emphasised by Perez's detailed and careful study. The head (fig. 43), which with the thorax has been formed by the eversion of the cephalic and thoracic imaginal discs from their sacs, is relatively large: two small tubercles (*an.*) mark the bases of the antennae. The proboscis is enclosed in a large flat sheath, which at this period appears to be distinctly divided into labral (fig 42, *lbr.*) and labial (*lb.*) portions. In a short time the parts of the proboscis are seen to develop in these sheaths. The femoral and tibial segments of the legs are closely adpressed and lie within a single sheath. The wings (*w.*) appear as sac-like appendages, and, as the nymphal sheath of the wing does not grow beyond a certain size, the wing develops in a slightly convoluted fashion by means of a fold which appears in the costal margin a short distance from the apex of the wing.

With a constant temperature of about 35° C., or even less, the exclusion of the imago may take place between the third and fourth day after pupation, but it is more usually four or five days as the larvae, when about to pupate, leave the hotter portion of the mass in which they have been feeding and pupate in the outer cooler portions or in near-by crevices or soil, etc.: this outward migration may be a provision for the more easy emergence of the excluded fly from the larval nidus[1]. In some cases the pupal stage lasts several weeks. I have never succeeded in keeping pupae through the winter, as I have been able to keep the pupae of *Stomoxys calcitrans* and other forms, although attempts were repeatedly made. As I have found no record of other investigators who have succeeded in keeping living pupae of *M. domestica* through the winter under natural conditions, it is fair to assume that the winter is never passed in the pupal state.

[1] I have found puparia at a distance of two feet from the manure heap in which the larvae fed and at a depth of nine inches.

EMERGENCE OF ADULT FLY.

When about to emerge, the fly pushes off the anterior end of the pupal case in dorsal and ventral portions by means of the inflated frontal sac or ptilinum, which may be seen extruded in front of the head above the bases of the antennae. The splitting of the anterior end of the pupal case is quite regular, a circular split is formed in the sixth segment and two lateral splits are formed in a line below the remains of the anterior spiracular processes of the larva, see fig. 41. The fly levers itself up out of the barrel-like pupa and leaves the nymphal sheath. With the help of the frontal sac, which it alternately inflates and deflates, it makes its way to the exterior of the heap and crawls about while its wings unfold and attain their ultimate texture, the chitinous exoskeleton hardening at the same time; when these processes are complete, the perfect insect sets out on its career.

In cases where the larvae have entered earth to pupate, as they frequently are accustomed to do, or where the breeding material has been covered with earth, the emerging flies are able, with the assistance of the ptilinum, to crawl through the overlying soil. Stiles and McMiller (1911) showed that if faecal material containing fly larvae is buried under sand to a depth of forty-eight to seventy-two inches the larvae will crawl up and pupate near the surface and emerge.

SUMMARY OF DURATION OF THE DEVELOPMENT.

If the foregoing results which were obtained at Manchester during the summers of 1905 to 1909 be summarised, it will be found that the shortest time occupied by *M. domestica* in passing through the different stages of its development was as follows: egg, eight hours; first larval stage, twenty hours; second larval stage, twenty-four hours; third larval stage, three days; pupa, three days, giving a total time of eight days four hours as the shortest period which could be obtained under experimental conditions for the duration of the whole life-cycle. It is unlikely, however, that in the hottest weather the development would be accomplished in less than nine days. These minimum periods are the result of a large series of experiments in which it was a more common occurrence to have a batch of larvae developing in ten, fifteen or twenty days.

The internal temperature of the breeding places, such as manure heaps, would tend to produce a shorter period of development than might be obtained with the same air temperature under experimental conditions where smaller quantities of food material are used, and this fact should be remembered in generalising from experimental results. It is difficult to experimentally imitate, except by incubation, such natural conditions as occur in a manure heap or privy midden, where, owing to the larger amount of material conserving and probably producing heat, a higher constant temperature is maintained. All experimental results, except those of incubation, tend to give a long rather than a short rate of development. In many cases where the average temperature was 20° C., but the food material rather dry, the developmental period was about three weeks, and where the temperature was low and the food became dry it extended to as much as six weeks, the greater time being spent in the pupal state, which was sometimes of three or four weeks duration.

The minimum period obtained by Newstead (1907) in his observations in Liverpool was ten to fourteen days from the deposition of the egg to the emergence of the adult.

Griffith (1908), experimenting in the south of England at Hove, obtained as his shortest times for the life-cycle of the house-fly four and a half days to six days from egg to pupa and three and a half days from pupa to adult fly, giving a minimum of eight days.

Reference has been previously made to the observations, in the United States, of Packard and Howard. The former, working at Salem in Massachusetts, gave as the minima, or the probable minima: egg, twenty-four hours; first larval stage, twenty-four hours; second larval stage, twenty-four hours; third larval stage, three days; pupal stage, five days. The shortest time for the development, therefore, was ten days.

The shortest time obtained by Howard in August 1895 at Washington was ten days, as already shown.

Herms (1911) states that in Berkeley, (California), where the temperature is rarely above 80° F. with a mean temperature of 48° F. during the winter, the development is completed usually in from fourteen to eighteen days, less often in twelve days.

In India, at Benares, Smith (1907) gives the rate of development of *M. domestica* in horse-manure as eight days.

FACTORS GOVERNING THE DEVELOPMENT.

Temperature.

The rate of development depends primarily on the temperature of the substance on which the larvae are feeding. This temperature may be reduced from without, for example, by the temperature of the surrounding air, or increased from within by such a heat-engendering agency as fermentation. The influence of a high temperature was shown in my experiments, in which batches of larvae were reared in horse-manure kept in a moist condition in an incubator at a constant temperature of 35° C. At this temperature the development was completed in eight to nine days. Griffith (*l.c.*) found that the life-history was completed in the same time on rearing larvae in an incubator at a high temperature of about 22°—23° C. I found that a higher temperature of 40° C. was too great for the larvae; they were, as it were, cooked and perished at such a temperature.

The observations of Smith in India, which have been quoted, indicate the shortening effect on the life-history of a high temperature under natural conditions.

In the rearing experiments which were carried out in the open air, and under natural conditions with regard to variability of temperature, I found that larvae reared in horse-manure at an average, though not a constant, daily temperature of 22·5° C., occupied from fourteen to twenty days in their development according to the air temperature.

The effect of the temperature on the development is a direct one and affects all stages from the egg to the pupa. To test this question fully a series of experiments in which the developmental stages overlapped each other, some being in the egg stage while others were in the larval and pupal stage, were plotted out on a chart together with the mean temperatures, and it was clearly shown that wherever a continuous high temperature occurred there was a corresponding shortening of the particular stage of development and *vice versa*, when the temperature was low the time of all the stages was lengthened; this included the three larval stages which were short or long according to the temperature.

Griffith (*l.c.*) also found this to be the case in his experiments and his observations on the subject are of interest. He states: "When the larvae were kept in proper warmth until they formed pupae, the flies came out even in a rather cold temperature and quickly in the warm. When the larvae were kept cool and the pupae warm, some flies came out but all were small; it has been a regular occurrence that cold surroundings and plenty of food have produced small flies, if any. Such small flies are incapable of reproduction; and herein is seen the wisdom of flies being able to lay more than one batch of eggs."

Character of larval food.

The effect of the character of the food, on which the rate of development also depends, is well shown by a comparison of the times of the developmental periods in two of the experiments where the average daily temperature was practically the same, namely, 19·3° C. and 20·5° C. In the former experiment, in which human faeces were used, the development was completed in twenty days, and in ·the latter, in which bananas were used, the development occupied twenty-seven days; the time was rather lengthened in both cases by the fact that the larval food was rather dry, but equally dry in both experiments as they were kept together; had more moisture been present the times would probably have been correspondingly shortened.

Desiccation.

It was experimentally proved that, when larvae were reared in batches on the same kind of food material with conditions as regards temperature the same, the developmental period was longer for those larvae which were subject to dry conditions than for those subject to moist conditions. In an experiment at an average temperature of 22° C. larvae reared on horse-manure which was kept in a rather dry condition took thirty days to complete the development, and another batch at the same temperature, but on horse-manure which was kept moist, the development was completed in thirteen days. Under similar conditions, with regard to temperature, the rate of development is directly proportional to the condition of the food as regards moisture. Dry conditions not only retarded development in some of my experiments to five and six weeks, but also tended to produce flies of sub-normal size. Moisture is necessary for the development, and if the food becomes too dry the result is fatal, as the larvae perish.

Fermentation.

A fourth and a most important factor affecting development and one intimately connected with the previous factors—temperature, character of food, and moisture—is that of fermentation, to which reference has already been made. This process appears to take place in the substances on which the larvae best subsist. Whether the suitability of the food is determined by the nature of its fermentation is a point which I was unable to determine, but which I am inclined to believe. I feel certain, however, that the calorific property of fermentation is the most important part of this process on account of its direct relation to the time of development; the endogenous heat of excremental products and decaying substances acting either in addition to, or independently of, the temperature of the surrounding air is of great advantage in accelerating the rate of development.

SEXUAL MATURITY.

On account of the importance of knowing how long a period elapsed from the time when the fly emerged until it became sexually mature and deposited eggs, I made attempts to carry flies through to the second generation. The difficulty of keeping flies alive in captivity has already been referred to in discussing the longevity of *M. domestica*. In one experiment, however, I succeeded in obtaining flies of the second generation. It was found that the flies became sexually mature in ten to fourteen days after their emergence from the pupal state and, four days after copulation, they began to deposit their eggs, that is, from the fourteenth day onwards to the time of their emergence[1]. As this experiment was an important one, it will be of general interest to give the details as recorded in my notes.

Experiment series No. B.2.07

Larvae reared on human excrement which was purposely allowed to dry up; the time of development of this series was twenty days.

21 August, 1907. First fly emerged.
28   ,,   All flies had emerged.

From this date onward they were given fresh horse-manure daily.

The maximum and minimum temperatures of the succeeding days were as follows (two other readings were taken in addition):

28 Aug.	Max. sun,	104° F.,	shade,	97° F.		
29 ,,	,,	,,	74 F.,	,,	72 F.	
30 ,,	,,	,,	117 F.,	,,	104 F.,	Min. 50° F.
31 ,,	,,	,,	—	,,	75 F.,	,, 57 F.
2 Sept.	,,	,,	117 F.,	,,	89 F.,	,, 47 F.
3 ,,	,,	,,	90 F.,	,,	84 F.,	,, 50 F.
4 ,,	,,	,,	79 F.,	,,	74 F.,	,, 42 F.

On Sept. 4th the following note appears:

"Flies from B.2.07 seem very lively and healthy; two were observed copulating. If the average date of emergence is taken, namely Aug. 25th, they are on the average 10 days old, so they evidently become sexually mature about this age, as I have not observed any similar proceedings (copulation) earlier."

On Sept. 6th a note is made that the abdomens of the females of this series are markedly distended with eggs.

On Sept. 9th (Temp.: Max. sun, 108° F., shade, 96° F., Min. 55° F.), larvae were found in the manure given to this series of flies on Sept. 4th upon which day the eggs were laid.

It should be pointed out that every batch of fresh manure was very carefully and thoroughly examined for eggs before being given to the captive flies to prevent any error.

NUMBER OF GENERATIONS.

As it has been shown that, under favourable conditions as regards temperature, etc., flies may deposit eggs about the fourteenth day after emergence from the pupal state, it may be

[1] Dr L. O. Howard informed me verbally (Jan. 1914) that in a series of experiments carried out at Washington, D.C., U.S.A., in which observations on this

justifiably assumed that, given a spell of very hot weather in which development could be accomplished in about nine days, the progeny of a fly would be laying eggs in about twenty-three or twenty-four days after the eggs from which they were hatched had been deposited by the parent fly. Remembering that a single fly may deposit from 100 to 150 eggs in a single batch and that during its life-time it may deposit from four to six batches, it is not difficult to account for the enormous swarms of flies which occur in certain localities during the hot summer months.

Howard (1911) has calculated that by September 10th the progeny of a single over-wintered fly which deposited its eggs on April 15th would number 5,598,720,000,000,000—if they all lived; but they do not all live, nor do all the eggs hatch. Nevertheless, these calculations serve a useful purpose by indicating in a graphic manner the potential fecundity of the house-fly.

pre-oviposition period were made, the period was found to be several days less than my own observations indicated; this will undoubtedly be found to be the case and a pre-oviposition period of seven days or even less is conceivable.

CHAPTER VIII

THE EXTERNAL FEATURES OF THE FULL-GROWN LARVA

THE external appearance of the acephalous muscid larva or "maggot" (fig. 40) is well-known. It is conically cylindrical. The body tapers off gradually to the anterior end from the middle region. The posterior moiety is cylindrical, and except for the terminal segment the segments are almost equal in diameter. The posterior end is obliquely truncate. The cuticular integument is divided by a number of rings; this ringed condition is brought about by the insertion of the segmentally-arranged somatic muscles the serial repetition of which can be clearly understood by reference to fig. 48. The average length of the full-grown larva of *M. domestica* is 12 mm.

The question as to the number of segments which constitute the body of the muscid larva is a debated subject. I have, however, taken as my criterion the arrangement of the somatic musculature. Newport (1839) considered that the body of the larva of *Musca vomitoria* consisted of fourteen segments, but if the anterior portion of the third segment, that is, my first post-oral segment, is included, there were fifteen, to which view he appeared to be inclined. Counting the anterior segment or "head" as the first, Weismann (1863 and 1864) considers that the body is composed of twelve segments. Brauer (1883) is of the opinion that there are twelve segments, but that the last segment is made up of two; Lowne follows this view in his description of the blow-fly larva and considers that there are fifteen post-oral segments. I am unable to accept Lowne's view. Counting the problematical cephalic segment, for which I shall use Henneguy's (1904) term

"Pseudo-cephalon," as the first segment, I believe that it is succeeded by twelve post-oral segments, making thirteen body-segments, in all, which is the usual number for dipterous larvae, as Schiner (1862) has also pointed out. My study of the somatic musculature, as I shall show, indicates the duplicate nature of the apparent first post-oral segment, so that the apparent second post-oral segment (IV), that is, the segment posterior to the anterior spiracular processes, is really the third post-oral segment or fourth body-segment.

The cephalic segment cannot be considered as homologous with the remaining twelve segments, which are true segments of the body, as shown by their musculature and innervation. This segment (fig. 44, I), for which Henneguy's term " pseudo-cephalon " is very suitable, probably represents a much reduced and degenerate cephalic segment, its present form being best suited to the animal's mode of life.

We may consider the greater part of the cephalic segment of the larva as having been permanently retracted within the head ; this is shown by the position of the pharyngeal skeleton, to the whole of which the name " cephalo-pharyngeal skeleton " has been given. All that now is left of the cephalic segment consists of a pair of oral lobes, whose homology with the maxillae is very problematical, and at present is not safely tenable.

In an interesting and very suggestive paper Becker (1910) has attempted to trace the gradual development of the "acephalic" condition of the dipterous larva, such as we find in the muscid larvae, from a larva such as *Simulium* which possesses a distinct head. He shows a gradual drawing in of the cephalic segments and the chitinous sclerites covering the same in a series represented by the larvae of *Chironomus, Stratiomys* and *Atherix* until in the muscid larva the cephalic sclerites are completely withdrawn into the anterior end of the larva and form the cephalo-pharyngeal plates.

On the dorsal side the oral lobes are united posteriorly. Each bears two conical sensory tubercles (*o.t.*), which are situated, the one dorsally, and the other anterior to this and almost at the apex of the oral lobe. Becker (*l.c.*) adopts Weismann's nomenclature in his description of the larval head and calls the upper of these sensory tubercles the antennae and the lower the " *maxillentaster.*"

The ventral and ventro-lateral surfaces of the oral lobes are traversed by a number of channels, which will be described later.

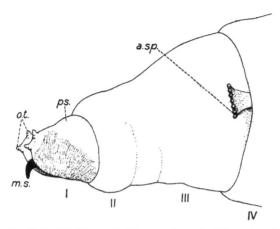

FIG. 44.   Lateral (left) aspect of the anterior end of the mature larva.
I—IV. Body-segments.   *a.sp.* Anterior spiracular process showing seven spiracular papillae.   *m.s.* Mandibular sclerite.   *o.t.* "Optic" tubercles.   *ps.* Pseudocephalon.

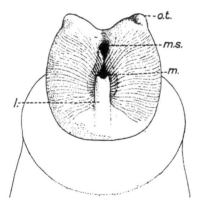

FIG. 45.   Ventral aspect of the Pseudo-cephalon and second body-segment of the mature larva showing the two oral lobes traversed by the food channels.
l. Lingual-like process.   *m.* Mouth.

The post-cephalic segment, which is composed of the first and second post-oral segments and represents the second and third segments of the body, is conical in shape.   The first post-oral segment (II), to which Lowne gave the name "Newport's segment,"

is limited posteriorly by a definite constriction and is covered
with minute spines. The second post-oral segment bears laterally
at its posterior border the anterior spiracular processes (*a.sp.*).
The remaining segments of the body—four to twelve—are on
the whole similar in shape. At the anterior edge of the ventral
side of each of the sixth to twelfth body-segments there is a
crescentic pad (fig. 40, *sp.*) bearing minute and recurved spines;
these are locomotory pads by means of which the larva moves
forwards and backwards. It is important to note that these pads
are situated on the anterior border of the ventral side of each
segment as they do not appear to have been carefully placed in
some of the previous figures of this species. In addition to these
spiniferous pads there are two additional pads of a similar nature,
one on the posterior border of the ventral side of the twelfth
body-segment, and the other posterior to the anus.

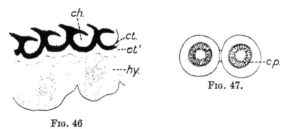

FIG. 46

FIG. 46. Longitudinal section through the surface of one of the oral lobes of
mature larva to show the food-channels.
ch. Food-channel.   *ct.* Outer layer of cuticular integument.   *ct′.* Inner layer of
the same.   *hy.* Hypodermis.

FIG. 47. Transverse section through two of the papillae of the anterior spiracular
process to show the clear central lumen.
c.p. The cuticular processes.

The terminal or thirteenth body-segment is obliquely truncate,
but the truncate surface, which occupies more than half the posterior end of the larva, is not very concave as in the blow-fly larva.
It bears in the centre the two posterior spiracles (fig. 61, *p.sp.*),
which are described in detail with the tracheal system. On the
ventral side of the terminal segment are two prominent anal lobes
(fig. 40, *an.l.*), which are important agents in locomotion.

The cuticular integument is thin and rather transparent, so
that in the younger larvae many of the internal organs can be

seen through it. In older larvae the fat-body assumes large proportions and gives the larva a creamy appearance, obscuring the internal organs. The cuticle (fig. 46) is composed of an outer rather thin layer of chitin (*ct.*), which is continuous with the chitinous intima of the tracheae and proctodaeal regions of the alimentary tract. Below this layer there is a thicker layer of chitin (*ct'.*), which does not stain so deeply. In places this lower layer is penetrated by the insertions of the muscles. The cuticle lies on a layer of stellate hypodermal cells (*hy.*), which are well innervated, and attain a large size in the posterior segments of the body.

CHAPTER IX

THE INTERNAL STRUCTURE OF THE FULL-GROWN LARVA

MUSCULAR SYSTEM.

THE muscular system of the larva consists of a segmental series of regularly repeated cutaneous muscles, forming an almost continuous sheath beneath the skin, together with a set of muscles in the anterior segments of the body which control the cephalopharyngeal sclerites and pharynx. In addition to this there are a set of cardiac muscles and the muscles of the alimentary tract.

I have been unable to find a detailed description of the muscular system of the muscid larva, and I do not think that Lowne's excuse for dismissing the cutaneous muscles of the blow-fly larva with a very brief statement, because "the details possess little or no interest," was justified, considering how little is known about the muscular systems of insect larvae, and constant reference to the classic work of Lyonet (1762) on the caterpillar is not sufficient to satisfy the inquiring student of to-day. The muscular system of the larva, therefore, will be described in some detail.

Muscles of the body-wall (fig. 48).

The cutaneous muscles are repeated fairly regularly from segments (by segments, I mean body-segments) four to twelve and a detailed description of the muscles of one of these segments will serve for the rest. The muscles, though continuous in most cases from segment to segment, are attached to the body-wall at the junction of the segments. The most prominent muscles are the dorso-lateral oblique recti muscles. In segments six to twelve

Fig. 48. Muscular system of the body-wall of the right side. The straight dorsal line is the median dorsal line of the body, and the curved ventral line is the median ventral line.

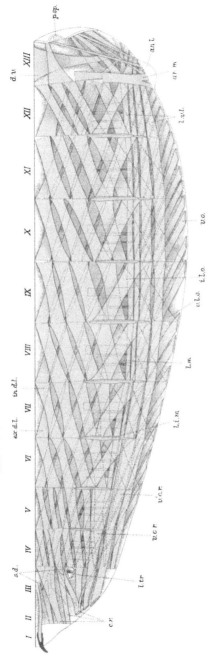

I—XIII. Body segments. an.l. Anal lobe. an.m. Anal muscle. c.r. Cephalic retractor muscle. d.v. Dorso-ventral muscle of the terminal segment. ex.d.l. External dorso-lateral oblique recti muscles. i.l.o. Internal lateral oblique muscle. in.d.l. Internal dorso-lateral oblique recti muscles. l.i.m. Lateral intersegmental muscle. l.m. Lateral muscles. l.tr. Branch of lateral tracheal trunk communicating with the anterior spiracular process. l.v.l. Longitudinal ventro-lateral muscles. p.sp. Posterior spiracle. s.d. Stomal dilators. v.c.r., v'.c.r. Ventral cephalic retractor muscles. v.l.o. Ventro-lateral oblique muscle. v.o. Ventral oblique muscle.

there are four pairs each of external (*ex.d.l.*), and internal dorso-lateral oblique recti (*in.d.l.*) muscles; in segments four and five there are five pairs of external and six pairs of internal dorso-lateral oblique muscles. Ventral to these muscles are four pairs of longitudinal ventro-lateral muscles (*l.v.l.*); the muscle bands of the two more ventral pairs are double the width of those of the two more lateral pairs. In the fifth segment there is only one of the more lateral pairs of the longitudinal ventro-lateral muscles present, and in the fourth segment only the two more ventral pairs remain. In addition to these muscles there are two other pairs of oblique recti muscles; these are, a pair of ventro-lateral oblique muscles (*v.l.o.*) and a pair of internal lateral oblique muscles (*i.l.o.*); both of these are absent in the segments anterior to the sixth. The foregoing muscles, namely the dorso-lateral oblique, the internal lateral oblique, the ventro-lateral oblique and the longitudinal ventro-lateral, by their contraction, bring together the intersegmental rings and so contract the body of the larva.

Attached externally to the anterior ends of the longitudinal ventro-lateral muscles are a number of pairs of ventral oblique muscles (*v.o.*); they vary in number from two to eight pairs in each segment. The number increases posteriorly from two pairs in segment four to four pairs in segment five, five pairs in segment seven, seven pairs in segment ten, eight pairs in segment eleven; the number of pairs then decreases to six or seven pairs in segment twelve. The more ventral pairs of these muscles are not attached at their posterior ends to the intersegmental ring but to the ventral wall of the segment and no doubt assist in bringing forward the ventral spiniferous pads. In segments four to twelve there are three pairs of lateral muscles (*l.m.*) situated next to the hypodermis and attached in a dorso-ventral position; these will assist in drawing the dorsal and ventral regions of the segments together and so increase the length of the larva. Between segments four and five and the remaining segments to twelve there is, on the intersegmental ring, a pair of lateral intersegmental muscles (*l.i.m.*); these by their contraction bring about a decrease in the size of the intersegmental ring and so assist the lateral muscles in increasing the length of the larva.

The muscles of the last segment (XIII) are not regularly arranged

as in the preceding segments; they consist of three main groups:
(1) the recti muscles, which assist in contracting the segments;
(2) the anal muscles (*an.m.*), which are attached ventrally to the
anal lobes (*an.l.*); and (3) the dorso-ventral muscles (*d.v.*), which
by their contraction assist in lengthening the segment. In addition
to these there are certain small muscles in relation with the posterior spiracles.

In the second and third segments the recti muscles are reduced
to four pairs and the attachment of the two lateral and external
pairs of muscles has led me to regard the apparently single first
post-oral segment as consisting of two segments; it is not a single
post-cephalic or pro-thoracic segment as it has been called. There
is quite a distinct internal division and the external constriction
has been already noticed. This view does not necessarily alter the
homology of the third segment, which may still be regarded as
pro-thoracic if this is desirable. The segment which I regard as
the second body-segment may be a rudiment of the cephalic region
which has been almost lost, and this loss, or, as I prefer to regard
it, this withdrawal of the head, only serves to make any discussion
as to the homologies of these anterior segments with those of
the adult extremely difficult, and, I believe, at present valueless.
Further, comparative studies of the larvae of the calyptrate
muscidae are necessary before we can arrive at any definite conclusions concerning the composition of the bodies of these larval
forms.

The cephalo-pharyngeal muscles.

These muscles (fig. 49) consist of four sets: (1) the cephalic
retractor muscles, which by their contraction draw the anterior end
of the larva and the pharyngeal mass inwards; (2) the protractor and
depressor muscles of the pharyngeal mass; (3) the muscles controlling the mandibular, dentate, and hypostomal sclerites; and
(4) the internal pharyngeal muscles.

There are four chief pairs of cephalic retractor muscles, of
which the two ventral pairs are by far the largest. The more
ventral of these two pairs (*v'.c.r.*) arises on the ventral side from
the posterior end of the sixth segment, internal to the ventro-
lateral longitudinal muscles; the other pair (*v.c.r.*), which is double,

arises more laterally from the posterior end of the fifth segment. The remaining pairs of cephalic retractors arise from the posterior end of the third segment. All the cephalic retractor muscles are

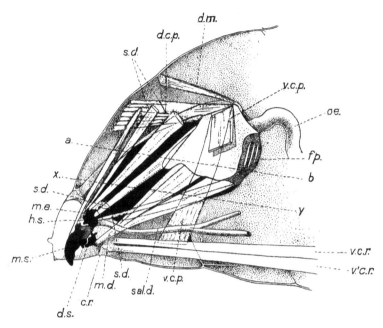

Fig. 49. Muscles of the cephalo-pharyngeal sclerites of the mature larva seen from the left side. The muscles of the body-wall have been omitted with the exception of the large cephalic retractor muscles.

a.b., *x.y.* Levels and direction of the oblique sections shown in Figs. 50 and 51. *c.r.* Cephalic ring. *d.c.p.* Dorsal cephalic protractor muscle. *d.m.* Right pharyngeal depressor muscle. *d.s.* Dentate sclerite. *f.p.* Chitinous floor of the posterior region of the pharynx showing the bases of the T-ribs. *h.s.* Hypostomal sclerite. *m.d.* Mandibular depressor muscle. *m.e.* Mandibular extensor muscle. *m.s.* Mandibular sclerite. *s.d.* Stomal dilator muscles. *sal.d.* Common salivary duct. *v.c.p.* Ventral cephalic protractor muscles. *v.c.r.* and *v'.c.r.* Ventral cephalic retractor muscles.

inserted anteriorly into a ring, the cephalic ring (*c.r.*), on the anterior border of the second segment, the first post-oral segment.

There are two pairs of cephalo-pharyngeal protractor muscles, a dorsal (*d.c.p.*) and a ventral pair (*v.c.p.*). Both are rather broad fan-shaped muscles inserted by their broad ends in the middle of the third segment, slightly to the sides of the dorsal and ventral median lines respectively. The dorsal and ventral muscles of each side are inserted together on the dorso-lateral region of the posterior end of the pharyngeal mass. The pair of depressor muscles (*d.m.*) which are situated dorsally, are attached by their broader ends to the intersegmental ring between segments three and four. They are inserted on the posterior end of the dorsal side of the pharyngeal mass; by their contraction the posterior end of the pharyngeal mass is raised, the result being that the sclerites articulated to its anterior end are depressed.

There remain six pairs of muscles controlling the mandibular, dentate and hypostomal sclerites, one pair controlling the two foremost sclerites and four pairs controlling the hypostomal sclerite. The mandibular extensor muscles (*m.e.*) are attached to the bodywall in the third segment on each side of the median line and between the dorsal cephalo-pharyngeal protractors. They are inserted on the dorsal side of the mandibular sclerite (*m.s.*); by their contraction they elevate the sclerite. This sclerite is depressed by the contraction of a pair of muscles which control the dentate sclerites (*d.s.*), the latter fitting into a notch on the ventral side of the mandibular sclerite. The mandibular depressor muscle (*m.d.*) is attached to the posterior ventral process of the lateral pharyngeal sclerite by the three bands into which the posterior portion of the muscle is divided; the anterior and single end of the muscle is inserted on the ventral process of the dentate sclerite.

Four pairs of muscles (*s.d.*) are inserted on the hypostomal sclerite (*h.s.*). Two more dorsal pairs are attached to the intersegmental ring between segments three and four as shown in fig. 48. The two more ventral pairs are attached to the lateral pharyngeal sclerites, one being attached to the ventral side of the posterior dorsal process and the other to the ventral process beneath the mandibular depressor. These muscles, which I call the stomal dilators, are inserted on the sides of the hypostomal sclerite. Their function is, I believe, to open and close the anterior pharyngeal aperture and so control the flow of fluid food into the pharynx and of the salivary secretion; the lowest pair of muscles may be more directly concerned with the latter function.

The pharyngeal apparatus is controlled, as in the adult fly, by a series of muscles. In the larval stadium, however, where so large an amount of food is required for the growth and building up of the future insect, there is a greater development and elaboration of the pharyngeal apparatus, including the muscles. In the greater anterior region of the pharynx that is, the part lying within the pharyngeal sclerites (fig. 51), the muscular system consists of two bands of oblique muscles (*o.ph.*) arranged in pairs. The muscles are attached dorsally to the inside dorsal edges of the lateral plates (*l.p.*) and ventrally to the roof of the pharynx (*r.ph.*),

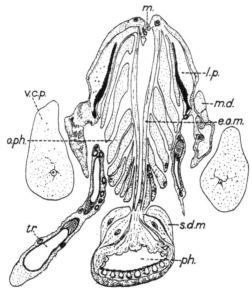

FIG. 50. Oblique section through the pharyngeal mass of the larva in the direction and at the level shown by the line *a.b.* in Fig. 49. (Camera lucida drawing.)

e.o.m. Elongate oblique pharyngeal muscle.   *l.p.* Lateral pharyngeal sclerite. *m.* Accommodating membrane.   *m.d.* Mandibular depressor muscle. *o.ph.* Oblique pharyngeal muscle.   *ph.* Pharynx.   *s.d.m.* Semicircular dorsal pharyngeal muscles.   *tr.* Trachea.   *v.c.p.* Ventral cephalic protractor muscle.

the ventral attachment being more posterior than the dorsal. The posterior region of the pharynx, which is between the lateral plates and the oesophagus (fig. 50), is controlled by two sets of muscles. Two pairs of elongate oblique muscles (*e.o.m.*) are attached dorsally to the dorsal edges of the lateral plates (*l.p*) and inserted ventrally on the roof of the pharynx; these muscles assist the previously described oblique pharyngeal muscles in raising and depressing the roof of the pharynx. They are assisted in enlarging and contracting the lumen of the posterior part of the pharynx by a number of semicircular dorsal muscles (*s.d.m.*), which by their contraction make the floor of the pharynx more concave, and it is these muscles, I believe, that are chiefly concerned in the maintenance of the peristaltic contractions of the pharynx, by means of which the fluid food, which has been sucked into the mouth by the pumping action of the pharynx, is carried on to the oesophagus.

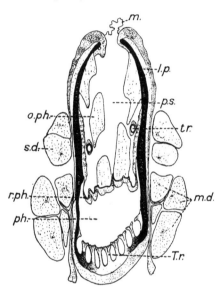

FIG. 51. Oblique section through the pharyngeal mass of the larva at the level shown by the line *x.y.* in Fig. 49. (Camera lucida drawing.)

p.s. Pharyngeal sinus.   *r.ph.* Roof of pharynx.   *T.r.* T-ribs of the floor of pharynx. Other lettering as in Figs. 49 and 50.

The similarity between the pharyngeal apparatus of the fly, that is, of the fulcrum, and that of the larva, is very striking, both with regard to the form of the skeletal structures and the musculature. If the pharynx of the larva were regarded as being homologous to that of the fly it would further support the view that the head of the larva had been permanently withdrawn into the succeeding anterior body-segments. These structures, however, may be merely analogous; the similarity of structure may have been brought about by similarity of function. Both larva and adult subsist on fluids which are sucked into the mouth and pumped into the oesophagus.

The series of muscular actions which takes place during locomotion appears to be as follows. By the contraction of the pharyngeal protractors the anterior end of the larva is extended, the mandibular sclerite being extended at the same time by the contraction of the mandibular extensor muscles. The mandibular sclerite is now depressed by the contraction of the mandibular depressors, and anchors the anterior end of the larva to the substance through which it is moving. A series of segmental linear contractions now takes place, initiated by the large cephalic retractor muscles, and carried on posteriorly from segment to segment by the dorso-lateral oblique, the internal lateral oblique, the longitudinal ventro-lateral oblique and ventral oblique muscles. Each segment as it comes forward takes a firm grip ventrally by means of the spiniferous pad. By the time the last spiniferous pad has become stationary the mandibular sclerite has left its anchorage, and by the contraction of the lateral and intersegmental muscles, which takes place from before backwards, the lengths of the segments of the larva are increased serially and the anterior end begins to move forward again, when the whole process is repeated.

NERVOUS SYSTEM.

The central nervous system of the larva (fig. 52) has attained what would appear to be the limit of ganglionic concentration and fusion. The boat-shaped ganglionic mass (fig. 53), which lies partly in the fifth segment, but the greater portion in the sixth segment, is a compound ganglion and represents the fusion of eleven pairs of ganglia similar to that which Leuckart (1858) describes in the first larval stage of *Melophagus ovinus*, but which, however, has not undergone so great a degree of concentration as in *M. domestica*. This ganglionic mass, which, for convenience and brevity, I shall call the ganglion (Lowne's "neuroblast") does not exhibit externally any signs of segmentation, the interstices between the component ganglia being filled up with the cortical tissue, whose outer wall forms a plain surface. In horizontal and sagittal sections, however, the component ganglia can be recognised and their limits are more clearly defined. The ganglion is surrounded by a thick ganglionic capsular sheath which is richly supplied with tracheae, and appears to be continuous with the outer sheath of the peripheral nerves. Two pairs of large tracheae (fig. 53) are found entering the ganglionic sheath, an anterior pair (*tr'.*) which runs in between the cerebral lobes, and a lateral pair (*tr".*) entering the ganglion beneath these lobes. In the young larva the cortical layer of cells is proportionately much thicker. The cortical tissue is made up of cells of varying sizes, but which can be grouped in two classes—smaller cortical cells and larger ganglionic cells. Most of the ganglionic cells appear to be unipolar, but there are many of a bipolar and multipolar nature present; they stain readily and possess fairly large nuclei. These ganglionic cells are arranged segmentally, and occur near the origin of the nerves. In the posterior region of the ganglion, where the nerves arise in close proximity, the ganglion cells are very numerous,

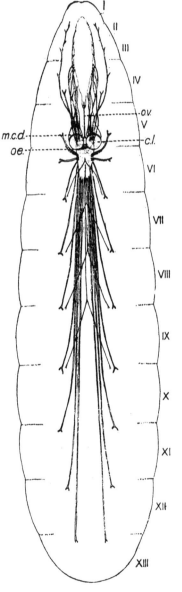

FIG. 52. Nervous system of the mature larva. The dorsal accessory nerves are shown by single black lines, and the outline of the pharyngeal mass is indicated by the dotted line.

I—XIII. Body segments of the larva. *c.l.* Cerebral lobes. *m.c.d.* Major cephalic imaginal discs. *oe.* Oesophagus. *o.v.* Anterior (oesophageal branch) of visceral nervous system.

relatively few of the cortical cells being found. A further demar-
cation of the component ganglia is brought about by median and
vertical strands of the ganglionic sheath-tissue, which perforate
the compound ganglion and occur as vertical strands along its
median line. Tracheae also penetrate the ganglion with these
strands of capsular tissue.

On the dorsal side of the anterior end of the ganglion is
situated a pair of spherical structures (*c.l.*), which may be termed
the "cerebral lobes." They are united in the median line dorsal

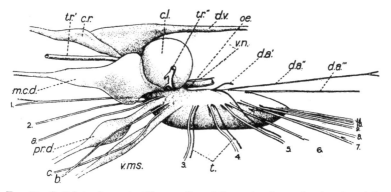

Fig. 53. Left lateral aspect of the ganglion of the mature larva showing the origin
of the nerves, position of the imaginal discs, and anterior end of the dorsal
vessel.

1—11. Eleven segmental nerves.   *a.*, *b.* and *c.* Nerves arising from the bases of the
stalks of the prothoracic and ventral mesothoracic imaginal discs.   *c.l.* Cerebral lobe.   *c.r.* Problematical cellular structure (Weismann's "ring").

d.a'., *d.a''.*, *d.a'''.* Dorsal accessory nerves.   *d.v.* Dorsal vessel.   *m.c.d.* Major
cephalic imaginal discs.   *oe.* Oesophagus.   *pr.d.* Prothoracic imaginal
disc.   *t.* Fine tracheae which arise in association with the segmental nerves,
others arise with some of the more posterior nerves, but for the sake of clearness they are not included in the figure.   *tr'.*, *tr''.* Tracheae entering the
ganglion.   *v.m.s.* Ventral mesothoracic imaginal disc.   *v.n.* Visceral
nerve.

to the foramen traversed by the oesophagus (*oe.*). These cerebral
lobes are chiefly of an imaginal character, and contain the funda-
ments of the supra-oesophageal ganglia and also of the optic
ganglia of the future fly (fig. 65). Each is surrounded by a thin
membranous sheath (*sh.*) and is connected with the major cephalic
imaginal discs by the optic stalk (*o.s.*).

The nerves arising from the ganglion may be divided into three
groups, according to their origin. Eleven pairs of nerves (fig. 53,

1—11) corresponding to the eleven pairs of ganglia arise, two from the anterior end and nine from the sides of the ganglion. Three pairs of nerves (a, b and c) arise laterally from the stalks of the prothoracic and mesothoracic imaginal discs. In the median dorsal line of the posterior half of the ganglion a single pair (d.a'.) and two median unpaired (d.a''., d.a'''.) nerves have their origin; these are accessory nerves.

The first pair of the two anterior pairs of nerves runs forward and innervates the posterior region of the pharyngeal mass; the anterior region of the latter is supplied by the second pair of nerves. These nerves also innervate the anterior segments of the body. The first (a) of the three pairs of nerves which arise from the stalks of the imaginal discs runs to the anterior end supplying the protractor and retractor muscles of the pharyngeal mass. The second (b) of these three pairs of nerves innervates the muscles of the body-wall of the third and fourth segments; the latter segment is also innervated by the third (c) of the three pairs of nerves. The succeeding nine pairs of lateral nerves are segmentally distributed, and innervate the muscles of the body-wall of segments five to thirteen. Each nerve bifurcates on reaching the muscles, and these branches further subdivide into very fine nerves.

The nerves which arise dorsally, and which I have called the accessory nerves, are interesting. The first pair (d.a'.) which arises about mid-way along the dorsal side of the ganglion, accompanies the pair of nerves supplying the seventh segment. The second (d.a''.), which is an unpaired nerve, bifurcates in the seventh segment, and the resulting nerves proceed to the body-wall in association with the nerves supplying the eighth segment. The third and posterior dorsal accessory nerve (d.a'''.) bifurcates in the seventh segment. Each of the resulting nerves undergoes a second bifurcation; the dextral nerve, bifurcating in the eighth segment, accompanies the nerves supplying the ninth segment; the sinistral nerve bifurcates between segments eight and nine, and the resulting nerves proceed to the tenth segment. None of the remaining lateral nerves appear to be accompanied by an accessory nerve, of which there are four pairs only. The ganglionic sheath is penetrated by tracheae, some of which arise from the ganglion in association with the nerves which they accompany to the body-wall. Two of

these tracheae are shown (fig. 53, *t.*).   Similar fine tracheae arise
with the three posterior pairs of lateral nerves, and on account of
their similarity to accessory nerves I at first mistook them for such,
even when dissecting with a magnification of sixty-five diameters,
until my serial sections showed their real nature. Without the aid
of sections it is impossible to distinguish these fine unbranching
tracheae from accessory nerves. I have mentioned this fact as
showing the necessity of supplementing the one method of study
by the other.

The *visceral or stomatogastric nervous system* (fig. 54) consists
of a small central ganglion (*c.g.*) lying on the dorsal side of the
oesophagus, immediately behind the transverse commissure of the

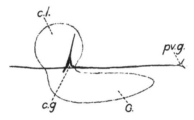

Fig. 54. Visceral or stomatogastric nervous system of the mature larva. The
position of the ganglion (*G.*) with the cerebral lobes (*c.l.*) is shown by means
of the dotted outline.

c.g. Central visceral ganglion.   *pv.g.* Proventricular or posterior ganglion.

cerebral lobes from the bases of which two fine nerves are given
off to join a fine nerve from the ganglion, which runs dorsally
towards the anterior end of the dorsal vessel. A fine nerve from
the ganglion runs forward on the dorsal side of the oesophagus
towards the pharynx. A posterior nerve (fig. 53, *v.n.*) runs from
the ganglion along the dorsal side of the oesophagus to the neck
of the proventriculus, where it forms a small posterior ganglion
(fig. 54, *pv.g.*), from which fine nerve-fibres arise and run over the
anterior end of the proventriculus.

SENSORY ORGANS.

The chief visible sensory organs which the larva possesses are the two pairs of conical tubercles (fig. 44, *o.t.*), which have been described already on the oral lobes. In section each consists of an external transparent sheath of the outer cuticular layer; beneath this and surrounded by a chitinous ring are the distal cuticularised extremities of a number of elongate fusiform cells grouped together to form a bulb. These are nerve-end cells and their proximal extremities are continuous with nerve-fibres by means of which they are connected to the ganglion. Judging from their structure these organs appear to be of an optical nature, and this is the usual view which is held with regard to their function. They would appear merely to distinguish light and darkness, which, for such cryptophagous larva, is no doubt all that is necessary. The negative heliotropism of the larva of the blow-fly has been experimentally proved by Loeb (1890), and my own observations confirm the same for the larvae of *M. domestica*.

The hypodermal cells are well innervated and the body-wall appears to be highly sensitive.

THE ALIMENTARY SYSTEM.

The alimentary tract increases in length at each of the larval ecdyses, and in the mature larva (fig. 55), its length is several times greater than the length of the larva. The great length of the alimentary tract of the larva compared with that of the fly is probably accounted for by the fact that a large digestive area is necessary for the rapid building up of the tissues from fluid food which takes place during the larval life. It is divisible into the same regions as the alimentary tract of the mature insect, but it differs from the latter in several respects. These regions are part of the original stomodaeal, mesenteric and proctodaeal regions of which the mesenteric is by far the longest in this larva. The regions of the alimentary tract which are derived from the stomodaeum and proctodaeum are lined with chitin of varying thickness which is attached during life to the epithelial cells, but is shed when the larva undergoes ecdysis. The mesenteron does not appear to be lined with chitin as it is in some insects, in which cases the chitinous intima usually lies loose in the lumen; it is, however, in the larva of *M. domestica*, usually lined with a lining of a mucous character. The whole alimentary tract is covered by a muscular sheath of varying thickness.

The *mouth* (fig. 45, *m.*) opens on the ventral side between the oral lobes. The ventral and ventro-lateral sides of the oral lobes are traversed by a series of small channels (fig. 46, *ch.*), which are made more effective by the fact that one side of the channel is raised and over-hangs the other so as to parti-ally convert the channels into tubes rather comparable to the pseudo-tracheae of the oral lobes of the fly, to which they have a similar function : the liquid food runs along these channels to the mouth. Distally, many of the channels unite; the resulting channels all converge and run into the mouth. The anterior border of the oral aperture is

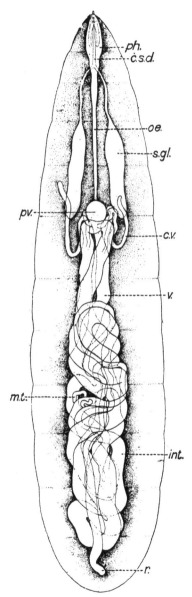

Fig. 55. Alimentary system of mature larva. The course of the ventriculus and intestine as they lie in the larva is shown by the dotted lines. The origins only of the Malpighian tubes are shown.

c.s.d. Common salivary duct. *c.v.* Caecum of ventriculus. *int.* Intestine. *m.t.* Malpighian tubule. *oe.* Oesophagus. *ph.* Pharynx. *pv.* Proventriculus. *r.* Rectum. *s.gl.* Salivary gland. *v.* Ventriculus.

occupied by the mandibular sclerite (fig. 44, *m.s.*), and the posterior border is bounded by a lingual-like process (fig. 45, *l.*) that is bilobed at its anterior extremity.

Cephalo-pharyngeal sclerites (fig. 56). The sclerites associated with the cephalo-pharyngeal region are rather similar to those of the second larval instar; they are, however, of a more solid and of a thicker character. Between the oral lobes is seen the median

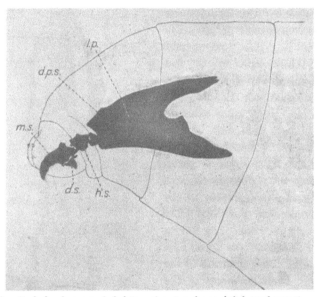

Fɪɢ. 56.   Cephalo-pharyngeal skeleton of mature larva, left lateral aspect.

d.p.s. Dorsal pharyngeal sclerite.   *d.s.* Dentate sclerite.   *h.s.* Hypostomal sclerite.   *l.p.* Lateral pharyngeal sclerite or plate, deeply incised posteriorly to form dorsal and ventral processes.   *m.s.* Mandibular sclerite.

uncinate mandibular sclerite (*m.s.*). The homology of this sclerite is obscure. Lowne regarded it as being the labrum; some authors consider that it represents the fused mandibles. As we know at present so little of the comparative embryology of these larvae it will be best to retain the name by which it is generally known. The basal extremity of the mandibular sclerite is broad, and at each side a dentate sclerite (*d.s.*) is articulated by means of a notch in the side of the mandibular sclerite, the function of which has been shown already in describing the muscles. The mandibular sclerite articulates posteriorly with the hypostomal sclerite (*h.s.*). This consists of two irregularly-shaped lateral portions united by a ventral bar of chitin; it is anterior to this bar of chitin that the salivary duct opens into the front of the pharynx. The sides of the hypostomal sclerite articulate with two processes on the anterior edge of the large pharyngeal sclerites (*l.p.*)[1].

The pharyngeal sclerites or plates recall the shape of the fulcrum of the adult fly. Each is wider posteriorly than anteriorly

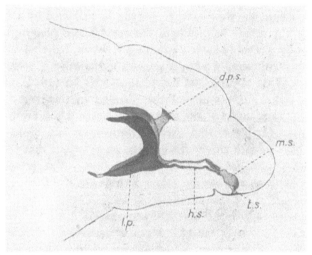

FIG. 57.   Cephalo-pharyngeal skeleton of the first larval instar; the outline of the pharyngeal mass is shown in dotted lines.

t.s. T-shaped sclerite of the left oral lobe.   Other lettering as in Fig. 56.

and the posterior end is deeply incised; at the base of this incision the nerves and tracheae which supply the interior of the pharynx enter. These sclerites vary in thickness, as will be seen in the figures of the sections of the pharynx. They are united dorsally at the anterior end by a dorsal sclerite, the dorso-pharyngeal sclerite (*d.p.s.*), and ventrally they are continuous with the floor of the pharynx.

[1] I have previously referred to these sclerites as the "lateral pharyngeal sclerites." As a result of some correspondence with Townsend concerning the terminology of these cephalo-pharyngeal sclerites we have agreed to call these the pharyngeal sclerites; they are the chief sclerites of the larval pharynx. The other terms used in this description have also been agreed upon by us as a result of a comparative study of the muscid larva.

The *pharynx* (figs. 50, 51) is similar in certain respects to that portion of the pharynx of the fly which lies within the fulcrum.

The whole length of the floor of the pharynx is traversed by a series of eight grooves separated by bifurcating ribs which are T-shaped in section (fig. 51, *T.r.*), and are called the "T-ribs" by Holmgren (1904); they form a series of eight tubular grooves. Holmgren believes that they may have been derived from a condition similar to that found in the pharynx of the larva of *Phalacrocera*, where the floor of the pharynx is traversed by a number of deep but closed longitudinal fissures. These pharyngeal ribs and grooves probably have a straining function, but they may also be of use in allowing a certain amount of the salivary secretion to flow backwards towards the oesophagus. Keilin (1912), who has made a comparative study of the pharynx in the larvae of the *Diptera Cyclorrhapha*, shows that all the larvae which are parasitic on animals and plants, and also the carnivorous and bloodsucking larvae, have no ribs on the floor of the pharynx but that such ribs are always present in saprophagous larvae. The presence or absence of these ribs can be taken to indicate the feeding habits of the larvae.

The musculature and action of the pharynx has been described. On the dorsal side of the pharyngeal mass, and attached laterally to the layer of cells covering the lateral sclerites, there is a loose membrane (fig. 50, *m.*), whose function, I believe, is to accommodate the blood contained in the pharyngeal sinus (fig. 51, *p.s.*) when the roof of the pharynx is raised. Posteriorly, the floor of the pharynx curves dorsally and opens into the oesophagus.

The *oesophagus* (fig. 49, *oe.*) is a muscular tube beginning at the posterior end of the pharyngeal mass. It describes a dorsal curve when the larva is contracted, and then runs in a straight line through the oesophageal foramen between the cerebral lobes of the ganglionic mass and dorsal to the ganglion to the posterior region of the sixth larval segment, where it terminates and opens into the proventriculus. It is of a uniform width throughout and is lined by a layer of flat epithelial cells (fig. 58, *oe.ep.*) whose internal faces are lined by a chitinous sheath (*ch.i.*) which is thrown into a number of folds. There is nothing of the nature of a ventral diverticulum forming a crop such as Lowne describes in the larva of the blow-fly.

The *proventriculus* (fig. 55, *pv.*) varies slightly in shape according to the state of contraction of the alimentary tract; in the normal

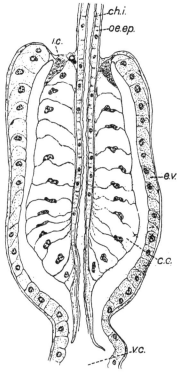

Fig. 58. Longitudinal section of the proventriculus of the mature larva. (Camera lucida drawing.)

c.c. Large cells forming the central hollow core of the proventriculus.   *ch.i.* Chitinous intima of the oesophagus.   *e.v.* Epithelial cells continuous with and similar in character to those of the ventriculus.   *i.c.* Ring of imaginal cells.   *oe.ep.* Oesophageal epithelial cells.   *v.c.* Lumen of ventriculus.

condition it is cylindrically ovoid and its axis is parallel with that of the body. As will be seen from the figure (fig. 58), it is rather similar to the proventriculus of the imago in general structure. The oesophageal epithelium penetrates a central core which is composed of large clear cells (*c.c.*); its lumen, being oesophageal, is lined with chitin. This core is surrounded by an outer sheath, the cells (*e.v.*) of which are continuous with those of the ventriculus. At the junction of the central core with the outer sheath of cells there is a ring of small more deeply-staining cells (*i.c.*). This ring was regarded by Kowalevski (1887) as the rudiment of the stomodaeum of the nymph, but Lowne is of the opinion that it develops in the nymph into the proventriculus of the imago. I believe that it forms a portion, at least, of the proventriculus of the imago, as it exhibits a very close resemblance to the ring of cells in this region figured in the section of the proventriculus of the imago (fig. 13).

The *mesenteron* of the mature larva is of very great length, and is not divisible into the two regions of ventriculus and small proximal intestine as in the imago, but appears to have the same character throughout; hence Lowne calls it the "chyle-stomach," which term, or ventriculus (fig. 55, *v.*), may be used to designate the whole region from the proventriculus to the point at which the malpighian tubes arise. It is very much convoluted and twisted upon itself. The course of the ventriculus is almost constant, and can be better understood from the accompanying figure than from any detailed description. At the anterior end four tubular caeca (*c.v.*) arise. Their walls consist of large cells whose inner faces project into the lumen of the glands; these glands were not present in the imago. The epithelium of the ventriculus (fig. 67) is composed of large cells (*e.v.*), which project into the lumen of the alimentary tract; they possess large nuclei and the sides of the cells facing the lumen have a distinctly striated appearance, which is absent in those epithelial cells covered with a chitinous intima. This striated appearance may be related in some way to the production of the mucous intima which is generally present in the ventriculus, and which appears to take the place of the loose chitinous intima or peritrophic membrane which occurs in this region in numerous insects, and which has been studied in detail by Vignon (1901) and others. Below the epithelial cells a number of small cells (*g.c.*) are found, which may be either gland cells or young epithelial cells. In addition to these cells small groups of deeply-staining fusiform cells (*i.c.*) are found below the epithelium. These, I believe, are embryonic cells from which the mesenteron of the imago arises. The malpighian tubes arise in the tenth segment at the junction of the ventriculus and the intestine.

The *intestine* (fig. 55, *int.*) is narrower than the ventriculus and runs forwards as far as the eighth segment, where it bends below the visceral mass and runs posteriorly, to become dorsal again behind the tenth segment, from whence it runs backwards, turning ventrally behind the visceral mass to become the rectum. The epithelium is thrown into a number of folds and is covered with a chitinous intima.

The *rectum* (*r.*) is very short and muscular, and the chitinous intima is fairly thick and continuous with the outer cuticular layer of the chitinous integument. It is almost vertical and opens by the anus on the ventral side of the terminal larval segment between the two swollen anal lobes.

Salivary glands.

There is a pair of large tubular salivary glands (*s.gl.*) lying laterally in segments five and six. Anteriorly each is continued as a tubular duct; the two ducts approach each other and join beneath the pharyngeal mass to form a single median duct (fig. 49, *sal.d.*) which runs forward and opens into the pharynx on the ventral side as already described. The glands are composed of large cells (fig. 68), which project into the lumen of the gland; they stain deeply and have large active nuclei. The salivary secretion, apart from the digestive properties which it has, is no doubt of great importance in making the food more liquid, as is also the case in the imago, and so rendering it more easy for absorption.

The malpighian tubes. (Fig. 55, *m.t.*)

These arise at the junction of the ventriculus and intestine in the tenth segment. A short distance from their origin they bifurcate and the resulting four tubules have a convoluted course, being mingled to a great extent with the adipose tissue. They are similar in appearance and histologically to those of the imago, consisting of large cells, of which only two can be seen usually in section; they consequently give the tubules a moniliform appearance. In the mature larva these cells appear to break down to form small deeply-staining spherical bodies. This histological degeneration begins at the distal ends of the tubules, which in the mature larva usually have the appearance shown in fig. 55 (*m.t.*); all the stages of degeneration can be traced out. This process may be a means of getting rid of the remaining larval excretory products.

THE RESPIRATORY SYSTEM.

The tracheal system (fig. 59) of the adult larva consists of two longitudinal tracheal trunks united by anterior and posterior commissures, and communicating with the exterior by anterior and posterior spiracles; the latter are situated in the middle of the oblique caudal end, and the anterior spiracles, which are not present in the first larval instar, are situated laterally at the posterior border of the third body-segment.

I believe that the anterior spiracles (*a.sp.*) are true functional spiracles, though for some time I was inclined to share Lowne's opinion that they were not functional. This latter view was due to the fact that it was difficult to understand how these spiracles could obtain air when they are immersed, as they usually are, in the moist fermenting materials on which the animal feeds. A careful microscopic examination of their structure, however, strengthens my belief that they are able, if necessary, to take in air; the occasions when this is possible are probably not infrequent. Each of the anterior spiracular processes consists of a fan-shaped body (fig. 44, *a.sp.*) bearing six to eight small papiliform processes. The papillae (fig. 47) open to the exterior by a small pore which leads into a cavity having a clear lumen surrounded by branched cuticular processes, whose function is probably to prevent solid particles from penetrating the spiracular channel. The body of the fan-shaped spiracular process is filled with a fine reticulum of the chitinous intima, which Meijere (1902) calls the "felted-chamber" (*Filzkammer*); through this meshwork the air can pass to the longitudinal tracheal trunk.

The posterior spiracles (fig. 61, *p.sp.* and fig. 62) are **D**-shaped with the corners rounded off and their flat faces are opposed. Each consists of a chitinous ring having, internal to the flat side, a small

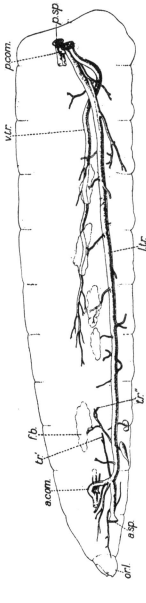

FIG. 59. The longitudinal lateral tracheal trunk of the left side seen latero-dorsally showing the origin of the tracheal branches; small portions only of the right trunk are shown.

a.com. Anterior tracheal commissure.   a.sp. Anterior spiracular process.   f.b. Fat-body.   or.l. Oral lobe.
l.tr. Longitudinal lateral tracheal trunk.   p.com. Posterior commissure.   p.sp. Posterior spiracle.
tr'. Trachea entering ganglion anteriorly.   tr''. Trachea entering ganglion laterally.   v.tr. Visceral
tracheal trunk.

pierced knob. Each chitinous ring encloses three sinuous slits, guarded by inwardly-directed fine dendritic processes; through these slits the air enters the small spiracular atrium, which is situated behind the spiracle. The spiracular atria communicate directly with the longitudinal tracheal trunks.

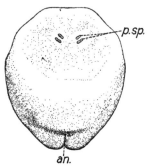

Fig. 60. Posterior end of larva in the second stage (2nd instar).
an. Anus.   *p.sp.* Posterior spiracle.

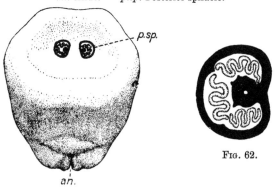

Fig. 61.

Fig. 61. Posterior end of mature larva (3rd instar).
Fig. 62. Posterior spiracle of mature larva of *Musca domestica*.

The origin and course of the branches of each of the longitudinal tracheal trunks (fig. 59, *l.tr.*) are the same, so that of the left side will be described only. Immediately behind the spiracular atria the short posterior tracheal commissure (*p.com.*) connects the two trunks. In the younger larvae this commissure is situated more anteriorly, but in the adult it is situated so far back and so close to the spiracles that its presence might easily be overlooked.

On the outer side of the tracheal trunk a large branch arises; this, the visceral branch (*v. tr.*), bends ventrally to the lateral trunk, and thus becoming internal to it enters the convoluted visceral mass with its fellow of the other side.

The visceral branches extend anteriorly as far as the seventh segment. In the twelfth and thirteenth segments the lateral tracheal trunk has a double appearance. A dorsal and a ventral branch arise in most of the segments, the dorsal branch chiefly supplies the fat-body, and the ventral branch supplies the viscera; both give off branches to the muscular body-wall. The anterior commissure (*a. com.*) is situated in the fourth segment. It crosses the oesophagus immediately behind the pharyngeal mass. On the internal side of the portion of the lateral tracheal trunk, that is anterior to the commissure, a branch arises, and running ventral to the pharyngeal mass it supplies the anterior end of the larva and the oral lobes. A branch that supplies the muscles of this region is given off external to the origin of the anterior commissure. Internal to the origin of the commissure two tracheae arise; the anterior branch enters and supplies the pharyngeal mass, and the posterior branch (*tr'.*) enters the ganglion ventral to the cerebral lobes. In the fifth segment another internal tracheal branch (*tr".*) enters the ganglion. These tracheae which supply the ganglion appear to run chiefly in the peripheral regions, where they divide into a number of branches, the fate of some of these being interesting. These branches are extremely fine, and they arise, as I have previously mentioned, in association with a number of the segmental nerves with which they run to the body-wall.

The Vascular System and Body Cavity.

The relations and structure of the vascular system of the larva are on the whole similar to those of the fly; there are, however, a number of modifications.

The *dorsal vessel*, which includes the so-called "heart," is a simple muscular tube lying on the dorsal side immediately beneath the skin, and extending from the posterior tracheal commissure to the level of the cerebral lobes of the compound ganglion in the fifth segment. Its wall is composed of fine striated muscle-fibres arranged transversely and longitudinally, but chiefly in the latter direction. The swollen posterior region (fig. 63), which is called the heart, lies in the last three or four segments, its anterior limit being hard to define. It consists of three distinguishable chambers, which, however, are not divided by septa. Three pairs of ostia (*os.*), each provided with a pair of internal valves (*v.*), are situated laterally, and place the cardiac cavity in communication with the pericardium, in which this portion of the dorsal vessel lies. There are three pairs of alar muscles controlling the action of this posterior cardiac region of the dorsal vessel. Lowne describes other openings in the wall of the "heart" of the blow-fly larva, but I have been unable to find others than those already described in this larva; it has three pairs only.

The dorsal aorta is the anterior continuation of the dorsal vessel, which gradually diminishes in diameter. When it reaches the fifth segment and lies above the ganglion, it terminates in a peculiar cellular structure (fig. 53, *c.r.*), which in the blow-fly has a circular shape and was called by Weismann the "ring." In the larva of *M. domestica* it has not so pronounced a ring-like appearance, but is more elliptically compressed and rather ∧-shaped. The cells of which it is composed have a very characteristic appearance, and are rather similar to a small group of cells lying on the neck of the proventriculus and at the anterior end of the dorsal vessel of the fly. From the lower sides of this cellular structure (*c.r.*) the outer sheaths of the major cephalic imaginal discs depend, and extend anteriorly to the pharyngeal mass, enclosing between them the anterior portion of the great ventral blood sinus.

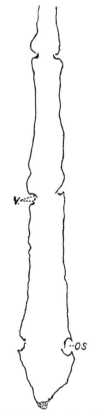

Fig. 63. Horizontal section of posterior or "cardiac" region of the dorsal vessel. (From camera lucida drawings.)

os. Ostium. *v.* Valvular flaps guarding the same.

The pericardium lies in the four posterior segments of the body, and is delimited ventrally from the general body-cavity by a double row of large characteristic pericardial cells. These cells have a fine homogeneous structure and are readily distinguished from the adjacent adipose tissue cells, whose size they do not attain. The pericardial cavity contains a profuse supply of fine tracheal vessels which indicates a respiratory function. A similar condition occurs in the blow-fly larva, and Imms (1907) has described a rich pericardial tracheal supply in the larva of *Anopheles maculipennis*, as also Vaney (1902) and Dell (1905) in the larva of *Psychoda punctata*.

The adipose tissue-cells (fig. 66, *f.c.*) form the very prominent "fat-body." They are arranged in folded cellular laminae that lie chiefly in the dorso-lateral regions of the body, and in section have the appearance shown in the figure. The cells have a similar structure to those of the adult fly; they are very large with reticular protoplasm containing fat globules, and there may be more than one nucleus in a single cell. As in the fly, the fat-body is closely connected with the tracheal system by means of a very rich supply of tracheae.

Two chief blood-sinuses can be distinguished—the pericardial sinus, which has already been described, lying in the dorsal region in the four posterior segments, and the great ventral sinus. The latter lies between the outer sheaths of the major cephalic imaginal discs and extends anteriorly into and about the pharynx; posteriorly, it encloses the ganglion and the convoluted visceral mass above which it opens into the pericardial sinus between the pericardial cells.

The blood which fills the heart and sinuses and so bathes the organs is an almost colourless, quickly coagulable fluid, containing colourless, nucleated, amoeboid corpuscles and small globules of a fatty character.

THE IMAGINAL DISCS.

As in other cyclorrhaphic Diptera, the imaginal discs of some of which have been described by Weismann (1864), Kunckel d'Herculais (1875–78) and Lowne, the imago is developed from the larva by means of these imaginal rudiments, which are gradually formed during the later portion of the larval life. They do not all appear at the same time, for whereas some may be in a well-developed state early in the third larval instar, others do not appear until the larva reaches its resting period or even later. The imaginal discs appear to be hypodermal imaginations, though their origin is difficult to trace in all cases; in many

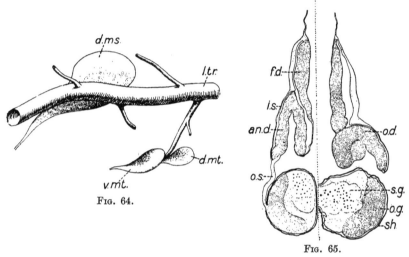

Fig. 64.

Fig. 65.

FIG. 64. Internal aspect of the posterior thoracic imaginal discs of the right side.

d.ms. Dorsal mesothoracic or alar imaginal disc.   *d.mt.* Dorsal metathoracic
    disc.   *l.tr.* Lateral tracheal trunk of the right side of larva.   *v.mt.* Ventral
    metathoracic imaginal disc.

FIG. 65. Longitudinal sections through the major cephalic imaginal discs of
    mature larva to show the position of the individual imaginal rudiments. The
    dextral section is more dorsal than the sinistral. (Camera lucida drawings.)

an.d. Imaginal disc of the antenna.   *f.d.* Facial imaginal disc.   *i.s.* Sheath
    of imaginal rudiments.   *o.d.* Optic imaginal disc.   *o.g.* Imaginal disc of
    the optic ganglionic structures.   *o.s.* Optic stalks.   *s.g.* Fundament of the
    imaginal supra-oesophageal ganglionic.   *sh.* Sheath of cerebral lobe.

instances they are connected with the hypodermis by means of a stalk of varying thickness. The imaginal disc or rudiment may consist of a simple or of a folded lamina of deeply-staining columnar embryonic cells, as in the wing discs, or of a number of concentric rings of these cells, as in the antennal and crural discs. They are usually closely connected with the tracheae and in some cases are innervated by fine nerves. Although the imaginal discs of *M. domestica* are similar in some respects to those of the blow-fly, as described by Lowne, there are several important differences, chief of which is the position of the imaginal discs of the metathoracic legs.

During the resting period of the larva the cephalic and thoracic discs can be distinguished, but the abdominal discs are small and not so obvious except in sections.

The cephalic discs.

The chief cephalic discs are contained in what at first appears to be a pair of cone-shaped structures in front of each of the cerebral lobes of the ganglion (fig. 53, *m.c.d.*); the cone, however, is not complete. The outer sheath of each of these major cephalic imaginal rudiments is continued dorsally, and joins the cellular structure mentioned previously (see fig. 66), thus enclosing a triangular space which is a portion of the ventral sinus. These sheaths are continued anteriorly and are connected to the pharyngeal mass, and it is through this connecting strand of tissue that the discs are everted to form the greater part of the head of the nymph. Immediately in front of the cerebral lobe is the so-called optic disc (fig. 65, *o.d.*), which in its earlier stages is cup-shaped, but later it assumes a conical form, having a cup-shaped base adjacent to the cerebral lobe. The optic disc is connected to the cerebral lobe laterally by a stalk of tissue, the optic stalk (*o.s.*), which becomes hollow later, and it is through this stalk that the optic ganglion and associated structures contained in the cerebral lobe appear to evaginate when the final metamorphosis and eversion of the imaginal rudiments takes place. The optic discs form the whole of the lateral regions of the head of the fly.

The remaining portion of the head-capsule of the fly is formed from two other parts of imaginal rudiments, the antennal and facial discs. The antennal disc (*an.d.*) lies in front, and internal to, the optic discs. Each consists of an elongate conical structure, in which at a later stage the individual antennal joints can be distinguished. The facial discs (*f.d.*) are anterior to the antennal discs and extend to the anterior end of the conical structure containing these three pairs of major cephalic discs, which will form the cephalic capsule.

In addition to these two other pairs of cephalic discs are found. A pair of small flask-shaped maxillary rudiments are situated one at the base of each of the oral lobes; a second pair of imaginal discs, similar in shape to the maxillary discs, is found adjacent to the hypostomal sclerite; the latter, I believe, are the labial rudiments, and will form almost the whole of the proboscis of the fly.

The thoracic discs.

In *M. domestica* there are five pairs of thoracic discs. The prothoracic imaginal discs (figs. 53 and 66, *pr.d.*) are attached to

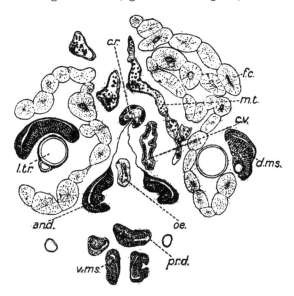

Fig. 66.   Transverse section of mature larva anterior to the ganglion and cerebral lobes to show the position of certain of the imaginal discs. The body-wall and muscles have been omitted. The folded character of the adipose tissue laminae can be seen in this section, and also the degenerating anterior portions of the malpighian tubules (*m.t.*). (Camera lucida drawing.)

an.d. Antennal disc.   *c.r.* Problematical cellular structure (Weismann's "ring"). *c.v.* Caecum of ventriculus.   *d.ms.* Dorsal mesothoracic (alar) imaginal disc. *f.c.* Adipose tissue cell.   *l.tr.* Lateral tracheal trunk.   *m.t.* Malpighian tubule cut rather longitudinally.   *oe.* Oesophagus.   *pr.d.* Prothoracic imaginal disc.   *v.ms.* Ventral mesothoracic imaginal disc.

the anterior end of the ganglion and slope obliquely forwards; the distal end of each is attached to the body-wall on the ventral side between segments three and four. These discs develop into the prothoracic legs, and probably also into the much-reduced prothoracic segment, as I was unable to discover any other rudiments corresponding to the dorsal imaginal discs of the mesothoracic and metathoracic segments. Arising from the sides of the ganglion immediately behind the attachment of the prothoracic rudiment are the imaginal rudiments of the mesothoracic legs and sternal region (*v.ms.*); the distal stalks of this pair of imaginal discs are attached to the body-wall at the posterior border of the fourth segment. The dorsal mesothoracic imaginal discs, from

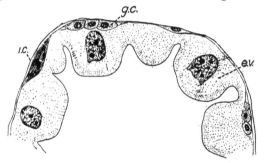

FIG. 67. Transverse section of a portion of the ventriculus of mature larva. (Camera lucida drawing.)

e.v. Epithelial cell of ventriculus showing large active nucleus and striated peripheral region of cell.   *g.c.* Probable gland cells.   *i.c.* Group of imaginal cells.

which originate the mesonotal region and the wings, may be termed the alar or wing discs. They form a pair of flattened pyriform sacs (fig. 64, *d.ms.*), lying one on each side of the ventral side of the fifth segment and slightly external to the lateral tracheal trunk (fig. 66, *d.ms.*), to a ventral branch of which each is attached. The metathoracic discs consist of two pairs of small pyriform masses (fig. 64) lying immediately behind the alar discs in the intersegmental line. They are attached to a ventral branch of the lateral tracheal trunk. The anterior rudiment (*v.mt.*) is the larger, and forms the imaginal metathoracic leg and sternal region; in the blow-fly and in *Volucella* it is interesting to note that this pair of imaginal discs is situated further forward, and is in association with the corresponding prothoracic and mesothoracic ventral discs. The smaller and more posterior disc (*d.mt.*) will develop into the remaining portion of the much reduced metathoracic segment, including the halteres.

Reference has already been made to other imaginal rudiments which occur in the abdominal region as regular patches of embryonic cells. The abdominal segments develop from numerous segmentally arranged plates of a similar nature which are found during the early pupal stage.

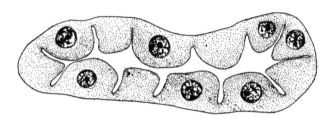

Fig. 68.   Transverse section of one of the salivary glands of the mature larva.
(Camera lucida drawing.)

During pupation the imaginal rudiments increase in size and are not destroyed by the phagocytes in histolysis, as is the case with most of the larval structures.   The cephalic discs are evaginated by the eversion of their sacs by way of the anterior end of the larva, a cord of cells attached to the dorsal wall of the anterior end of the pharynx marking the path of eversion.   A similar process takes place in the case of the thoracic imaginal discs, which, by their eversion, build up the whole of the skeletal case of the thorax and its dorsal and ventral appendages, the wings, halteres and legs.

PART III

THE NATURAL ENEMIES AND PARASITES OF THE HOUSE-FLY

CHAPTER X

ARACHNIDS AND MYRIAPODS

CHERNES NODOSUS SCHRANK.

THERE are frequently found attached to the legs of the house-fly small scorpion or lobster-like creatures which are Arachnids, belonging to the order *Pseudo-scorpionidea*; the term "chelifers" is also applied to them on account of the large pair of chelate appendages which they bear. The species which is usually found attached to *M. domestica* is *Chernes nodosus* Schrank (fig. 69). It is very widely distributed, and my observations agree with those of Pickard-Cambridge (1892), who has described the group.

The species is 2·5 mm. in length and Pickard-Cambridge's description of it is as follows:

"Cephalothorax and palpi yellowish red-brown, the former rather duller than the latter. Abdominal segments yellowish-brown; legs paler. The caput and first segment of the thorax are of equal width (from back to front); the second segment of the thorax is very narrow. The surface of the cephalothorax and abdominal segments is very finely shagreened, the latter granulose on the sides. The hairs on this part as well as on the palpi and abdomen are simple, but obtuse. The palpi are rather sharp and strong. The axillary joint is considerably and somewhat sub-conically protuberant above as well as protuberant near its base underneath. The humeral joint at its widest part, behind, is

considerably less broad than long; the cubital joint is very tumid
on its inner side; the bulb of the pincers is distinctly longer, to
the base of the first claw, than its width behind; and the claws
are slightly curved and equal to the bulb in length."

They appear to be commoner in some years than in others.
Godfrey (1909) says: "The ordinary habitat of *Ch. nodosus*, as
Mr Wallis Kew has pointed out to me, appears to be among refuse,
that is, accumulations of decaying vegetation, manure heaps,
frames and hot-beds in gardens. He refers to its occurrence in a

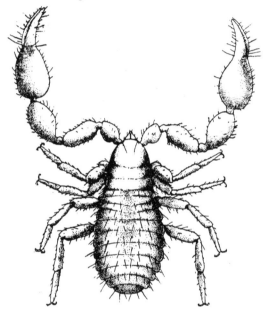

Fig. 69. *Chernes nodosus*, Schr. × 30.

manure-heap in the open air at Lille, and draws my attention to
its abundance in a melon-frame near Hastings in 1898, where it
was found by Mr W. R. Butterfield." In view of these facts it is
not difficult to understand its frequent occurrence on the legs of
flies, which may have been on the rubbish heaps either for the
purpose of laying eggs, or, what is more likely, because they have
recently emerged from pupae in those places and in crawling about,
during the process of drying their wings, etc., their legs were
seized by the *C. nodosus*.

The inter-relation of the *Chernes* and *M. domestica*, however, is one·of no little complexity; much has been written and many diverse views are held concerning it. An interesting historical account of the occurrence of these Arachnids on various insects has been given by Kew (1901). Three views are held in explanation of the association and they are briefly these : First, that the *Chernes*, by clinging passively to the fly, uses it as a means of transmission and distribution; second, that the Arachnid is predaceous; and, third, that it is parasitic on the fly. Owing to the unfortunate absence of convincing experimental proof in favour of either of the last two opinions, it is practically impossible to give any definite opinion as to the validity of these views; nevertheless they are worthy of examination.

The dispersal theory was held by Pickard-Cambridge and Moniez (1894). Whether the other views are held or not there is no doubt that such an association, even if it were only accidental, would result in a wider distribution of the species of *Chernes*, as the flies are constantly visiting fresh places suitable as a habitat for the same. Except in one or two recorded cases, the Arachnids are always attached to the legs of the fly, the chitin of which is hard and could not be pierced, a fact which is held in support of this theory as the only explanation of the association.

The parasitic and predaceous views are closely related. The Pseudo-scorpionidea feed upon small insects, which they seize with their chelae. It is suggested by some that the *Chernes* seizes the legs of the fly without realising the size of the latter. Notwithstanding its size, however, they remain attached until the fly dies and then feed upon the body. In some cases as many as ten of the Arachnids have been found on a single fly, and if the movements of the insect are impeded by the presence of a number of the *Chernes*, it will be easily understood that the life of the fly will be curtailed thereby. Pseudo-scorpionidea have been observed feeding on the mites that infest certain species of Coleoptera, and it has been suggested that they associated with the flies for the same purpose, although I do not know of any recorded case of a fly infested with mites carrying *Chernes* also. If this were the case the *Chernes* would be a friend and not a foe of the fly, as Hickson (1905) has pointed out.

There are few records to support the view that the *Chernes* is parasitic on the fly. Donovan (1797) mentions the occurrence of a Pseudo-scorpionid on the body of a blow-fly, and Kirby and Spence (1826) refer to their being occasionally parasitic on flies, especially the blow-fly, under the wings of which they fix themselves. It is probable that the *Chernes* seldom reaches such a position of comparative security on the thorax of the fly; should it succeed in doing so, however, it would become parasitic in the true sense of the word. As I have previously pointed out, little experimental evidence is at present available and further investigation is necessary before it is possible to maintain more than a tentative opinion with regard to this association between the *Chernes* and the fly. It is obvious that the association will result in the distribution of the Pseudo-scorpionid, but whether this is merely incidental and the real meaning lies in a parasitic or predaceous intention on the part of the Arachnid, as some of the observations appear to indicate, further experiments alone will show[1].

MITES OR ACARINA BORNE BY HOUSE-FLIES.

Most careful observers and even casual observers have noticed that house-flies are occasionally infested with small reddish mites which are attached to their bodies in various positions.

As early as 1735 de Geer observed small reddish Acari in large numbers on the head and neck of *M. domestica*. They ran about actively when touched. The body of this mite was oval in shape, completely chitinised, and polished; the dorsal side was convex and the ventral side flat. Linnaeus (1758) called this mite *Acarus muscarum* from de Geer's description, and Geoffroy (1764) found what appears to be the same, or an allied species of mite, which he called the "brown fly-mite." Murray (1877) describes a form, *Trombidium parasiticum*, which is a minute blood-red mite parasitic on the house-fly. He says: "In this country they do not seem so prevalent, but Mr Riley mentions that in North America,

---

[1] I have since endeavoured to throw some light on this question by keeping *Chernes* and flies in small vials. In no case, however, did I observe the *Chernes* feeding on the living fly, although they would feed occasionally on the dead flies.

in some seasons, scarcely a fly can be caught that is not infested with a number of them clinging tenaciously round the base of the wings." As it only possessed six legs it was doubtless a larval form. This species was named *Atoma parasiticum* and later *Astoma parasiticum* by Latreille[1]. Mr A. D. Michael informed me that the genus was founded on *Trombidium parasiticum* of de Geer. They were really larval Trombidiidae and *Atoma* was founded on larval characters; probably any larval *Trombidium* came under the specific name.

FIG. 70. *M. domestica* infested with *Trombidium* mites (X). Magnified nearly six times.

Howard (1911) quotes the following on the authority of Banks: "Latreille based a new genus and species on mites from the house-fly and he called it *Atomus parasiticum*. This is the young of one of the Harvest Mites of the family Trombidiidae but the adult has not been reared and is still unrecognised in Europe. Riley found these harvest mites on house-flies in Missouri, in some years so abundantly, he says, that scarcely a fly could be caught that was not infested with some of them clinging tenaciously

[1] *Magazin Encyclopedique*, Vol. IV, p. 15, 1795.

at the base of the wings. Later he succeeded in rearing the adult, and described it as *Trombidium muscarum*. In recent years Oudemans has described *Trombidium muscae* from larval mites found on house-flies in Holland. All these forms are minute, six-legged, red mites which cling to the body of the fly and with their thread-like mandibles suck up the juices of the host. When ready to transform they leave the fly and cast their skins, the mature mite being a free-living, hairy, scarlet creature about 1·5 mm. long. The adults are usually found in the spring and early summer, while the larvae are usually found in the autumn on house-flies and other insects." I have illustrated (fig. 70) a fly caught in Ottawa in September, 1909, which is infested with this species of mite.

Howard states further that "mites of the genus *Pigmeophorus*, of the family Tarsonemidae, have also been taken on house-flies. They cling to the abdomen of the fly, but it is not certain whether they feed on the insect or use it simply as a means of transportation. The hypopus or migratorial nymphal stage of several species of *Tyroglyphus* has been found on house-flies. This hypopus attaches itself by means of suckers to the body of any insect that may be convenient. The mites do not feed on the fly but when the fly reaches a place similar to that inhabited by the mites the latter drop off, cast their skins and start new colonies."

Anyone who has collected Diptera as they have emerged from such breeding places as hot-beds, rubbish and manure heaps will have noticed the frequently large number of these insects which are to be found carrying immature forms of the Acari. These are being transported merely by the flies in the majority of cases. Mr Michael informed me that he used to call such flies "the emigrant waggons"—a very descriptive term. Many of these mites belong to the group Gamasidae—the super-family Gamasoidea of Banks (1905). These mites have usually a hard coriaceous integument. In shape they are flat and broad and have rather stout legs. Sometimes immature forms of these mites swarm on flies emerging from rubbish heaps. Banks holds the opinion that they are not parasitic, but that the insect is only used as a means of transportation. It is difficult to decide whether this is so in all cases. I have illustrated (fig. 71) a specimen of the Lesser House-fly, *F. canicularis*, caught in a room; on the underside of the fly's abdomen a number of immature Gamasids are attached, apparently by their stomal regions. Mr Michael, to whom I submitted these mites, said that it was extremely difficult to identify immature Gamasids owing to the scarcity of our knowledge as to their life-histories, but he stated that they were very like *Dinychella asperata* Berlese.

These specimens may be truly parasitic, as I am inclined to believe, since many Acari are parasitic in the immature state, although the adults may not be so; on the other hand this form of attachment may be employed as a means of maintaining a more secure hold of the transporting insect.

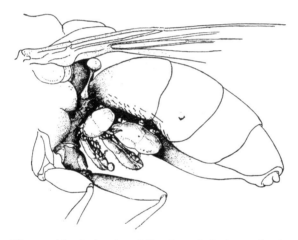

Fig. 71.   Thoraco-abdominal region of *Fannia canicularis*, ♀, showing Gamasids attached to the ventral side of the abdomen.

Ewing (1913) describes a new species of Gamasid mite *Macrocheles muscae* which is parasitic on *M. domestica*, always attaching itself, according to the author, in a definite place, namely, at the base of the abdomen on its ventral side, the anterior end of the larva being directed forwards. It feeds on the host. Its colour is dark *yellowish* brown; length 0·97 mm., width 0·62 mm. It has been found in the States of Oregon and New York.

Hamer (1909) records Gamasid mites as particularly affecting *Muscina stabulans*, especially in early June.

Berlese (1912) has reared what he considers to be the *Acarus muscarum* of Linnaeus from the Stable-fly, *Muscina stabulans*, and finds that the adult belongs to the genus *Histiostoma*. He also illustrates two Acarids attached to *M. domestica*; attached to the right anterior tibia is a larval *Trombidium* and attached to the left hind tibia is a migratory *Holostaspis marginatus* Herm., a species which is accustomed to attach itself to coprophagous insects.

By the transference of the hypopal or migratorial stage of those species of mites which are destructive to cheese and other foods, house-flies are frequently responsible for infecting such foods with mites. While their food is abundant the adult mites reproduce rapidly, the young mites developing into adults in a very short time. Should the food supply become exhausted or other unfavourable conditions supervene the almost fully grown mites develop the hard protective shells characteristic of the hypopal or migratorial stage. Thus protected they attach themselves to house-flies or other flies and trust that the inquisitive wanderings of their transporting host will carry them to pastures new.

Centipedes and Spiders.

The carnivorous habits of the centipedes are well-known but the peculiar genus *Scutigera* contains a number of species which feed upon insects, including the house-fly, when the opportunity occurs.

In the southern and eastern regions of the United States a species, *Scutigera forceps* Raf., occurs very frequently in houses where, according to Howard, its food consists principally of household insects such as *M. domestica*, clothes moths and small cockroaches. It is a small fragile-looking animal with unusually long legs and feeds at night, its long legs apparently being of great service to it in capturing its prey.

Kunckel d'Herculais (1911) records the occurrence of *Scutigera coleoptera* L. in France where it occurred especially in privies, hunting flies by night, its chief prey being the latrine fly *Fannia scalaris*, which breeds abundantly in such places (see p. 193).

The genus *Scutigera* is very widely distributed, *S. smithii*

Newport occurs in Australia and, like the aforementioned species, has been found to hunt the house-fly by night[1].

The relations which exist between spiders and flies are so universally recognised that it is difficult to make an original observation on the subject and perhaps hardly necessary in an account of this nature. House spiders and garden spiders will devour whatever house-flies "walk into their parlours," and the little jumping spider *Salticus scenicus* may be frequently observed leaping upon flies almost twice its size.

---

[1] Recently, I captured *Scutigera forceps* in Toronto (Canada) and fed it on flies.

CHAPTER XI

PROBABLY the most important of all the enemies of the house-fly is the parasitic fungus *Empusa muscae*. Towards the end of the summer large numbers of flies may be found attached in a rigid condition to the ceiling, walls or window-panes. They have an extremely life-like appearance, and it is not until one examines them closely or has touched them that their inanimate, so far as the life of the fly is concerned, condition is discovered. These flies have been killed by the fungus *Empusa muscae* Cohn, and in the later stages of the disease its fungal nature is recognised by the fact that a white ring of fungal spores may be seen around the fly on the substratum to which it is attached (fig. 72). The abdomen of the fly is swollen considerably, and white masses of sporogenous fungal hyphae may be seen projecting for a short distance from the body of the fly, between the segments, giving the abdomen a transversely striped black and white appearance.

The majority of flies which die in the late autumn—and it is then that most of the flies which have been present during the summer months perish—are killed by this fungus. Its occurrence, therefore, is of no little economic value, especially if it were possible to artificially cultivate it and destroy the flies in the early summer instead of being compelled to wait until the autumn for the natural course of events.

Empusa muscae belongs to the group Entomophthoreae, the members of which confine their attacks to insects, and in many cases, as in the case of the present species, are productive of great mortality among the individuals of the species of insect attacked. In England it may be found from the beginning of July to the end of October and usually occurs indoors.   Howard states that the epidemic usually ceases in Washington, U.S.A., in December. Its distribution almost coincides with that of the house-fly and it is the only species of *Empusa* which has, as yet, been recorded from the southern hemisphere.   While it is uncommon out-of-doors, I have found specimens of *Musca domestica* killed by it out-of-doors in Canada, at Ottawa.   It has been recorded out-of-doors in England where it was found attacking a species of Syrphid,

Fig. 72.   *M. domestica* killed by *Empusa muscae*, showing discharged spores (conidia).   (Photo by H. T. Güssow.)

Melanostomum scalare Fabr. on Esher Common[1].   Thaxter (1888) also mentions two cases of its occurrence out-of-doors in the United States, in both of which cases it had attacked species of *Syrphidae*.   The same author states that *Empusa muscae* is probably the only species which occurs on flowers attractive to insects, but he only observed it on flowers of *Solidago* and certain Umbelliferae.   He also records two other species of *Empusa* attacking the house-fly, namely, *E. sphaero-sperma* (Fres.) Thaxter and

[1] *Trans. Ent. Soc. London*, Proceedings, p. 57, 1908.

E. americana Thaxter.   *E. americana* attacks blow-flies and other flies similar in size to the house-fly and is frequently found out-of-doors.

E. muscae, besides occurring on *M. domestica* has been found on several species of *Syrphidae* and also on *Lucilia caesar* and *Calliphora vomitoria*.

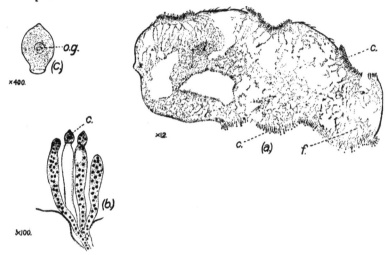

Fig. 73.   (a) Longitudinal (sagittal) section of abdomen of *M. domestica*, which has been killed by *Empusa muscae*, showing the feltwork of fungal hyphae filling the inside of the abdominal cavity and the production of conidia in the intersegmental regions.   × 12.   *c.* Conidiophores producing conidia.   *f.* Fungal hyphae.

(b) Four conidiophores showing the formation of conidia (*c*).   × 100 (approx.).

(c) Conidium of *Empusa muscae*.   × 400.   *o.g.* Oil globule.

The development of the house-fly fungus *Empusa muscae* was very carefully studied by Brefeld (1871). An *Empusa* spore which has fallen on a fly rests among the hairs covering the insect's body and there adheres. A small germinating hypha develops, which pierces the chitin, and after entering the body of the victim penetrates the fat-body. In this situation, which remains the chief centre of development, it gives rise to small spherical structures which germinate in the same manner as yeast cells, forming gemmae. These separate as they are formed, and falling into the blood sinus are carried throughout the whole of the body of the fly. It was probably these bodies that Cohn (1855) found, and he explained their presence as being due to spontaneous generation; he believed that the fly first became diseased and that the fungus followed in consequence. After a period of two or three days the fly's body will be found to be completely penetrated by the fungus, which destroys all the internal tissues and organs. The whole body is filled with the gemmae, which germinate and produce ramifying hyphae (fig. 73, *a*).

The latter pierce the softer portions of the body-wall between the segments and produce the short, stout conidiophores (*c*.), which

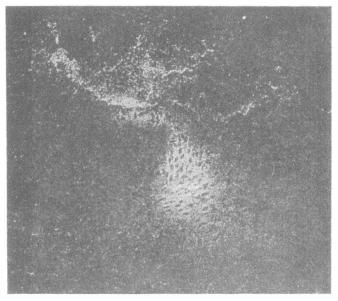

Fig. 74. Discharged conidia of *Empusa muscae* upon which *M. domestica* has been feeding as indicated by the proboscis marks. (Photo by H. T. Güssow.)

are closely packed together in a palisade-like mass to form a compact white cushion of conidiophores, which is the transverse white ring that one finds between each of the segments of a diseased, and consequently deceased, fly. A conidium now develops (fig. 73, *b*) by the constriction of the apical region of the conidiophore. When it is ripe the conidium (fig. 73, *c*) is usually bell-shaped, measuring 25–30 μ in length; it generally contains a single oil-globule (*o.g.*). In a remarkable manner it is now shot

11—2

off from the conidiophore, usually for a distance of about a centi-
metre, although I have seen spores discharged to a distance of
70 mm. In this way the ring or halo of white spores, which are
seen around the dead fly, are formed. My friend Mr H. T. Güssow
has confirmed in conjunction with me the external infection by
the conidiospores. He has also found that the conidiospores may
be taken up through the mouth, and fig. 74 shows the proboscis marks of flies which have been feeding on the discharged
conidiospores.

In some cases, although I find that it is not an invariable rule
as some would suggest, the fly, when dead, is attached by its
extended proboscis to the substratum. Giard (1879) found that
blow-flies killed by *Entomophthora calliphora* were attached by
the posterior end of the body. If the conidia, having been shot
off, do not encounter another fly, they have the power of producing
a small conidiophore, upon which another conidium is in turn
developed and discharged. If this is unsuccessful in reaching a
fly a third conidium may be produced, and so on. By this peculiar
arrangement the conidia may eventually travel some distance, and
it is no doubt a great factor in the wide distribution of the fungus,
once it occurs. On the fly itself short conidiophores may be found
producing secondary conidia.

Reproduction by conidia appears to be the only form of generation, as we are still uncertain as to the occurrence of a resting-
spore stage in this species. Winter (1881) states that he found
resting-spores in specimens of *M. domestica* occurring indoors ;
they also produced conidia which he identified as *E. muscae.*
These azygospores measured 30–50 μ in diameter, and were produced laterally or terminally from hyphae within the infected fly.
Giard (*l.c.*) describes resting-spores which were produced externally
and on specimens found in cool situations. Brefeld, however, is of
the opinion that *E. muscae* does not produce resting-spores. The
question of the production of resting-spores needs further investigation, and it is one of some importance. In the absence of
confirmatory evidence it is extremely difficult to understand how
the gap in the history of the *Empusa*, between the late autumn of
one year and the summer of the next, is filled. A number of suggestions have been made, many of which cannot be accepted ; for

example, Brefeld believes that the *Empusa* is continued over the winter in warmer regions, migrating northwards with the flies on the return of the summer! In the case of *Entomophthora calliphora*, Giard believes that the cycle is completed by the corpses of the blow-flies falling to the ground, when the spores might germinate in the spring and give rise to conidia which infect the larvae. Olive (1906) studied the species of *Empusa* which attacks a species of *Sciara* (Diptera) and found the larvae infected. He accordingly thinks that the disease may be carried over the winter by those individuals which breed during that period in stables and other favourable places. As I have shown, *M. domestica*, under such favourable conditions as warmth and supply of suitable larval food, is able to breed during the winter months, although it is not a normal occurrence so far as I have been able to discover. If, then, these winter-produced larvae could become infected they might assist in carrying over the fungus from one year to the next, and thus carry on the infection to the early summer broods of flies. This suggestion and the possible occurrence of a resting-spore stage or the infection by conidiospores surviving from the previous year, appear to me to be the probable means by which the disease may be carried over from one " fly-season " to the next.

Until this gap in the life-history is filled it will not be possible to determine experimentally the practical value of this fungus as a means of destroying the house-fly. No one, so far as I am aware, has yet succeeded in rearing the fungus on artificial media although many attempts have been made. Nor has it been possible in my own experience to infect flies from flies of a previous year which have been killed by the fungus. Investigations are now being conducted, according to Bernstein (1910), on this parasitic fungus in connection with the enquiry of the Local Government Board and the results will be awaited with interest[1].

[1] Since the above was written two interesting papers on this subject have appeared.

Güssow (1913) has given the results of a careful investigation on the life-history of *E. muscae*. He did not succeed, however, in carrying cultures of the fungus from spore to spore outside the fly and failed to find resting-spores. His observations indicate that flies may become infected by ingesting the spores into the alimentary tract.

Buchanan (1913) studied chiefly the possibility of bacterial dissemination by the spores of *E. muscae* and found that the spores were able to take up, and to carry when discharged from the bodies of the flies, bacteria of the colon group.   He was unable to cultivate the fungus artificially.   From the fact that in the body of the diseased fly the fungal hyphae closely invest the chorion of the eggs he believes that the larvae emerging from such eggs might become infected with the fungus and thus transmit the infection to the next generation.   I should point out, however, that there are two strong objections to this possibility.   First, once a fly is attacked by the fungus it would be incapable of ovipositing by the time the disease had reached a stage where the ovaries have become infected with the fungus. Secondly, even did the first objection not exist, the emergence of the larva from the egg is of such a nature as to permit the larva to escape infection, especially as the fungal hyphae do not penetrate the chorion of the egg.

Buchanan records the fact that Morgan has succeeded in obtaining an artificial culture of the fungus (see *Brit. Med. Journ.* Jan. 4th, 1913).

CHAPTER XII

INSECT AND VERTEBRATE ENEMIES

PARASITIC INSECTS.

LIKE other insects, the house-fly is subject to the attack of a variety of parasitic enemies and in the course of some investigations in 1908 on the house-fly in Illinois, Girault and Sanders (1909, 1910) found a number of Chalcidoid parasites of *M. domestica* and its near relatives. They all appeared to belong to the family

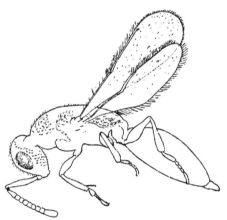

FIG. 75. *Spalangia* sp., a parasite reared from *M. domestica.* × 20.

Pteromalidae and three genera, namely, *Spalangia* (fig. 75), *Nasonia* and *Muscidifurax* were discovered attacking *M. domestica* and other Muscids, the last genus being previously undescribed.

In the first of the two papers mentioned the parasite *Nasonia brevicornis*, a new species, is described. This parasite, the authors state, is " stolid and serious, little heeding external influences and disturbances, quietly, persistently giving its whole attention to reproduction." It attacks *M. domestica* in the pupal stage. The female is a minute insect measuring from 1 mm. to 2·30 mm. in length and is of a metallic dark brassy-green colour; the eyes are garnet coloured. The male is about one-third smaller than the female, varying in length from 0·60 mm. to 2·00 mm. It is lighter in colour, more brassy in appearance, metallic and green; the eyes are sometimes a brilliant carmine. The wings of the male appear to be functionless as they have never been seen to fly and although the female is able to fly both sexes appear to prefer to crawl and are able to crawl quickly. They reproduce very rapidly like most Pteromalid parasites. Maggots and puparia of *M. domestica* were placed in a breeding jar with females of *Nasonia brevicornis* on September 9, 1908; reproduction of the parasites occurred, and on September 26, 1908, males and females of the Pteromalid parasites emerged. The average life-cycle of *Nasonia brevicornis* under natural temperatures is 22½ days; the duration of the development is longer in the spring than in the summer. The parasite hibernates in the larval stage in the puparium of its host and transforms to a pupa in the spring; a number of species of Pteromalids have this habit.

Another parasite of this group which these authors have found attacking the house-fly is *Pachycrepoideus dubius*. It was reared in experiments with *Nasonia brevicornis* and was obtained from the pupae.

The same authors also describe a third Pteromalid parasite attacking *Musca domestica* and its coprophagous allies. This species they have named *Muscidifurax raptor*; it is a minute black insect with clear wings and is somewhat solitary in its habits. The female apparently lays from thirty to forty eggs which are deposited in the pupae, in the interior of which the last annual brood passes the winter as a full-grown larva. The average duration of the summer broods was between nineteen and twenty days.

Richardson (1913) has described a new species, *Spalangia muscidarum*, which was bred from the pupae of *M. domestica* and of *Stomoxys calcitrans*[1].

---

[1] More recently Richardson (1913) has described the habits and development of *Spalangia muscidarum*. The highest proportion of parasitised house-fly puparia

There is no doubt, judging from the numerical abundance of the above parasites which the authors indicate in their papers, and from my own observations in the case of another species of Pteromalid, that these parasites are of importance in holding *M. domestica* in check where the parasites are sufficiently abundant.

In addition to the many Chalcid parasites of *M. domestica*, a number of parasites belonging to the family Cynipidae have been reared from the house-fly. Most of the Cynipidae are minute gall-forming insects, causing some of the well-known galls on plants such as the oak galls. One sub-family, the Figitines,

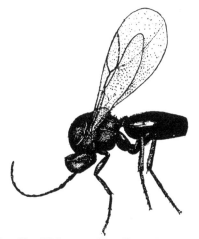

FIG. 76. *Figites* sp. Muscid parasite. × 28.

however, are parasitic on the larvae of insects, having been obtained from dipterous, neuropterous and coleopterous larvae, and they do not attack plants. These insects are described by Kieffer

found was 9 *Spalangia* larvae and pupae from 22 puparia. He found two generations, and a third more or less irregularly, each year in Massachusetts. The winter is probably passed in the pupal stage and the parasites emerge in the spring. The second and strongest generation emerges during late July or August, while a third may appear late in September or early October.

Bishopp (1913) found *S. muscidarum* a parasite of *M. domestica*, *Stomoxys calcitrans* and *Haematobia serrata* in Texas.

Pinkus (1913) studied the life-history and habits of *Spalangia muscidarum* as a parasite of *S. calcitrans* at Dallas, Texas. The parasitism of *M. domestica* was also studied and he discusses the possibility of the artificial propagation of the parasites.

(1902) in his monograph on the Cynipidae. The members of the genus *Figites* (fig. 76) are parasitic on dipterous larvae which live in various kinds of excrement, and in consequence the minute adult flies may be caught frequenting human, horse, cow and other kinds of excrement for the purpose of depositing their eggs on the dipterous larvae which may be found there. The species most commonly attacked belong to the genera of the dipterous families which have coprophagous larvae, such as the *Muscidae*, *Anthomyidae, Sarcophagidae* and *Scatophagidae*. The commonest species, *Figites scutellarius* Rossi, has been reared by Förster from *M. domestica* according to Dalla-Torre. Other species which have been reared from *M. domestica* are *F. striolatus* and *F. anthomyiarum*. Further investigations will no doubt indicate that a number of the species of *Figites* which have already been described are able to parasitise *M. domestica*.

PREDACEOUS INSECTS.

As Howard has pointed out, it is a remarkable fact that the larvae of *M. domestica* are not destroyed to a greater extent than our observations would appear to indicate by the numerous species of predaceous insects which feed upon soft-bodied insects. To a small extent the ground-beetles (Carabidae) and rove-beetles (Staphylinidae) may sometimes be found feeding or induced to feed upon the larvae of *M. domestica*, but not to the extent that they are accustomed to feed upon other soft-bodied larvae such as many lepidopterous, coleopterous and other dipterous larvae.

Packard (1874) records the occurrence of what was probably a Dermestid beetle, which he figures; this was found by him in a pupa of *M. domestica*. Berg (1898) states that in South America a species of beetle, *Trox suberosus* F., locally known as " Champi," is an indirect destructor of the common fly.

Ants are frequently responsible for the destruction of *M. domestica* in the egg, larval and adult stage. Wheeler (1910) records ants feeding on house-flies and their larvae. He informs me that he has seen the Fire Ant *Solenopsis geminata* Fab. at Quiraguá, Guatemala, feeding upon the larvae of *M. domestica* which they extracted from human excrement. Howard (1911) refers to the observations of a number of persons on the destruction of *M. domestica* by ants. Capt. P. L. Jones found it impossible to rear *M. domestica* in the Philippine Islands unless the eggs and larvae (in manure) were protected from ants. Stallman (1912) reports the destruction of fly larvae by red ants in Arizona. It would appear, however, that usually ants do not prey upon *M. domestica* to an extent sufficient to make the results of their predatory habits appreciable.

The common species of wasps (*Vespa* spp.) have been frequently observed by myself and others preying upon flies which they occasionally destroy in large numbers. A Canadian engineer, who had been engaged in railroad construction in northern Ontario, described to me the mystification caused by the presence of large numbers of the wings of flies on the table of their cabin. On investigation, it was found that a large species of wasp was catching the flies in the cabin; after capturing a fly the wasp took it to a beam immediately over the table and there cut off the wings before eating it or carrying it away.

The large hairy Robber-flies of the dipterous family *Asilidae* frequently catch flies. I have observed and captured *Laphria canis* Will. in the act of catching and eating *M. domestica*[1].

In some parts of India it is the custom, I have been informed by residents in that country, to employ a species of *Mantis*, one of the predatory "praying insects," to destroy the house-flies.

Compere (1912) has found earwigs destroying the maggots of stable-flies in fresh manure in southern China and believes that they are an important factor in the control of house-flies in Hongkong. It is hardly likely, however, that they would exert so great an influence as is suggested.

The destruction of the larvae of *M. domestica* by the larvae of the allied muscid *Muscina stabulans* and of *Hydrotaea dentipes*, is recorded by Portchinsky (1913). He found that the larvae of *M. stabulans*, having completed the second stage, follow and attack the larvae of *M. domestica* and soon exterminate all that happen to be living with them. The larva of *M. stabulans* that had killed

---

[1] Since the above was written I have described (Hewitt, 1914) the predaceous habits of the common yellow dung fly *Scatophaga stercoraria* L. which was found destroying *M. domestica* and other Muscidae.

the larva of *M. domestica* was never observed to devour it alone, a number of other individuals usually joined in the feast. Nor were the larvae of *M. stabulans* ever seen to devour each other. Portchinsky also found that they would bore into and eat the bodies of dead flies. The habits of *M. stabulans* are considered later (pp. 207–210).

Vertebrate Enemies.

That the house-fly has a number of vertebrate enemies, exclusive of man, is a matter of common observation. Cats will sometimes sit in a window and catch flies. Rats have also been observed destroying flies. Birds will destroy *M. domestica* in both the adult and larval stages though not in preference to other species of insects.

Poultry will feed upon the larvae and pupae of *M. domestica* which they may find in the stable-yard and are sometimes of some service in this respect.

Lizards, toads and frogs will capture the adult flies whenever an opportunity occurs, but their influence in the matter of control is too slight to be noticeable.

CHAPTER XIII

PROTOZOAL PARASITES

HERPETOMONAS MUSCAE-DOMESTICAE BURNETT.

THE Herpetomonas of the house-fly, *Herpetomonas muscae-domesticae* Burnett, has been known for many years as a flagellate parasite of the alimentary tract of *M. domestica*, but the discovery of the relation to certain diseases of a number of species of Trypanosomes and allied flagellates has been responsible for a considerable addition to our knowledge of the life-history of this and other species of flagellates inhabiting the alimentary tracts of insects.

In 1878 Stein figured a flagellate which he called *Cercomonas muscae-domesticae*, identifying it with the *Bodo muscae-domesticae* described by Burnett and the *Cercomonas muscarum* of Leidy. For this form Kent (1880–81) instituted a new genus *Herpetomonas*. When the haemo-flagellates were being studied some years later, Prowazek (1904) described with great detail the development of this species. In the previous year Léger (1908–1909) had given a short account of the species. Patton (1908–1909) has also described the life-history of *H. muscae-domesticae*, and his account has been confirmed by Porter (1909), Mackinnon (1910) and Wenyon (1911 and 1913) who in his later paper gives a careful account of the cytology of the flagellate.

The full-grown flagellate (fig. 77, VIII) measures 30–50 μ in length. The body is flattened and lancet-shaped, the posterior end being pointed and the anterior end bluntly rounded. The alveolar endoplasm contains two nuclear structures. In the centre is the large "trophonucleus" (*tr.*); it contains granules of chromatin, but is sometimes difficult to see. Near the anterior end the deeply staining rod-shaped "kinetonucleus" (blepharoplast of many authors) (*k.*) lies, usually, in a transverse position. The single stout flagellum, which is a little longer than the body of the flagellate, arises from the anterior end, near the kinetonucleus. Prowazek describes the flagellum as being of a double nature and having a double origin; this, which is a mistaken interpretation, is repeated by Lingard and Jennings (1906). Prowazek (1913) has corrected what he considers misinterpretations of his original statements concerning *H. muscae-domesticae.* By statistical methods he has found that the greater number of the flagellates examined in Rovigno are biflagellate.

This mistake concerning the double nature of the flagellum was pointed out by Léger and Patton, and their ideas have since been confirmed by Porter, Mackinnon and Wenyon, who have studied this and other species of *Herpetomonas.* In some flies Porter found that practically every *H. muscae-domesticae* which was seen exhibited the so-called "double-flagellum." The appearance of a double-flagellum represents the beginning of the longitudinal division of the flagellate (VI). Patton (1908) figures a stage in *H. lygaei* with a double-flagellum, and Léger (1902) and Porter found the same appearance in *H. jaculum* preparatory to division. From the figures that these authors give it may be understood how this mistake has arisen. Through this misinterpretation Prowazek was led to consider that the parasite was of a bipolar type, in which the body has been doubled on itself so that the two ends came together and the flagellum remained distinct. The flagellum, according to Léger, is continued into the cytoplasm as a thin thread, which stains with difficulty, and terminates in a double granule above the kinetonucleus; this double granule is no doubt the "diplosome" of Prowazek. According to the latter author another deeply staining double thread (*s.t.*), that appears to be spirally coiled, runs backwards from the kinetonucleus and terminates posteriorly in a distinct granule, shown in fig. 77, VIII. Wenyon (1913) believes that a cytopharynx is present, using as an argument the presence of bipolar bodies which he believes are bacteria taken up by the Herpetomonas.

The flagellates congregate in the proventriculus or in the posterior region of the intestine, where they become united by

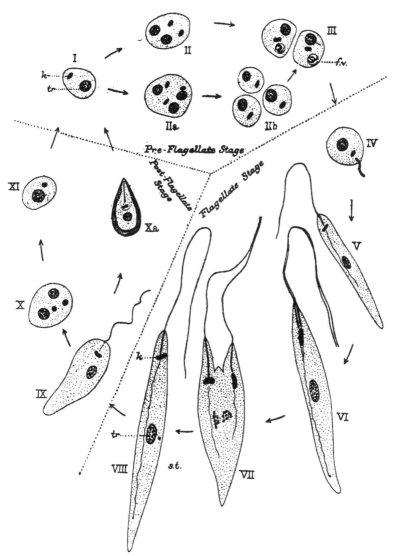

Fig. 77. Diagram of the life-cycle of *Herpetomonas muscae-domesticae* Burnett. Arrangement chiefly after Patton; figures after Léger, Patton, and Prowazek. I—III. Pre-flagellate stage. IV—VIII. Flagellate stage: V. Young flagellate. VI. Flagellate beginning to divide, flagellum having already divided. VII. Advanced stage of division. VIII. Adult flagellate. IX—XI. Post-flagellate stage: IX. Degeneration of flagellum. Xa. Post-flagellate stage completed by formation of gelatinous covering, containing double row of granular bodies (Prowazek). *f.v.* Flagellar vacuole. *k.* Kinetonucleus. *s.t.* Spiral chromophilous thread. *tr.* Trophonucleus.

their anterior ends to form rosettes. Prowazek states that in the rosette condition the living portion of the flagellate resides, as it were, in the long tail-like process.

Patton divides the life-cycle of *H. muscae-domesticae* into three stages—the pre-flagellate, flagellate, and post-flagellate. The last two are common, but the first stage is not common, and Prowazek appears to have overlooked it. For convenience I have described the flagellate stage first, and the process of division in this stage is simple longitudinal fusion. The nuclei divide independently, and the kinetonucleus usually precedes the trophonucleus. The latter undergoes a primitive type of mitosis, in which Prowazek recognised eight chromosomes (VII). The flagellum divides longitudinally, and each of the two halves of the kinetonucleus appropriates one of the halves with its basal granule.

The pre-flagellate stage, which Patton (1909) describes, usually occurs in the masses which lie within the peritrophic membrane[1]. They are round or slightly oval bodies (I), their average breadth being 5·5 μ. The protoplasm is granular and contains a trophonucleus and kinetonucleus. Division takes place by simple longitudinal division or multiple segmentation, and in this manner a large number of individuals are formed (II *b* and III). These develop into the flagellate stage: a vacuole, the flagellate vacuole (III, *f.v.*) appears between the kinetonucleus and the rounded end of the pre-flagellate form, and in it the flagellum appears as a single coiled thread, which is extended when the vacuole has approached the surface.

The flagellate form has already been described, and in the concluding portion of the flagellate stage, which, according to Prowazek, is found in starved flies, these forms are found collecting in the rectal region, and attaching themselves by their flagellar ends in rows to gut epithelium. The more external ones begin to shorten, during which process the flagella degenerate (IX) and are shed. Thus a palisade of parasites is formed, the outer ones being rounded and devoid of flagella, and some of them may be found dividing (X). Léger (1902) terms these the "formes gregariennes," and maintains that the existence of these "gregarine" forms is a

[1] I assume that Patton refers to this membrane by the term "peritricheal membrane."

powerful argument in favour of the flagellate origin of the Sporozoa, which he had previously suggested, and which Butschli had put forward in 1884. After the degeneration of the flagellum a thickened gelatinous covering is formed, containing a double row of granular bodies (X*a*), and these cysts are regarded by Patton as the post-flagellate stage. Wenyon (*l.c.*) finds that nuclear multiplication may occur during encystment. Dunkerly (1911) describing the life-history of a flagellate under the name of *Leptomonas muscae-domesticae*, to which reference is made later, records the occurrence in the rectum of flies of small oval bodies similar in appearance to these cysts; they were found near the rectal glands, and flies containing these cysts contained no flagellate forms.

The cysts pass out with the faeces of the fly and dropping on the moist window pane or on food are taken up by the proboscides of other flies which are thus infected.

Prowazek describes dimorphic forms of the flagellate stage, which he regards as sexually differentiated forms, but Patton, in a letter to me, says that he is unable to find any of these complicated sexual stages. According to Prowazek, one of these forms is slightly larger than the other, and has a greater affinity for stain. The dimorphic forms conjugate; their cell substance and nuclei fuse, and a resting-stage cyst is formed, but the subsequent stages have not been followed. He further states that the sexually differentiated forms may force their way into the ovaries where they undergo autogamy and infect the subsequent brood. Mackinnon in her study of *Herpetomonas* was unable to find, after a careful search, any infection of the ova. She never found any larval infection in *M. domestica*, but it was commonly found in *Scatophaga* and *Fannia*. It is suggested that the infection of the adult fly is probably fresh and is independent of that of the larva.

In Madras Patton found that 100 per cent. of the flies were infected with the flagellate; Prowazek found it in 8 per cent. of the flies at Rovigno. In the cold season in the plains (India) Lingard and Jennings (*l.c.*) found the flagellate in less than 1 per cent. of the flies examined; in the hills (Himalayas), at an elevation of 7500 feet, the flagellates were most numerous during the hottest season of the year, and gradually decreased in number to October and November, when none were discovered. Wenyon (*l.c.*) found that the majority of house-flies at Aleppo, Syria, contained *H. muscae-domesticae*. Although films of the contents of the alimentary tracts of a large number of house-flies were made and examined by me at Manchester (England) I was unable to find with any certainty these flagellates, nor was Dunkerly (*l.c.*) more successful.

One of the chief points of interest in connection with this flagellate is its similarity to the parasite of Kala-azar. This resemblance prompted Rogers (1905) to suggest that the latter parasite was a *Herpetomonas* which Patton has since not only conclusively proved to be the case but has also shown that the bed-bug can act as its carrier.

CRITHIDIA ˙MUSCAE-DOMESTICAE WERNER.

Werner (1906) described this parasite from the alimentary tract of *M. domestica*, where he stated that it occurred in the alimentary tracts of four out of eighty-two flies examined.

It measures 10–13 μ in length, the length of the body being 5–7 μ and the flagellum 2–6 μ. As in other members of the genus *Crithidia*, which is closely allied to *Herpetomonas*, the breadth of the body is great compared with the length, and the kinetonucleus and trophonucleus are rather close together. A short, staining, rod-like body lies between the kinetonucleus and the base of the flagellum. The flagellum is single. Dividing forms undergoing longitudinal division were frequently found. The kinetonucleus appears to divide first, followed in succession by the flagellum and the trophonucleus. Forms undergoing division and showing a single trophonucleus and double kinetonucleus and flagellum were also found. Cases occurred in which the fission began at the non-flagellate end of the body. No conjugating forms were found, nor any wandering into the ovaries.

Lingard and Jennings (*l.c.*) describe certain flagellates of a flag-shaped or rhomboidal nature, which I am strongly of the opinion are species of *Crithidia* and not species of *Herpetomonas*. Closely following Prowazek's account of *H. muscae-domesticae* they describe and figure all their forms as having two flagellae in the flagellate stage. If one allows for the rupture of the flagellum from the bodies of the organism in making the film, some of their figures are not unlike those of *Crithidia gerridis*, parasitic in the alimentary tract of an Indian water-bug, *Gerris fossarum* Fabr., and described by Patton (1908).

Rosenbusch (1910) describes this species of *Crithidia* as being very numerous. The flagellates were found covering considerable patches of the peritrophic membrane of the intestine to which they were attached by their anterior ends. They showed irregular flagellate movements. In shape they were flattened laterally, the end of the body gradually tapering off to a blunt extremity. The pointed anterior end terminates in the free flagellum. Rosenbusch gives the following as the dimensions of his flagellates: length, 8–25 μ; width, 1–2 μ; length of flagellum, 12 μ. Post-flagellate stages were also found; these were shortened and rounded and possessed no flagellum.

It should be pointed out that there is in the minds of a number of those who have studied these flagellates a doubt as to whether these flagellates, which have been described as *Crithidia*, are really distinct parasites; they may be stages in the life-history of *Herpetomonas*. Patton and Strickland (1908) call attention to the marked resemblance of Werner's parasites to the post-flagellate stages of *Herpetomonas muscae-domesticae*. In living specimens Patton has seen the flagellates of the house-fly collecting in masses in the rectum of the insect where the typical long forms shorten, divide and eventually round up, and these authors rightly point out that Werner's contention that the difference in size between his parasites (*Crithidia*) and the flagellates of *H. muscae-domesticae* is an argument in favour of their being unconnected forms will not hold good in view of the marked dissimilarity in form found in the different stages of the life-history of *H. muscae-domesticae*. Dunkerly (*l.c.*) has described as a distinct flagellate *Leptomonas muscae-domesticae*, to which I shall refer shortly. Here again a non-specialist of the flagellates is in doubt as to its relationship. Further investigation will alone clear up these uncertain and disputed points; in the meantime I am recording these flagellates under the names and in the manner in which they have been described. They may be distinct or developmental forms of two species or of only one.

Leptomonas muscae-domesticae. In a paper on some stages in the life-history of what he considers to be a distinct flagellate, Dunkerly (*l.c.*) attempts to show that *Crithidia* is not a valid genus and cannot be applied as a generic name to any form, as it has simply been the name given to two stages of a *Leptomonas.* He describes a Leptomonad which actively divides in the intestine or in the malpighian tubules of the fly and also very active slender forms which often show an undulating membrane. Encystment while attached to the rectal wall in large numbers probably takes place and the cysts may be passed out with the faeces and thus infect fresh flies. Porter (1911) calls attention to the faulty reasoning contained in Dunkerly's paper, as she considers it, and the lack of evidence. Wenyon (*l.c.*) believes that the *Leptomonas* of Roubaud and others is merely a not actively dividing *Herpeto-monas muscae-domesticae* and that both may pass through a transition into flagellates of a trypanosome type.

Patton (1909) refers in a critical paper to the discovery which he made of another flagellate in the malpighian tubules of *M. domestica* but I am unaware of any further reference to it.

CHAPTER XIV

THE PARASITIC NEMATODE: *HABRONEMA MUSCAE* CARTER

CARTER (1861) appears to be the first to have described a parasitic worm in *M. domestica*. He describes a bi-sexual nematode infesting this insect in Bombay and found that: "Every third fly contains from two to twenty or more of these worms, which are chiefly congregated in, and confined to, the proboscis, though occasionally found among the soft tissues of the head and posterior part of the abdomen." His description of this nematode, to which he gave the name *Filaria muscae*, is as follows: "Linear, cylindrical, faintly striated transversely, gradually diminishing towards the head, which is obtuse and furnished with four papillae at a little distance from the mouth, two above and two below; diminishing also towards the tail, which is short and terminated by a dilated round extremity covered with short spines. Mouth in the centre of the anterior extremity. Anal orifice at the root of the tail." He gives the length as being one-eleventh of an inch and the breadth as one three hundred and thirteenth of an inch. In his description of his figures of the worm he calls what is evidently the anterior region of the intestine the "liver." Leidy (1874) found from one to three specimens of *F. muscae* in about one fly in five. He stated that this parasitic worm was one-tenth of an inch long and occurred in the proboscis. Ercolani (1874) describes the discovery of a nematode in the proboscides of flies. Von Linstow (1875) describes a small nematode, which he calls *Filaria stomoxeos*, from the head of *Stomoxys calcitrans*; this larva measured 1·6 to 2 mm. in length. Ransom (1913) points out that this may be the larva of *Habronema microstoma*. Harrington (1883) refers to a paper read by Taylor before the Montreal meeting of the American Association for the Advancement of Science on "The House-fly as a Carrier of Contagion." This observer, when dissecting a house-fly, noticed a minute thread-worm emerging from the ruptured proboscis measuring eight-hundredths of an inch in length. Incidentally, the same investigator reported the results of feeding flies on the spores of the red rusts of grasses (*Tricholoma*) and found these ingested. Generali (1886) describes a nematode from the common fly, which he calls *Nematodum spec.* It is highly probable, as my friend Dr A. E. Shipley suggested to me, that Generali's nematode and the *F. muscae* of Carter are identical. Diesing (1861) created the genus Habronema for the *Filaria muscae* of Carter, and his description is practically a translation of Carter's original description. Piana (1896) describes a nematode from the proboscis of *M. domestica*, which, in the occurrence of the male and female genital organs in the same individual, he says, resembles Carter's nematode. He finds that at certain seasons of the year and in certain localities it is very rare, while at others it may occur in 20–30 per cent. of the flies. The larva, after fixation, measured 2·68 mm. in length and 0·08 mm. in breadth. It was cylindrical and gently tapering off at the extremities, with the mouth terminal.

In many hundreds of flies which I dissected when making the morphological studies which have been described in the earlier chapters of this book, only two specimens of this nematode were found, but as I did not seek it specially it is very possible that specimens may have been unnoticed. The specimens which I found measured 2 mm. in length and agreed entirely with the descriptions of *Habronema muscae* (fig. 78). Both specimens were found in the head between the optic ganglia and the cephalic air sacs.

Recently, a very complete and unusually interesting study of *Habronema muscae* (Carter) has been made by Ransom (1911, 1913). This author has cleared up the mystery which hitherto surrounded the parentage of this nematode and has discovered that the adult is a parasite of the horse, the house-fly acting as a carrier of the larval form. The following account is taken from Ransom's description of the occurrence, structure and life-history of *H. muscae*.

One hundred and thirty-seven flies were examined by him for the presence of *Habronema*; 39 of these, or 28 per cent. were found to be infested. While Carter found as many as 20 larval nematodes in a single fly, Ransom never found more than 8. Out of 43 flies of which a record was kept 25 had but one parasite, 6 had 2 parasites, 6 had 3 parasites, 3 had 4 parasites, 2 had 5 parasites, and 1 had 8 parasites.

A record was kept of the location of the parasites in the case of 37 flies. In 17 cases the head was infested, in 8 cases the thorax, and in 19 cases the abdomen. In 12 cases the head only was infested, in 4 cases the thorax only, and in 15 cases the abdomen only. In one case 6 larvae were found in the head; the largest number found in the abdomen was 5 (in one case). Encysted larvae were found in the abdomen in 6 flies, in 5 of which no worms were present in either the head or thorax. All of the flies in which encysted larvae were found had recently emerged.

One hundred and thirty-seven pupae were examined and a record kept and larvae of *Habronema* were found in 23, that is in 17 per cent. The author points out that some of the younger stages of the parasites may have escaped detection. Encysted worms were found in 11 out of 23 pupae. The location of the larvae in the pupae was as follows : abdomen infested nine times, head once, thorax once. Only one nema-

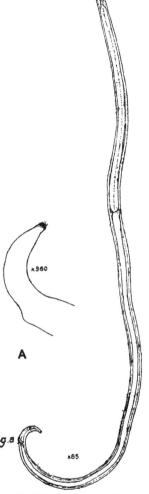

Fig. 78. *Habronema muscae*(Carter). Full grown larva. × 85. *g.a.* Genito-anal aperture.
A Caudal end of *Habronema muscae*. × 360.

tode was found in 170 larvae of *M. domestica*; this specimen was encysted.

The adults of *Habronema muscae* were found in the stomach of the horse together with the last larval stage, which is the last stage in the fly and the first in the horse. The adult male of *H. muscae* varies in length from 8 to 14 mm., and from 250 to 300 μ in maximum width. The female varies in length from 13 to 22 mm., and in width in the region of the vulva from 250 to 400 μ. The cuticle is faintly marked with transverse striations. The head is armed with two lateral and four submedian papillae. The tail of the male curves ventrally. The spicules are very dissimilar; the left spicule is long and slender measuring about 2·5 mm. in length, the right spicule is shorter and thicker measuring about 500 μ in length. In the female the vulva is situated about one-third of the length of the body from the anterior end.

The eggs in the uterus of *H. muscae* measure about 40–50 μ long by 10–12 μ wide. The embryos pass out of the body of the horse in the faeces. It has not been determined whether they undergo further development before entering the larvae of *M. domestica* or whether their entry is forcible or with the food; the latter method of entry seems the more probable. The earliest definitely known larval stage of *H. muscae* found in the fly occurred in a pupa from a culture of horse faeces. This specimen measured 450 μ in length. Second stage larvae were found enclosed in a cyst. The sixth larval stage, which is the final larval stage, was found in both flies and horses; the length varies from 2·6 to 3·2 mm., and the width at the widest part from 55 to 70 μ. The larval stage is reached about the time the flies emerge from the pupal state. The infection of the horse probably takes place by the swallowing of infected flies or of larvae which have escaped from flies. As the proboscis of the fly is a common location of the larval nematodes the escape of the larval nematodes on to the moist surface of the horse's lips is not unlikely.

Ransom points out that the presence of larval nematodes in flies is of interest to entomologists and sanitarians alike, as it may serve as a means of determining with some degree of accuracy what proportion of flies in a given locality find their breeding place in horse manure.

The occurrence of parasitic nematodes in the head of the fly is of further economic interest, for although *M. domestica* is not a blood-sucking species and the nematode previously described is not of the nature of the pathogenic *Filaria bancrofti*, there is no reason why the house-fly should not, under the necessary conditions, carry pathogenic nematodes which might easily get on to the food of man.

PART IV

OTHER SPECIES OF FLIES FREQUENTING HOUSES

CHAPTER XV

THE LESSER HOUSE-FLY *FANNIA CANICULARIS* L[1].
AND THE LATRINE FLY, *F. SCALARIS* FAB.

THE two species of flies *Fannia canicularis* L. and *F. scalaris* Fab. are, on account of their habits, of considerable economic importance in their relation to man. They belong to the dipterous family *Anthomyidae*, many of which resemble the house-fly in general appearance, on which account *F. canicularis* is very frequently mistaken by the uninitiated for *M. domestica* which are not full grown or "young" house-flies. They are characterized chiefly by the close approximation of the eyes of the male, the comparatively large squamae or lobes on the posterior sides of the bases of the wings, and the open first posterior or apical cell (5 *R.*) of the wing (cf. fig. 7). Most of the larvae feed upon decaying vegetable or animal substances.

Without close examination, the two species under examination are liable to be mistaken for the same species, but such an examination will serve to separate them. The abdomens of both species are conical, but the basal segments of the abdomen of *F. canicularis* are partially translucent, and the abdomen of *F. scalaris* is black overspread with bluish grey; each of the mid tibiae of the latter species bears a distinct tubercle which is not found in *F. canicularis* (fig. 81).

[1] Until recent years this species has always been referred to as *Homalomyia canicularis*, but by the rules of priority the generic name *Fannia* of Robineau Desvoidy, 1830, which was given in his *Essai sur les Myodaires*, will have to replace Bouche's genus *Homalomyia*, by which generic name these species have been previously designated but which genus was not created until 1834.

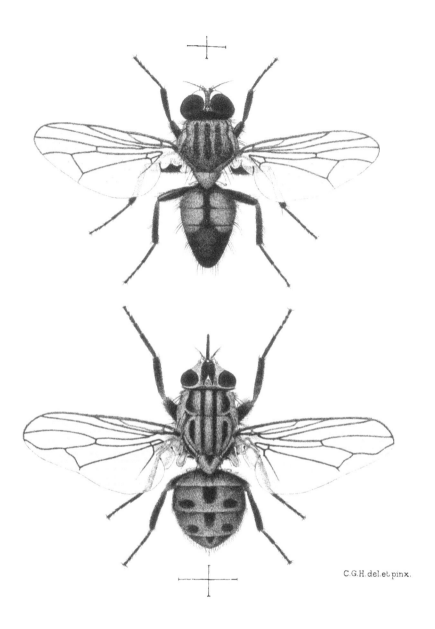

C.G.H. del.et pinx.

FIG. 79. Lesser House-fly, *Fannia canicularis* L. Male.

FIG. 80. Stable Fly, *Stomoxys calcitrans* L. Female.

THE LESSER HOUSE-FLY. *Fannia canicularis* L.

This species (fig. 79) is the less common of the two species of flies found in houses. Its occurrence and frequency are, however, very variable, and no valid explanation has been found so far in my investigations to account for this variability. *F. canicularis* is more abundant than *M. domestica* for a short time during the early part of the summer, usually in May and June. With the beginning of the hot weather the numbers of the latter increase enormously and replace the Lesser House-fly. In many cases which were observed the latter seemed to retreat in small numbers to the rooms of the house not devoted to cooking, and they may

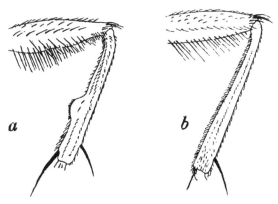

FIG. 81. Median joints of middle pair of legs (right); posterior aspect.
a. *Fannia scalaris*.   b. *F. canicularis*.

be frequently found flying in a characteristic, jerky and hovering manner around chandeliers, etc., in the living and bed rooms. In country houses, however, they frequently occur in numbers in the kitchens, as an examination of fly-traps and papers in such places indicates. An observation recorded by Austen (1911) illustrates the earlier occurrence of *F. canicularis* as compared with *M. domestica*. Out of a collection of more than 430 flies caught in a kitchen in Leeds, from May 19th to July 18th, only 48 specimens of *M. domestica* occurred, all the rest being *F. canicularis*, which outnumbered the former in the ratio of 8 or 9·5 to 1.

The numerical abundance of *F. canicularis* in comparison with the abundance of *M. domestica* varies considerably. In a collection of 4000 flies which I made in different situations, such as kitchens, restaurants, bed-rooms, etc., in 1907, this species formed 11·5 per cent. of the total number. In 1900 Howard found that in collections made in different cities of the United States only about 1 per cent. of a collection of over 23,000 flies made in rooms where food was exposed were *F. canicularis*, and over 98 per cent. were *M. domestica*. Hamer in 1908, in collections made in kitchens and "living rooms" of houses near depots for horse-manure in London, found that the percentage of *F. canicularis* varied from 17 per cent. to 24 per cent. Niven gives the results of collections made at six different stations in Manchester. The total number of flies caught was 8553, of which 8196 were *M. domestica*, 293 *F. canicularis*, and 64 were other species. Thus *F. canicularis* constituted 3·4 per cent. of the total of the fly population. Robertson (1909) gives the results of similar collections made in Birmingham where, of 24,572 flies caught, 91 per cent. were *M. domestica* and 4·7 per cent. were *F. canicularis*. From observations which I have made in many localities in different neighbourhoods, I do not think that this species would often form more than 25 per cent. of the total fly population. After *M. domestica*, however, it is the next fly of importance inhabiting houses, and well deserves the title of the Lesser House-fly. It is known in Germany as "die kleine Stubenfliege."

The male of *F. canicularis* differs from the female in some respects. In the male the eyes are close together, and the frontal region is consequently very narrow; the sides of this, which are the inner orbital regions, are silvery white, separated by a narrow black frontal stripe. In the female the space between the inner margins of the eyes is about one-third of the width of the head; the frons is brownish black, and the inner orbital regions are dark ashy grey. The bristle of the antenna of *F. canicularis* is bare; in *M. domestica*, it will be remembered, the bristle bears a row of setae on its upper and lower sides. The dorsal side of the thorax of the male is blackish grey with three rather indistinct longitudinal black lines. In the female it is of a lighter grey, and the three longitudinal stripes are consequently more distinct. The abdomen of the male *F. canicularis* is narrow and tapering compared with that of *M. domestica*. It is bronze-black in colour, and each of the three abdominal segments has a lateral translucent area, so that when it is seen against the light, as on a window-pane, three and sometimes four pairs of yellow translucent areas can be seen by the transmitted light. In the female the abdomen is short in proportion to its length and is pyriform in shape, greenish or brownish-grey in colour with a golden attachment. The average length of the species is 5·7 mm.

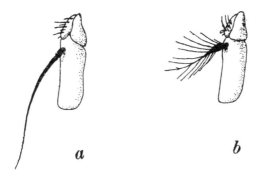

FIG. 82.   Antennae, (*a*) of *F. canicularis*, (*b*) of *M. domestica*.

Great disparity in the proportion of males to females is found in this species as it occurs in houses. Hamer showed, in 1909, that the males constitute from 75 to 85 per cent. of the total flies of this species caught in balloon fly-traps and on fly-papers. This, however, does not indicate a disparity in the proportion of males to females in the species, as I have found that the females are more commonly found out-of-doors, especially in the neighbourhood of the breeding-places.

The breeding habits of this species are somewhat similar to those of the house-fly, *M. domestica*. The larvae breed in decaying and fermenting vegetable and animal matter, and also in excrementous matter. In 1848 Heeger recorded it as living in the caterpillars of *Epischnia canella*[1]; Roth found them in

[1] Larvae of *F. canicularis* were found feeding on the body of a dead caterpillar (*Xylina* sp.) in a breeding cage in my laboratory on September 5th, 1913.

the nest of the humble bee, *Bombus terrestris,* and Schiner observed them in the bottom of a box in which a dormouse had been kept. Taschenberg also records the larvae as being found in snails, in old cheese and in pigeon-nests; he reared the flies from sugar-beet, and Brischke found the larvae in the stalks of rape. I have found them commonly in human excrement and in a variety of decaying vegetable substances, even in rotting grass (*cf. Stomoxys*). In England they may be found in the larval stages from May to October. Howard has reared them from human excrement during the same period in the United States, and in Canada I have observed the same period for the occurrence of the larval stages. Larvae of *F. canicularis* were found by Carter and Blacklock (1913) in a case of external myasis in a monkey (*Cercopithecus callitrichus*). They were removed, together with the larvae of *Muscina stabulans* and *Calliphora erythrocephala,* from the nasal and facial region and the right side of the body near the groin, but it is probable, as these authors point out, that they may have been derived from an external source. The eggs are white and cylindrically oval.

The larvae of *F. canicularis* (fig. 83) is wholly different from that of *M. domestica,* its body being provided with a number of appendages or spiniferous processes. These are arranged in three pairs of longitudinal series, and there are in addition two pairs of series of smaller processes.

The body is compressed dorso-ventrally, and the surface is roughened in character and in places spiniferous. It consists of twelve segments, of which the first, or pseudocephalic segment, is often withdrawn into the second or prothoracic segment, as shown in the figure. The posterior end of the body is very obliquely truncate. The full-grown larva measures 5 to 6 mm. in length. The three series of pairs of spiniferous flagelliform processes, or appendages, are arranged as follows: A dorsal series consisting of ten pairs of processes, commencing with an antenna-like pair of processes at the anterior border of the prothoracic segment (segment II) and slightly increasing in size posteriorly. A latero-dorsal series of ten pairs of processes which commence on segment III and is continued to the posterior end of the body. A latero-ventral series which commences on segment III and is continued posteriorly. These flagelliform processes are spiniferous, the spines being well developed at the bases of the processes and gradually decreasing in size distally. The twelfth or anal segment is provided with three pairs of these processes of unequal size; the most anterior pair is the longest on the body, and the intermediate pair is shorter.

There is a series of pairs of small, almost sessile branched appendages (fig. 85) situated near and slightly posterior to the bases of the latero-dorsal appendages. These were described by Kieffer. Each of these processes has three to four branches, and they carry a small nucleiform organ which Chevril (1909) has also described. He believes that this organ is of the nature of an exuvial gland and correspondent to Verson's gland.

On the ventral surface of the body, and extending posteriorly from segment III, there is to be found a series of pairs of small spiniferous papillae. Between these there is on each segment a transverse row of four groups of spines.

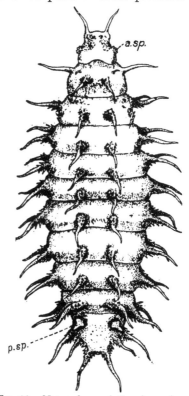

Fig. 83. Mature larva of *Fannia canicularis*, L.   × 17.

a.sp. Anterior spiracular processes.
p.sp. Posterior spiracular processes.

The anterior or prothoracic spiracular processes (fig. 84) have usually seven finger-like lobes, though the number may vary from five to eight, and between the second and third lobes there appears to be a small stigmatic organ. The posterior spiracular processes have a trilobed appearance, but a close examination reveals their four-lobed character shown in fig. 86; a stigmatic orifice is situated at the extremity of each lobe.

The spiny character of the flagelliform appendages and body of the larva causes particles of dirt to adhere readily to the bodies and appendages of the larvae. In consequence the larvae have a very dirty appearance, and their external features are almost hidden by the accumulated particles of dirt and filth adhering to them.

The larval period may extend over a week, or it may last for three or four weeks if the substances in which the larvae are feeding become rather dry. When fully grown it is covered fairly thickly with dirt, which is of great assistance in the formation of the pupal case, as this is formed of the larval skin.

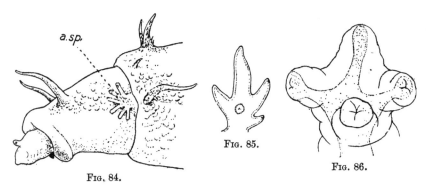

Fig. 84.   Fig. 85.   Fig. 86.

Fig. 84. *Fannia canicularis.* Lateral aspect of cephalic region of larva ; *a.sp.* anterior spiracular process.

Fig. 85. *F. canicularis.* Palmate and sessile dorsal appendage of larva.

Fig. 86. *F. canicularis.* Posterior spiracular process of larva.

In changing into the pupa, the cephalic region is retracted and the length of the larva is thereby decreased. The larval skin, with its covering of dirt particles, forms the co-arctate pupal case. Before pupating, the larva leaves the very moist substance in which it may have been living and seeks a dryer situation. The pupal period extends over a period of seven to twenty-one days, or longer, and it is not unlikely that larvae, which have developed very late in the season, pass the winter in the pupal state, as is the case in certain other species of Anthomyid flies. The adult fly emerges by pushing off the anterior segments of the pupal case.

THE LATRINE FLY.   *Fannia scalaris* Fab.

This species, which, on account of its most common breeding habits, may be called the Latrine-fly, is very common both in European countries and in North America. Owing to its general similarity, it is often confused with the Lesser House-fly, *Fannia canicularis*, but the chief differences have already been indicated.

In the male the frontal triangle on the head is black and is continued as a thin line to the vertex, being bordered on each side by a silvery white stripe. The antennae and palps are black. The thorax and scutellum are black and somewhat polished; the humeri are light-coloured. The abdomen is black, overspread with bluish-grey, and has a darker median stripe from which dark transverse bands arise, forming by their junction with the median stripe black triangular markings. The legs are black and the middle femur is swollen ventrally, bearing on its broader side a group of brush-like bristles, as will be seen from fig. 81. The middle tibia is provided, as shown, with a distinct tubercle near the distal end.

The colouring of the female is more distinctly grey, with a faint longitudinal striping on the thorax; the transverse markings on the abdomen are also indistinct. The head is grey with a wide frons.

F. scalaris is slightly larger than *F. canicularis*, measuring up to 6 mm. in length.

The habits of this species are somewhat similar to those of *F. canicularis*, but it prefers excrementous matter as a nidus for the eggs and is commonly found breeding in human excrement. It has been recorded breeding in human excrement by Schiner, Taschenberg, Howard and Newstead, and I have also bred it from this material in England and in Canada, both in privies where the excrement was found in a semi-liquid condition and on rubbish-tips or dumps, where it was mixed with ashes or clinkers. Swammerdam figured what would appear to be the larva of this species as breeding in human excrement. Taschenberg also refers to its breeding in mushrooms. In 1908 Dr David Sharp submitted the larva of this species to me for examination. He had found it in rotting fungus in the New Forest in September 1905 and noticed its similarity to Swammerdam's Latrine Larva.

The larvae emerge as early as eighteen hours after the deposition of the eggs, and become full-grown in six to twelve days.

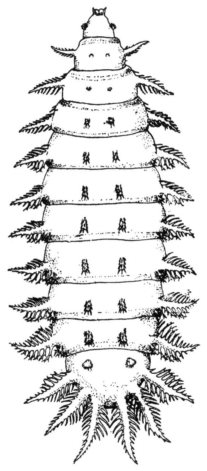

FIG. 87.   Larva of Latrine Fly, *Fannia scalaris* Fab.   × 12.

The shortest time which I have recorded for the pupal stage was nine days, which was in the month of August, but I believe that, under very favourable conditions, the pupal stage would be passed in a shorter time.

The larva of this species (fig. 87) has a general resemblance to that of *F. canicularis*, but a closer examination will reveal very marked differences and a number of distinguishing characters. In shape it is similar to the larva of *F. canicularis*, being compressed dorso-laterally. The appendages or processes, however, are very different. The pair of antenna-like processes at the anterior and upper edge of the prothoracic (second) segment are much shorter than those of *F. canicularis*, as will be seen from the figure, where they are shown dorsal to the oral lobes. On the dorsal side of the larva, from segment III to segment XI, is a series of nine pairs of short and somewhat thick processes of a very spiny character; the first two pairs being little more than spinous tubercles. As the processes of the third segment differ from the succeeding segment, they may be mentioned separately. There is a pair of latero-dorsal processes bearing spines. Ventral and slightly anterior to the base of each of these processes is

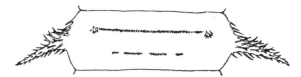

Fig. 88.   *Fannia scalaris*.   Larva.   Ventral aspect of segment VII.

a small spiniferous papilla. A short spinous latero-ventral appendage is situated slightly more posteriorly. Viewed from above, the larva is seen to be surrounded by a fringe of feather-like processes. Segments IV to XI are each provided with a pair of pinnate latero-dorsal processes which gradually increase in size posteriorly. Three pairs of these pinnate processes surround the obliquely truncate dorsal surface of the twelfth segment. Situated laterally and ventral to the series of pinnate processes is a series of latero-ventral processes which are spinous (as shown in fig. 88), but much less pinnate and shorter than the latero-dorsal series. The latero-ventral processes of segment XII are situated more ventrally than those of the preceding segments, and their usual place is taken by a small group of spines. Posterior to the base of each of the latero-dorsal processes of segments V to XI is a small branched process.

13—2

On the ventral side of the larva, extending from segments IV to XI, there is a series of pairs of small spiniferous papillae as shown in fig. 88, each of which is situated at the end of a transverse row of spines. Posterior to this transverse row of spines there is a shorter row of spines, divided into four groups. The anterior or prothoracic, spiracular processes are six to eight-lobed, the usual number of the lobes being seven. The posterior spiracular processes are very similar to those of *F. canicularis*. Vogler (1900), who has given a good description of this larva, illustrates the anterior spiracular processes with eight lobes, and his figure of one of the posterior spiracular processes is not very clear.

The feathery character of the processes of *F. scalaris* is probably associated with the fact that the larvae usually live in substances of a semi-liquid character, where such processes will be more advantageous than those of *F. canicularis* for life in such a medium. It may be of interest to note in this connection that the spiniferous and branched lateral appendages of the larvae of the genus *Fannia* were considered by Walsh (1870) and probably by other entomologists, to be "branchiae" or gills. Walsh (*l. c.*) stated: "The larvae......wallow in moist decaying matter, whether animal or vegetable; and as in such situations they would be sometimes stifled for want of air, if they breathed through the spiracles or breathing holes with which all air-breathing insects are supplied, nature has replaced the spiracles by lateral 'branchiae' or gills, by means of which they are able, after the manner of a fish, to extract the air from the fluids around them," and he compares them to the gills of the Ephemerid larvae.

Prior to pupation the larva leaves the moist situation for one of a drier character, and the pupation is similar to that of *F. canicularis*.

F. scalaris is more commonly found than *F. canicularis* as the cause of intestinal myiasis, and it also breeds more commonly in human excrement. These facts make its economic relation to man one of some importance.

CHAPTER XVI

THE STABLE FLY, *STOMOXYS CALCITRANS* LINN. (fig. 80)

OWING to a general resemblance which this species bears to *Musca domestica*, and to the fact that it is a blood-sucking species, it is frequently mistaken for *Musca domestica*, which is supposed to "bite" under certain conditions. This has led to the popular but obviously inaccurate idea that house-flies bite. The biting fly is usually *Stomoxys calcitrans*. It is naturally an out-door species and loves the sunlight, coming indoors usually on the approach of rain when the sky is dull, hence it has been named the "Storm-fly," and this fact, namely, the presence of *S. calcitrans* indoors during dull weather, has led to the popular misconception that house-flies bite during such meteorological conditions.

In England and Canada I have found this species common and widely distributed, occurring especially in the country from July to October. In the United States it appears to be abundant during the same period. During these months it may be often found in houses, although Hamer's observations (1908) appear to indicate that the presence of cowsheds in which they occur in large numbers, does not affect their numbers in houses. In England I have found *S. calcitrans* in large numbers in the windows of a country house in March and April, and it may be found frequently out-of-doors on a sunny day in May and throughout the ensuing summer months. It occurs occasionally in-doors in November and is commonly found in cowsheds and stables throughout the winter months. Its association with the cowsheds and stables has given it the name of stable fly, by which it is now generally known.

The recent investigations in Massachusetts, U.S.A., by Brues, Sheppard and Rosenau (see 1911 to 1913), on the relation of *S. calcitrans* to poliomyelitis or infantile paralysis, in which they experimentally demonstrated the ability of the fly to transmit the disease to healthy monkeys, which experiments were confirmed in Washington by Anderson and Frost (1912), have resulted in increased attention being paid to this species, which is now being studied by a number of investigators including myself. In later experiments Anderson and Frost (1913) failed to confirm their previous results[1].

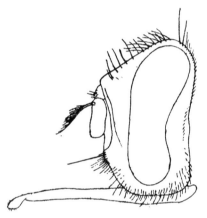

FIG. 89.   Head of *Stomoxys calcitrans* L.   Left lateral aspect.

Stomoxys calcitrans is slightly larger and more robust than *M. domestica*, measuring about 7 mm. in length. It can be readily distinguished by the awl-like proboscis which projects horizontally forward and slightly upwards from beneath the surface of the head (fig. 89). The bristles of the antennae bear setae on their upper sides only, as in the allied *Glossina*. The general colour is brownish or greyish with a greenish tinge; the dorsal side of the thorax has four dark longitudinal stripes, the outermost pair being interrupted. At the anterior end of the dorsal side of the thorax the medium light-coloured stripe has a golden appearance which is very distinct when the insect is seen against the light.

---

[1] In a more recent paper Sawyer and Herms (1913) describe a series of experiments in which they were unable, under varied conditions, to transmit poliomyelitis from monkey to monkey through the agency of *S. calcitrans*.

The abdomen is broad in proportion to its length, and each of the large second and third segments has a single median and two lateral brown spots; there is also a median spot on the fourth segment. The fourth longitudinal or median nervure (cf. fig. 7, *M*. 1 + 2) of the wing in *S. calcitrans* has not the pronounced angular bend found in the same nervure in *M. domestica*.

The life-history was first studied in anything like a complete manner by Newstead (1906) and I was able to confirm his observations during 1907, 1908 and 1912. Portchinsky added to our knowledge in 1910. In the United States Bishopp (1913) has recently made a valuable contribution to the knowledge of the insect's biology and life-history, and Mitzmain (1913) has given a summary of his investigations in the Philippine Islands which have extended over two years. These references are given as it is not intended to give an exhaustive account of this insect here. It may be added that morphological studies have been made by Newstead, Tulloch (1906) and Brain (1912). The following account of the life-history is taken from the foregoing accounts and my own notes.

Both sexes are able to suck blood. After emerging from the puparium the proboscis lies extended backwards, along the ventral side of the thorax. It soon bends forward and hardens, and in six to eight hours after emergence the fly may take its first meal. The feeding period lasts from two to twenty minutes. It has been found to feed chiefly on cattle, horses, dogs and man, being especially attentive to the ears of dogs. Bred flies may begin ovipositing on the ninth day, usually after they have had the third or fourth feed. This species breeds in the following substances, apparently with varying preferences in different countries: horse manure, cow manure, sheep dung, human excreta, the straw of oats, rice, barley and wheat, fermenting cut grass, decaying vegetable substances, including fungi, and in animal substances, no doubt in a state of decomposition.

The eggs may be laid singly or in batches of as many as seventy-two (Newstead). The eggs are white, cylindrically oval in shape, somewhat resembling a banana, and measuring 1 mm. in length. A groove widening at the anterior end runs along the side of the egg. Mitzmain states that the maximum number of eggs deposited at one period was ninety-four, and that the maximum number of eggs deposited by a single female may be placed at, at least, six hundred and thirty-two. Bishopp found that two feedings are usually necessary between the deposition of each lot of eggs.

There are three larval stages and the larvae are creamy white in colour and have a shiny translucent appearance; the young larvae are even more translucent. The adult larvae, which measure about 11 mm., are rather similar to those of *M. domestica,* but they can be distinguished by the character of the posterior spiracles. These (fig. 90, *A*) are wider apart than in *M. domestica* and are triangular in shape with rounded corners; each of the corners

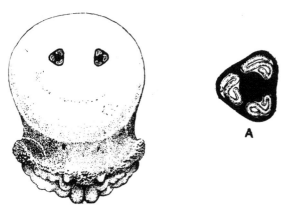

Fig. 90.   Posterior end of mature larva of *S. calcitrans.*
A Posterior spiracle of the same, enlarged.

subtends a space in which a sinuous aperture lies.   The centre of the spiracle is occupied by a circular plate of chitin.   The anterior spiracular processes are five-lobed.

The developmental stages, according to my own and the observations of those who have studied the life-history in temperate and tropical climates, are as follows: egg stage, twenty hours to four days; larval stage, seven to thirty days; pupal stage, five to twenty days.   The whole life-history, therefore, from the deposition of the eggs to the emergence of the adults, may vary from about thirteen days to seven or eight weeks.   The longest time which I observed in any of my experiments was a few days over ten weeks (Oct. 4th to Dec. 15th).   In temperate climates it is possible that the pupae may exist through the winter, and the larval stages may also be found in warm situations during the winter. In studying the longevity of the adults Bishopp found that they could be kept, when water and sugar syrup were supplied, for twenty-three days. Mitzmain found that a female fly can live a maximum of at least seventy-two days and a male a period of ninety-four days. The length of the life-history depends on the conditions with regard to temperature, moisture and nature of food; absence or presence of light also appears to influence the rate of development.

It is not intended to discuss the relation of *S. calcitrans* to disease; the reader is referred to the excellent accounts of Brues, Rosenau and Sheppard, and also those of Anderson and Frost[1], for an account of their observations and experiments on the relation of this fly to poliomyelitis. Its blood-sucking habits may occasionally enable it to be the vector of the anthrax bacillus causing anthrax in cattle and malignant pustule in man[2]. Schuberg and Kuhn (1911) have experimentally demonstrated that *S. calcitrans* can infect animals with trypanosomes and spirochaetes[3].

[1] See also Sawyer and Herms (1913).

[2] Mitzmain (1914) obtained positive results in the transmission of anthrax by *S. calcitrans*.

[3] Schuberg and Böing (1913) found that *S. calcitrans* transmitted fatal infection of *Streptococci* to rabbits not only immediately but after periods of 2 minutes to 24 hours had elapsed after the imbibition of the organism. *S. calcitrans* failed to convey anthrax infection to the goat, but did so in the case of one of two sheep experimented with.

CHAPTER XVII

THE BLOW-FLIES, *CALLIPHORA ERYTHROCEPHALA* MEIG.
AND *C. VOMITORIA* L., AND THE SHEEP MAGGOT OR
"GREEN BOTTLE" FLY, *LUCILIA CAESAR* L.

THE large blow-fly or "blue bottle," *C. erythrocephala*, is a
widely distributed and common species in Europe and North
America. In the past, but less commonly now, the name of the
other species, *C. vomitoria*, has been indiscriminately applied to
both species. *C. vomitoria* is, however, much less common than
C. erythrocephala. They can be distinguished by the fact that in
the latter species the genae are fulvous to golden yellow and are
beset with black hairs, whereas in *C. vomitoria* the genae are
black and the hairs are golden red.

Calliphora erythrocephala has been described in detail by
Lowne (1870, 1895). Its appearance, with its bluish-black thorax
and dark metallic blue abdomen, is sufficiently well known as to
render a description of the adult flies unnecessary. Its length
varies from 7 to 13 mm. The larvae are necrophagous, and the
flies deposit their eggs on any fresh, decaying or cooked meat, and
also upon dead insects ; Howard (1909) has found the fly on fresh
human faeces. On one occasion, when obtaining fresh food material
in the form of wild rabbits upon which to rear the larvae of *C.
erythrocephala*, I found the broken leg of a live rabbit, which had
been caught in a spring trap set the previous evening, a living
mass of small larvae, which were devouring the animal while it
was still alive. An enormous number of eggs are laid by a single
insect ; Portchinsky (Osten Sacken, 1887) found from 450 to 600
eggs, though I have not found so many. Fabre, in his *Souvenirs
entomologiques*, records *C. vomitoria* depositing 300 eggs in one
batch and more were subsequently deposited, and he believed, on
the evidence which he secured, that as many as 900 may be

deposited. He found the flies would emerge from pupae buried under 60 cm. of sand.

With an average mean temperature of 23° C. (73·5° F.) and using fresh rabbits as food for the larvae, the following were the shortest times in which I reared *C. erythrocephala*. The eggs hatched from ten to twenty hours after deposition. The larvae

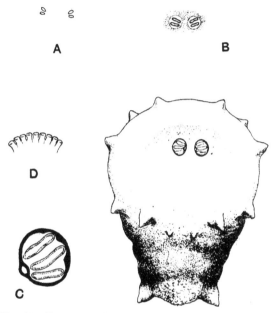

A   B

D

C

FIG. 91.   Posterior end of mature larva of *C. erythrocephala*.

A Posterior spiracles of first larval stage of *C. erythrocephala*, Mg.
B Posterior spiracles of second larval stage of *C. erythrocephala*.
C Posterior spiracle of mature larva of *C. erythrocephala*.
D Anterior spiracular process of mature larva of *C. erythrocephala*.

underwent the first ecdysis eighteen to twenty-four hours after hatching; the second moult took place twenty-four hours later, and the third larval stage lasted six days, the whole larval life being passed in seven and a half to eight days. Fourteen days were spent in the pupal state; thus the development was complete in twenty-two to twenty-three days. I have no doubt that this time could be shortened by the presence of a very plentiful supply of food, as an enormous amount, comparatively, is consumed.

The full-grown larva may measure as much as 18 mm. in length. There are two distinct mandibular sclerites. The anterior spiracular processes (fig. 91 D) are usually nine lobed. The posterior extremity is surrounded by six pairs of tubercles arranged as shown in fig. 91; there is also a pair of anal tubercles. The posterior spiracles (fig. 91 C) are circular in shape and contain three straight slit-like apertures. In the second larval instar there are only two slits in each of the posterior spiracles (fig. 91 B) and in the first larval instar each of the posterior spiracles (fig. 91 A) consists of a pair of small slit-like orifices.

C. erythrocephala is an out-door fly, but it frequently enters houses in search of material upon which to deposit its eggs, and also for shelter. From its habit of frequenting faeces, which may often be observed, especially in insanitary courtyards and similar places, it is not improbable that it occasionally may bear intestinal bacilli on its appendages or body or internally, and thus carry infection. Its flesh-seeking habits may also render it liable to carry the bacilli of anthrax should it have access to infected flesh, and such meat-seeking habits render the ingestion of the larvae into the human digestive system extremely possible. Buchanan (1907) recovered the bacillus of swine fever from blow-flies which were caught on the carcases of pigs which had died of swine fever. Further infection experiments with C. erythrocephala are discussed in a later chapter.

THE SHEEP MAGGOT OR "GREEN BOTTLE" FLY, LUCILIA CAESAR L.

This fly is an out-door species which is sometimes found in houses, into which it is generally driven by adverse weather conditions for the purpose of shelter. It is more commonly found in farm and country houses. Although it is not so large as C. erythrocephala, being more similar in size to M. domestica, it is frequently called a "blue bottle" and is often referred to as the "green bottle" fly. The colouring is more brilliant than that of C. erythrocephala, being of a burnished gold, sometimes bluish appearance and sometimes of a shining green colour.

The flies are usually found on dead animals and carrion; they also occur on the excrement of man and other animals. On all such substances the larvae are able to feed. In Europe the larvae feed on the matted wool and on the flesh of the backs of sheep, from which habit they are popularly known as "sheep maggots." In Canada, however, as a result of a careful inquiry I have only been able to discover a few cases of such a feeding habit for the larvae, which results in the production of large ulcerated areas on the backs of the sheep, causing severe loss of flesh and sometimes death. Banks (1912) states that Meinert has reared *Lucilia nobilis* from larvae taken from the ears of a sailor. An excellent account of the *Lucilia* flies has been given by MacDougall (1909), and Herms (1911) gives a full account of the habits of *L. caesar*. In England I have usually obtained it from the backs of sheep. Howard (1900) reared it from human excrement.

The larvae are very similar to those of *C. erythrocephala* except in size, and Portchinsky considered them otherwise indistinguishable. The full-grown larva measures 10 to 11 mm. in length. The larval life lasts about fourteen days and the pupal stage a similar length of time, but my experience in the field would lead me to believe that under favourable conditions development may be completed in a much shorter time.

PROTOCALLIPHORA GROENLANDICA.

In the careful investigation which he carried out Hamer (1908 and 1910) found this species, and occasionally *P. azurea*, in the neighbourhood of a glue and size factory, and also at a railway siding to which were brought, in addition to stable manure and other refuse, the bones for the glue factory. Hamer states that so numerous were the larvae in the sacks of bones that "the ground beneath some vans containing these sacks was found one day last summer to be covered with larvae so that from a little distance this portion of the yard surface had the appearance of snow." The adult flies were caught in the neighbourhood of the factory from the middle of June until mid-September and it was found that the adoption in the late summer of a system of destroying the larvae had some influence in lessening the fly nuisance in surrounding houses.

The larvae of *Protocalliphora* are similar to those of *Calliphora*.

CHAPTER XVIII

THE first of these two species, *Pollenia rudis*, is common in Europe and North America. On account of their habits and general appearance they are usually mistaken by the uninitiated for house-flies emerging from their winter sleep. Early in spring and sometimes during mild days in winter they may be found crawling sluggishly around, as is their habit. Out-of-doors one may find them on the snow around buildings or on the walls. In-doors they buzz lazily, frequently in considerable numbers, in the windows and especially on the windows of unoccupied rooms. I have found them out-of-doors on the snow at Ottawa as early as the middle of March. In England I have observed them entering the sun-lit window of a bed-room in a country house and swarming over the window panes. They appear to frequent especially country houses covered with creepers and vines. The name "Cluster-fly" has been given to them on account of their habit of congregating in large numbers in and about houses and other occupied buildings. They appear to prefer unoccupied rooms, no doubt on account of the absence of disturbance.

Howard (1911) refers to the observations of Dall in the United States, which confirm my own made on the habits of this fly in England and Canada. Dall states that this fly was a great nuisance in country houses near Geneva, New York State. They were a terror to housekeepers since they were found in all kinds of places such as in and on beds, wardrobes, behind pictures, etc. They would form large clusters about the ceilings of clean, dark rooms seldom used. It was stated that about the first of April they came out of the grass and flew up to the sunny side of houses, which they entered. They remained in evidence until some time in May and then disappeared until September. Large numbers were said to occur often under buildings, between the earth and the floor. An examination of these flies and a comparison with *Musca domestica* will show their distinct character. Their characteristic sluggish habits have already been mentioned. They are slightly larger in size and darker in colour and the thorax is sparsely covered with yellowish coloured hairs. When at rest the wings are folded more closely together over the back than is the case of the house-fly.

As in the case of many other common species of Diptera, we have little information in regard to the breeding habits of this species. Robineau-Desvoidy states that the flies of the genus *Pollenia* deposit their eggs on decomposing animal and vegetable matter; in this regard they resemble, in general, the other members of the Muscidae. Howard (1910) records the rearing of a single specimen of *P. rudis* from cow manure in Washington D.C. in December, and quotes J. S. Hine as reporting the rearing of numbers of Cluster flies in the summer of 1910 from cow manure in the pasture. Copeman, Howlett and Merriman (1911) record the occurrence of three specimens of *P. rudis* in a collection of 200 larvae and pupae obtained from fresh refuse on the Norwich Corporation tip at Postwick. Keilin describes *P. rudis* as parasitic on a species of earthworm *Allobophora chlorotica* Sav. He states that it pupates in the earth, the pupal stage lasting from thirty-five to forty-two days.

MUSCINA STABULANS FALL.

So similar is this fly to the true house-fly in general appearance, though slightly larger in size, that it is almost invariably mistaken by the untrained observer for a large house-fly. According to my own observations it usually occurs in and near houses in the early summer, about June, and generally about the same time that the Lesser House-fly, *Fannia canicularis*, is the predominant domestic species. Hamer (1910) found that the largest number of *M. stabulans* were captured in August and early September. In large collections which have been made in

North America and England of the flies occurring in houses, a number of specimens of this species are usually found (see p. 66). Cleland (1912) records it in New South Wales as occurring in houses on the window panes or on the table at meal time; in the neighbourhood of houses he states that it is found feeding in house refuse and round the garbage can.

It is larger than *M. domestica* and more robust in appearance. It varies from 7 to nearly 10 mm. in length. Apart from its greater size it may be distinguished from *M. domestica* by the fact that the median or fourth longitudinal vein (cf. fig. 7, *M.* 1 + 2) of the wing is only slightly curved ventrally instead of being bent upwards through a pronounced angle as in *M. domestica.*

Its general appearance is grey. The head is whitish-grey with a "shot" appearance. The frontal region of the male is velvety black and narrow; that of the female is blackish-brown, and is about a third of the width of the head. The bristle of the antenna bears setae on the upper and lower sides. The dorsal side of the thorax is grey and has four longitudinal black lines; the scutellum is grey. The abdomen, as also the thorax, is really black covered with grey; in places it is tinged with brown, which gives the abdomen a blotched appearance. The legs are rather slender, and are reddish-gold or dirty orange and black in colour.

The eggs are laid and the larvae feed upon various kinds of decaying or decomposing vegetable and animal substances. They have been reared from fungi, decaying fruit, such as apples and pears, cucumbers and miscellaneous vegetables. They sometimes attack growing vegetables, and I have reared them in considerable numbers with root maggots from radishes. In the latter case the eggs or young larvae may be introduced with the manure, but I have found them also attacking growing plants where no manure had been used. Aldrich has reared it in Idaho from rotten radishes. They breed in excrement, such as cow dung and human excrement from which Howard (1900) reared specimens. Cleland (1912) describes this species as occurring on human faeces and house refuse in Australia. In Europe they have been found feeding on caterpillars and larval bees and in Canada Fletcher (1900) records the species as parasitic on the noctuid caterpillar *Peridromia saucia* Hbn. in British Columbia. In the United States they have been reared from the pupae of the cotton-worm and the Gipsy moth; Riley was of the opinion that in the first case rotten pupae only were fed upon. In 1891 it was reared on the masses of larvae and pupae of the Elm-leaf beetle. Other observers record it as being reared from the pupae of such Hymenoptera as *Lophyrus*.

The most complete account of *M. stabulans* is that given by Portchinsky (1913). Reference has already been made to its habit of destroying the larvae of *M. domestica*. This author states that the larvae have been found not only in the excrement of man, cattle and horses, but also in raw and cooked meat, on carcases of different vertebrates such as mammals, birds and amphibians, on invertebrates such as insects, their larvae and pupae, in rotten bulbs and vegetables, in fungi and in old cheese, etc. Portchinsky in his detailed account of the life-history and habits of *M. stabulans* states that 160 eggs are deposited; they are spread singly or in lines over the larval food. He confirms Bouché's statement that the fly is able to pass through its developmental stages in about a month, and thus several generations a year may be produced.

The larva may reach a length of 11 mm. It is creamy white in colour. There are two closely approximate mandibles. The anterior spiracular processes are usually five-lobed (occasionally six) and are somewhat like hands from which the fingers have been amputated at the first joint. The posterior spiracles are separated by a space less than the diameter of each; they are rounded, and each encloses three triangular-shaped areas containing each a slit-like aperture. I have not been able as yet to study the complete life-history; Taschenberg (*t.c*) states that it occupies five or six weeks, but my observations would indicate that this period can be shortened.

The habits of this species which lead it to frequent excrement and food render it a potential disease-carrier, as Portchinsky also affirms. Banks (1912) records the passing by a child suffering from diarrhoea of a considerable number of the larvae of this species and refers to Laboulbene's record of the larvae of *M. stabulans* being vomited by a person suffering from bronchitis. Carter and Blacklock (1913) record the occurrence of the larvae in a case of external myiasis in a monkey, together with the larvae of

Fannia canicularis. The relation of this species to intestinal myiasis is referred to later.

It is interesting to note in conclusion that Hamer (1910) states that a remarkable point in regard to *Muscina* was the frequency with which it was infested with parasites. These were apparently Gamasid mites. In early June, out of 300 flies examined, 40 specimens were infested. So closely together were the parasites aggregated in some instances that no part of the fly's abdomen was visible, the mites forming a kind of chain mail. Berlese (1912) records the occurrence of the common mite *A. muscarum* (see p. 158) on this species of fly, and it is not unlikely that the aforementioned mites may have been the same species.

CHAPTER XIX

ALLIED MUSCID FLIES AND MISCELLANEOUS FLIES FOUND IN HOUSES

INDIAN HOUSE-FLY, *MUSCA DOMESTICA* SUB SPECIES *DETERMINATA* WALKER.

THIS Indian variety of the house-fly was described by Walker (1856) from the East Indies. His description is as follows: "Black with a hoary covering; head with a white covering; frontalia broad, black, narrower towards the feelers; eyes bare; palpi and feelers black; chest with four black stripes; abdomen cinereous, with a large tawny spot on each side at the base; legs black; wings slightly grey, with a tawny tinge at the base; praebrachial vein forming a very obtuse angle at its flexure, very slightly bent inward from thence to the tip; lower cross vein almost straight; alulae whitish, with pale yellow border; halteres tawny."

In appearance and size it is very similar to *M. domestica*. Its breeding habits are also similar. Aldridge (1904) states that at certain seasons of the year it is present in enormous numbers. The method of disposal of the night soil is to bury it in trenches about one foot or less in depth. From one-sixth of a cubic foot of soil taken from a trench at Meerut and placed in a cage, 4042 flies were hatched. Lieutenant Dwyer collected 500 from one cage covering three square feet of a trench at Mhow. Specimens in the British Museum collection were obtained from hospital kitchens, and Smith found them in a ward at Benares.

They have also been recorded from the N.W. Provinces, Kangar Valley (4500 feet), Dersa, and I have received specimens from Aden.

14—2

MUSCA ENTENIATA BIGOT.

This species of house-fly has a distribution somewhat similar to the last species and like it has a marked resemblance to *M. domestica* as Bigot's (1887) description indicates : " Front très étroit, les yeux, toutefois, séparés. Antennis et palpes noirs ; face et joues blanches ; thorax noir avec trois larges bandes longitudinales grises ; flancs grisâtres, écusson noir avec deux bandes semblables ; cuillerons et balanciers d'un jaunâtre très pâle ; abdomen fauve, avec une bande dorsale noire et quelques reflets blancs ; pieds noirs ; ailes hyalines ; cinquième nervure longitudinal (Rondin) coudée suivant un angle légèrement arrondi, ensuite un peu concave ; deuxième transversale (l'extrême) presque perpendiculaire, légèrement bisinueuse, soudée à la cinquieme longitudinale, à égale distance du coude et de la première nervure transversale (l'interne)."

M. enteniata measures 4 to 5 mm. in length. The British Museum collection contains specimens sent by Major F. Smith from Benares, with these notes : " Bred from human ordure ; hospital ward fly ; at an enteric stool ; bred from cow dung fuel cakes." I have received specimens from Suez and Aden, and it is recorded as breeding in human excrement in Khartoum (Balfour, 1908) and in stable refuse, as also are *M. domestica* and *M. corvina*. It will be seen, therefore, that its breeding habits are very similar to those of *M. domestica* and the sub-species *determinata*. It is interesting to note the choice of cow dung as a breeding place, especially in conjunction with the economic status of the cow in India.

MUSCA VETUSTISSIMA WALK.

This is a common species in Australia and can be readily distinguished from *M. domestica* by the silvery white-thoracic stripes. Cleland (1912), who reprints Walker's description of the species, writes of it as follows : " This fly is essentially an out-of-door species. Very rarely indeed is it found within the house and only occasionally in out-houses with open doors and windows. When found in these situations it is usually attracted by some source of food, such as the carcass of an animal, even though freshly killed. They have been found in abundance feeding on the pus of blood from the sores on the head of a sheep. As soon as one ventures outside, if the weather is warm, this species attaches itself to one's person. In some country situations that part of one's clothing sheltered from the wind may be covered with a dense mass of these insects. They are especially annoying by hovering around and finding their way into one's eyes, nose and mouth."

It is very common in the neighbourhood of Sydney, N.S.W., and in all the country districts in New South Wales. It is also common in Melbourne, around Adelaide, and occurs in Perth, West Australia.

The abundance of this fly and its habit of hovering round and alighting on the face of human beings makes it probable, as Cleland suggests, that if such eye diseases as epidemic conjunctivitis and trachoma are transmitted by flies, this species would be incriminated. It will readily eat dried blood. Cleland states that smears of anthrax blood were made on glass slides, and after several days were confined with one of these flies in a test tube. Some of the blood was eaten and anthrax bacilli were cultivated from the resulting faeces, and the fly itself, on being emulsified in saline solution and injected into a guinea pig, gave the animal anthrax.

THE ROOT MAGGOT FLY. *ANTHOMYIA RADICUM*
MEIGEN (fig. 92).

This member of the Anthomyidae has been found in houses, especially those in or near the country. It is not, however, a house-fly in the same sense as its congeners, *F. canicularis* and *F. scalaris,* and while its habits lead it in search of excrement it is not likely to be a serious factor as a vector of pathogenic organisms. It may occasionally be responsible for intestinal myiasis and a case recorded by Austen (1912) is referred to later.

In size and general appearance this species resembles *Fannia.* The female, which is shown in fig. 92, is olive grey in colour and has the upper frontal region of the head, that is, between the eyes and superior to the bases of the antennae an orange rufous colour. The male is darker in colour, the dorsal side of the thorax being blackish with three black longitudinal stripes; the frontal region is very narrow; the abdomen is grey with a dark median stripe. The average length of the body is 5 mm.

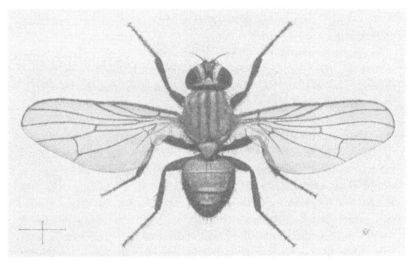

Fig. 92.   The Root Maggot Fly *Anthomyia radicum* Meig.   Female.   × 9.

The flies are common in the summer and may be found in the neighbourhood of manure.   In a study which I made (1907) of the life-history of this species it was found that the eggs were very frequently deposited on horse manure which served as a common breeding place for this fly.   The insect's popular name has been derived from the fact that the larvae also commonly feed upon the roots of certain cultivated cruciferous plants such as cabbages, radishes, etc.

The eggs hatch out from eighteen to thirty-six hours after deposition.   The first larval stadium lasts twenty-four hours, the second forty-eight hours, and five days later the larva changes into a pupa, the whole larval life occupying about eight days. The pupal stage lasts ten days, so that in warm weather the development may be completed in nineteen to twenty days.   The

full-grown larvae (fig. 93) measures 8 mm. in length, and may be
distinguished by the tubercles surrounding the caudal extremity.

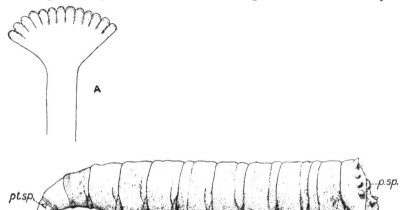

Fig. 93.   Mature larva of *Anthomyia radicum* Meig.   × 12.
p.sp. Posterior spiracle.   *pt.sp.* Anterior spiracular process, A same enlarged.

In this species there are six pairs of spinous tubercles surrounding
the posterior end, and a seventh pair is situated on the ventral
surface posterior to the anus.   The tubercles of the sixth pair,

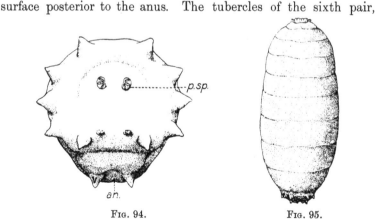

Fig. 94.   Fig. 95.
Fig. 94.   Posterior end of mature larva of *Anthomyia radicum* Meig.   *an.* Anus.
Fig. 95.   Pupa of *A. radicum.*   × 11.

counting from the dorsal side, are smaller than the rest, and are
bifid.   The arrangement of the tubercles can be seen in fig. 94.

The anterior spiracular processes (fig. 93 A) are yellow in colour and have thirteen lobes.

THE MOTH FLIES. *PSYCHODA* SPP.

There may be found frequently on window panes small, grey, moth-like flies known as "moth flies" or "owl midges." The wings of these minute insects are large and broad in proportion to the size of the body and are densely covered with hairs; when the insect is at rest the wings slope in a roof-like manner. The larvae of some species breed in excrement ; Newstead (1907) has found the larvae in human excrement; others occur in decaying vegetable substances. I have bred them from rotting potatoes, while certain species breed in water, especially when it is polluted with sewage. Such aquatic species have the spiracular apparatus modified to suit their changed life. Welch (1912) has recently described the life-history of a new species, *Psychoda albimaculata*, which was found breeding in the sewage and water of the experimental tanks of the Chicago sewage works. Although a form, *Phlebotomus*, which occurs in southern Europe, northern Africa and north India, has blood-sucking habits, most of the species are of little economic importance in their relation to man except in their *rôle* as scavengers.

Among other species of flies which occur in or visit houses may be mentioned the Cheese Maggot Fly, *Piophila casei* L., whose larva the well-known "cheese skipper" lives in cheese, ham and other animal substances, usually of a decomposing character. The life-history of this species may be passed in so short a time as three weeks, or it may be more prolonged. Occasionally the larvae may be more responsible for cases of intestinal myiasis. The flesh-flies, of which group *Sarcophaga carnaria* is a common species, occasionally occur in houses. They breed in excreta and decomposing animal matter, especially dead insects. The Yellow Dung Fly, *Scatophaga stercoraria* L., a rather large yellow fly which is commonly found on and breeding in cow dung and other excreta, is found in country houses on rare occasions. Reference has already been made (p. 171) to the useful character of this species as an enemy of *M. domestica.* Sometimes small shining black flies belonging to the genus *Scenopinus* are found in the windows of houses. They measure about one quarter of an inch in length and have been called "window flies" by Comstock. The larvae of these flies appear to be carnivorous in their habits.

Many other flies accidentally stray into houses, but so far as we have been able to ascertain by their feeding and breeding habits, the species which have been considered are the only flies which are of special economic interest. No general statement, however, can be made, for in spite of the advance in our knowledge as to the economic relationships of the Diptera which has taken place during the past few years, the information we now have available only serves to indicate how little we really know concerning this important order of insects.

PART V

THE RELATION OF HOUSE-FLIES TO DISEASE

CHAPTER XX

THE DISSEMINATION OF PATHOGENIC ORGANISMS BY FLIES

ALTHOUGH *M. domestica* is unable to act as a carrier of pathogenic micro-organisms in a manner similar to that of the mosquito, so far as we know at present, nevertheless its habits render it a very potent factor in the dissemination of disease by the mechanical transference of the disease germs. These habits are the constant frequenting and liking for substances used by man for food on the one hand and excremental products, purulent discharges, and moist surfaces on the other. Should these last contain pathogenic bacilli, the proboscis, body and legs of the fly are so densely setaceous (see fig. 96) that a great opportunity occurs, with a

FIG. 96. Tarsal joints of one of posterior pair of legs of *Musca domestica*. Lateral aspect to show densely setaceous character.

maximum amount of probability, for the transference of the organisms from the infected material to either articles of food or such moist places as the lips, eyes, etc. As I have already pointed out (1907), *M. domestica* is unable to pierce the skin, as certain persons have suggested. The structure of the proboscis will not permit the slightest piercing or pricking action, which fact eliminates such an inoculative method of infection. It is as a mechanical carrier, briefly, that *M. domestica*, and such allies as *F. canicularis*, etc., though to a less degree, may be responsible for the spread of infectious disease of a bacillary nature, and an account will now be given of the *rôle* which this insect plays in the dissemination of certain diseases[1].

While flies have been shown, as the evidence advanced later will indicate, to carry pathogenic and non-pathogenic organisms, especially the non-spore-bearing micro-organisms, on the exterior of their bodies, recent investigations, notably the excellent studies of Graham-Smith (1909–13), Torrey (1912) and of Cox, Lewis and Glynn (1912), would appear to indicate forcibly that the carriage of micro-organisms in the digestive tract and infection therefrom is even more important than mechanical transference on the appendages and exterior of the body. The longer duration of the life of the micro-organism carried in the digestive tract of a fly increases the possibility of infection either by "vomit" spots resulting from regurgitation or by faecal spots. This important aspect of the problem will be discussed shortly.

It should be pointed out that whereas in some of the diseases the epidemiological evidence adduced in support of transference of disease germs by flies is confirmed bacteriologically, in others only the former evidence exists. Should neither form of evidence be available in support of the idea that *M. domestica* plays a part in the dissemination of the infection of a particular disease, it is essential, nevertheless, that if such a method of transference is conceivable the possibility of this insect being able to carry the pathogenic organism should be realised. This possibility is governed by such factors as the presence of *M. domestica*; its access to the infected or infective material, this being attractive to the insect either because it is moist or because it will serve as food for itself or its progeny; and a certain power of resistance for a short time against desiccation on the part of the pathogenic organisms, although, as in the case of the typhoid bacillus, the absence of this factor is not fatal to the idea, as it may be overcome by the fact that the fly is able to carry the organisms in its digestive tract or to take on its appendages an amount sufficient to resist

---

[1] Though it should be unnecessary, I wish to explain, as I have been occasionally misunderstood by medical men and others, that *M. domestica* is not regarded as being the cause of any disease, but as a carrier of the infection.

desiccation for a short time. The last factor is the presence of suitable culture media, such as certain foods, or moist surfaces as the mouth, eyes, or wounds, for the reception of the organisms which have been carried in the digestive tract or on the body or appendages of the fly. If these conditions are satisfied the possibility of *M. domestica* or its allies playing a part in the transference of the infection should be carefully considered, and this suggestive or circumstantial evidence will be discussed in certain of the diseases which follow, in addition to the epidemiological and bacteriological evidence.

HISTORICAL.

There is no doubt that if a careful search were made in old writings many references would be found attributing unhealthy conditions to the presence of flies. The idea is a very old one, the scientific examination and experimental proof, however, is of comparatively recent date.

Mercurialis in 1577 (quoted by Nuttall), believed that flies carried the virus of plague from persons suffering from the disease to the food of healthy people. Riley (1910) refers to an early and remarkable statement of Kircher who, in his *Scrutinium Physico-medicum*, published in Rome in 1658, said: " There can be no doubt that flies feed on the internal secretions of the diseased and dying, then flying away, they deposit their excretions on the food in neighbouring dwellings, and persons who eat it are thus infected." Kircher attributes this theory to Mercurialis. Sydenham (1666) considered that if swarms of house-flies were abundant during the summer, the autumn would be unhealthy, although Holscher (1843) disagreed with this idea. Crawford (1808) believed that insects in a general way, especially house-flies, acted as disseminators of disease. Leidy (1871) declared his belief that house-flies were responsible for the spread of hospital gangrene and wound infection during the American Civil War and reaffirmed his convictions in later writings. In 1871 Lord Avebury called attention to the habits of flies which alight on decomposing substances and carry impurities especially the secretions of unhealthy wounds. Rather than regard them as dipterous angels dancing attendance in Hygeia, he said we should look upon them as winged sponges spreading hither and thither to carry out the foul behests of Contagion. Many others have called attention to the general facts and also their connection with specific diseases; the latter authorities will be mentioned in considering the various diseases. Of the former the following may be mentioned : Hewson (1871), Cobbold (1879), Megrun (1875), Laboulbene (1875), White (1880), Slater (1881), Grassi (1883), Taylor (1883), Moore (1893), Coplin (1899), Parker (1902), Martini (1904), Bergey (1907), Scott (1909), Skinner (1909) and others (see Bibliography).

METHODS BY WHICH BACTERIA ARE SPREAD BY FLIES.

In discussing the relation of house-flies to disease the various methods by which they are able to distribute the pathogenic and other bacteria must be borne in mind. It has generally been understood that the mechanical transference of bacteria and other micro-organisms on the exterior of the fly, that is on its body and appendages, was sufficient to account for the distribution of infection. In the experimental evidence which is quoted in the succeeding sections this assumption has generally been made by the investigators. Now while this is true in many instances, there is very strong evidence derived from careful experiments that the infection from the exterior of the fly is far from being the sole method of bacterial distribution.

Flies which frequent infectious matter not only inevitably contaminate the exterior of their body and their appendages, but by feeding upon such matter, which is one of the two reasons for their visiting it, they take the bacteria and other organisms which it contains into their alimentary tracts. In consequence the crop, the stomach and the intestines become the temporary resting places of the bacteria. In the gut of the fly the micro-organisms, especially those of a non-spore-producing kind, will naturally remain in a viable condition for a considerably greater length of time than on the exposed external surface of the insect, and the period during which infection may be distributed will be lengthened. In treating of the various pathogenic organisms I shall have occasion to quote many instances of the last fact.

From the gut of the fly the infection may be distributed in two ways, either from the mouth or from the anus. In describing the method of feeding, the habit which flies have of regurgitating the food and depositing the liquid in the form of " vomit spots " was mentioned. Such vomit spots may distribute micro-organisms taken up with the food. Also, during the act of feeding, micro-organisms previously absorbed when feeding on infected matter may be deposited by the proboscis. Infection from the anus may take place by means of infected faecal matter, so promiscuously deposited as "fly specks."

A most thorough and valuable series of experiments with a view to ascertaining the possibilities of the aforementioned methods of distributing infection has been conducted by Graham-Smith (1910–13) in connection with the Local Government Board's inquiry. The importance of this aspect of the problem warrants my quoting the results of his experiments somewhat fully.

The greater possibility of non-spore-bearing bacteria surviving in the gut of the fly than on the exterior and also probable infection by the proboscis were shown in his experiments with *Bacillus enteritidis* (Gaertner). It was shown that *B. enteritidis* may be present in the contents of the crops and intestines of flies for at least seven days after infection. Flies can infect plates over which they walk for some days (seven days in the experiment) in spite of the fact that the organisms can be seldom isolated from their legs (once in 32 cultures). When walking over the culture plates flies constantly place their proboscides on the medium and in most cases leave imprints on its surface. The colonies of bacteria develop round these marks. As Graham-Smith points out, the infection of the plate is probably due to inoculation by the proboscides of the flies.

The same investigator in another series of experiments utilised *Bacillus prodigiosus*, which was chosen on account of its being easily cultivated and identified on plate cultures. Further, it seemed likely that its results would give some information as to the behaviour of other non-spore-bearing organisms. Experiments on the duration of life of *B. prodigiosus* on the exterior and in the alimentary tract of flies showed that this organism may remain alive on the legs and wings for at least 18 hours after feeding.   In exceptional cases it may remain alive longer.   In the contents of the crop and intestine and on the proboscis it is present for four or five days, after which time its numbers gradually diminish and after 17 days the cultures yielded negative results.

A careful series of experiments with a view to determining whether *B. prodigiosus* multiplied in the crop appeared to indicate that no multiplication takes place. In this connection it may be remarked that Nicoll (1911), to whose experiments reference will be made later, finds that certain of the non-lactose fermenting bacteria appear to be capable of multiplying in the intestine of the fly.

The extent to which the persistence of *B. prodigiosus*, and no doubt other bacteria of like character, in the faecal deposits, that is, the infectivity of the flies in this respect, is affected by the character of the food, is shown by a series of experiments in which the faeces from infected flies which had been fed on milk, syrup and sputum were collected at various periods and cultures made. *B. prodigiosus* was not recovered after 48 hours from the faeces of flies fed on sputum.   The faeces of flies fed on milk contained the bacteria, that is, were infective, for seven days and those of flies fed on syrup for four days after feeding.

That infected flies will infect both liquid and solid food upon which they feed was shown by feeding flies artificially infected with *B. prodigiosus* and also *B. pyocyaneus* on milk and with the former organism on sugar.   It was shown that flies infected with these non-spore-producing organisms were able, by feeding upon it, to contaminate milk for ten or eleven days (in the case of *B. prodigiosus*).   That the contamination would appear to have been solely from the alimentary tract or proboscis was shown by the fact that the organism (*B. prodigiosus*) could not be cultivated from the limbs of the infected flies eleven days after infection, although the limbs were heavily contaminated on the second day. After feeding flies on syrup infected with *B. prodigiosus* it was shown that they could infect sugar, by being allowed to feed upon it subsequently, for at least two days.

Flies, when unusually hungry, will suck at the deposits of other flies which they may find, in the same way that they will suck at dried milk spots, and it was shown that clean, uninfected flies will infect themselves from the deposits, both faecal and vomit, of infected flies. The clean flies will sometimes infect themselves from the vomit or faeces deposited by infected flies several days after infection. In no case were the limbs of the flies infected.

The fly in the milk is unfortunately an occurrence altogether too common. The infection of milk by the immersion of infected flies was shown by Graham-Smith in his experiments, in which *B. prodigiosus, B. pyocyaneus* and a pink *Coccus* were used, to be possible for as long a period as 74 hours after the flies had fed upon the infected matter. The micro-organisms were usually to be found in the crops or intestinal contents of the flies during that period.

Experiments along similar lines were carried out with blowflies (*Calliphora erythrocephala*) which were able to produce gross infection with non-spore-bearing bacteria (*B. prodigiosus* and *B. pyocyaneus*) for six to nine days.

The foregoing experiments clearly indicate the manner in which and extent to which infected flies may infect milk and sugar, and no doubt other foods containing these ingredients, by feeding upon them. They also demonstrate the more serious nature of such infection compared with infection by means of the limbs or body owing to the infective organisms persisting for a much greater length of time in the alimentary tract of the fly.

Infection from Flies Bred in Infected Material.

In addition to the direct mode of transference of the typhoid bacillus, by mechanical means or by feeding on infected material, some experiments of Faichnie (1909) indicated what may be a still more important means of dissemination. He believed that one of the most important sources of fly infection was from flies bred in enteric excreta ; such insects thereby became carriers of bacilli for the remainder of their lives, spreading the infection chiefly by means of their excreta. They were, in fact, typhoid " carriers."

On August 12th, 1909, three ounces of faeces infected with *B. typhosus* were placed in a box of earth and covered with a wire cage into which about 30 flies were liberated.   These flies died in a day or two, but on August 26th, fourteen days later, a single fly hatched and twelve flies emerged on the day following.   The box of earth was now replaced by a sterilised earthenware plate, and the wire cage was changed for a bell-shaped mosquito net.   The flies were fed upon sugar and water.   On August 26th one fly, a day old, was chloroformed and transfixed with a red-hot needle: its body was flamed (i.e. singed) and it was put into a bottle of sterile salt solution.   After shaking up, 1 c.c. of this solution was put into McConkey broth which remained unchanged for 48 hours. The fly was then crushed with a sterile glass rod and a drop plated: this gave *B. typhosus*.   On August 27th four other flies were similarly treated; the control experiment in McConkey was negative but *B. typhosus* was obtained from the crushed flies.

Similar results were obtained on September 3rd from two six-day old flies and on September 6th from two nine-day old flies treated in the same manner.   On September 10th two flies, thirteen days old, were placed in a sterile bottle for 24 hours and then removed.   The bottle was washed out with salt solution, and from this *B. typhosus* was recovered.   The two flies were then crushed in salt solution, not having been flamed, and *B. typhosus* was obtained.   On September 13th a sixteen-day old fly was placed in a sterile bottle for half an hour and then removed; two drops of excrement were visible and from sterile salt solution which was added *B. typhosus* was obtained.   This fly was flamed and crushed and the bacillus was recovered.   *B. typhosus* was not recovered from another sixteen-day old fly similarly treated.

A second series of experiments was carried out with the faeces of a man suffering from paratyphoid fever (*B. paratyphosus* A.), the diagnosis having been made by a blood culture.

On August 22nd two ounces of liquid faeces infected with *B. paratyphosus* A. were put into a box of earth and about thirty flies were allowed to feed on it, etc.   In a day or two they died owing to the absence of water.   On September 1st one fly hatched out; on September 3rd, 12 flies were seen, and on that date the earth was replaced by a sterile plate as in the previous experiments. On September 1st one fly, one day old, was examined as before : the McConkey broth control was negative and after being flamed and crushed *B. paratyphosus* A. was obtained. Similar results were obtained from four flies examined on September 3rd. On September 10th, three flies each seven days old were placed in a sterile bottle, and from their excrement *B. paratyphosus* A. was recovered : the flies were then examined, the McConkey broth test was negative, and the bacillus was recovered from the crushed flies. On September 10th, a fly ten days old was examined but the bacillus was not recovered ; it was recovered, however, from two other flies also ten days old.

From these experiments, which Faichnie appears to have carefully carried out, it will be seen that out of thirteen flies bred from a typhoid stool six at least contained *B. typhosus* in their intestines, and the bacillus was recovered from the excrement and intestine of a fly sixteen days old. Similarly, from a paratyphoid stool at least four flies out of eleven contained *B. paratyphosus* A. in their intestines and in each series of experiments one fly only was found not to contain the bacillus.

He also found *B. typhosus* in flies from Lahore, once ; from Kamptee, twice ; from Nasirabad, once in flies from the bungalow of an officer who had enteric fever, and once from flies in the officer's mess there ; from Nowgong, twice, once in flies from the Royal Artillery Coffee-shop, and again in flies from the trenching ground ; making a total of nine times in three months. Except in the case of the flies from Nasirabad, two flies were always flamed before examination, and a control of the washed flies was taken before crushing, so that there is no doubt that the bacillus was actually in the interior of the fly most probably in the intestine.

Summarising the conclusions reached as the result of his first series of experiments (1909) carried out at Kamptee, he says: " Experience seems to show that infection conveyed by flies' legs, natural though it may appear to all from experiments carried out to prove its possibility, is not a common nor even a considerable cause of enteric fever. On the other hand infection by the excrement of flies bred in an infected material explains many conclusions previously difficult to accept. In a word, it is the breeding ground that constitutes the danger, not the ground where the flies breed."

In a series of experiments with *Bacillus pyocyaneus* Bacot (1911) appears to prove conclusively that this bacillus if ingested during the larval life of *M. domestica* is able to retain its existence during the complex changes accompanying metamorphosis and to continue its existence in the gut of the adult fly after emergence from the puparium. He points out that the flies in Faichnie's experiments may have re-infected themselves by feeding on the contaminated material. To avoid the possibility of this in his experiments the pupae were sterilised and placed in clean sand in clean tubes. Commenting on Bacot's experiments, Ledingham states in a footnote that he has successfully isolated *B. typhosus* from pupae of *M. domestica*, the larvae having fed on the organism. All chances of external infection from the exterior of the puparium were removed by the careful method he devised of examining the bacterial content of the pupal interior. The puparium was held lightly between the left thumb and forefinger so that its blunt extremity was free. The extremity was seared by means of a small searing iron, and at the same time flattened. It was then pierced by a fine capillary pipette controlled by a rubber teat. The pupal contents are stirred up by the extremity of the pipette and finally drawn up into the tube whence they are squirted on to culture plates.

In a later paper Ledingham (1911) has carried his investigations still further, and shows that although *B. typhosus* was supplied in considerable quantities to the larvae of *M. domestica*, all attempts to demonstrate the bacillus in the pupae or resulting adult flies were unsuccessful until recourse was had to the disinfection of the ova. After this preliminary disinfection both larvae and pupae gave pure growths of *B. typhosus*. The author's chief conclusion appears to be that the typhoid bacillus can lead only a very precarious existence in the interior of the larvae or pupae which possess apparently a well-defined bacterial flora of their own, as Nicoll (1911) has recently shown. In the experiments it was not really possible to determine whether the *B. typhosus*, though recoverable from the pupae, was in fact actively multiplying in the pupal interior or gradually dying out. There was some indication that the latter was the case, as the typhoid colonies recovered from the pupa in the one successful instance were extremely few in number, while the larvae which had been feeding on *B. typhosus* contained enormous numbers of the bacilli as both cultural and microscopical study demonstrated.

Graham-Smith (1911) carried out experiments on flies bred from artificially infected larvae. He used the blow-fly *Calliphora erythrocephala* and fed the larvae on meat infected with the anthrax bacillus. These experiments gave results of a positive character and are described in a later chapter on anthrax.

In a further report Graham-Smith (1912) finds that of the non-spore-producing organisms only those which are adapted to the conditions prevailing in the intestine of the larvae, such as Morgan's bacillus and certain non-lactose fermenting bacilli survive through the metamorphosis and are present in the adult flies. Such organisms as *Bacillus typhosus*, *B. enteritidis* and *B. prodigiosus* rarely survive.

More recently Tebbutt (1912) has found that such pathogenic organisms as *Bacillus dysenteriae* (type " Y ") cannot be recovered from pupae or flies reared from larvae to which these organisms have been administered. When larvae which have been reared from disinfected ova have been fed on *B. dysenteriae* (type " Y ") the organism is recovered in a small proportion of cases from the pupae and flies. Under similar conditions *B. typhosus* was not recovered. Further experiments appeared to indicate that the process of metamorphosis is accompanied by a considerable destruction of bacteria present in the larval stage. Tebbutt believes that there is but a remote possibility of flies becoming infected from the presence of pathogenic organisms in the breeding ground of the larvae.

Nevertheless, it would appear that the experiments which have been described demonstrate the possibility that flies, bred from larvae which have bred in infected matter, may carry infection internally, and the practical significance of these results surely needs no explanation.

CHAPTER XXI

THE CARRIAGE OF TYPHOID FEVER BY FLIES

OF all infectious diseases the conditions incidental to this disease are most favourable for the transference of infection by *M. domestica*, and it is no doubt on this account that the greatest attention has been paid to the *rôle* of house-flies in the dissemination of this disease. The chief favourable condition is that the typhoid bacillus occurs in the stools of typhoid and incipient typhoid and "carrier" cases. Human excrement attracts flies not only on account of its moisture but as suitable food for the larvae. The infected excrement is often very accessible to flies, especially in military camps, as will be shown shortly; the flies also frequent articles of food and not infrequently the moist lips of man. Such are the conditions most suitable for the transference of the bacilli, and it is on account of the frequent coincidence of these conditions that flies can play, and have played, such an important *rôle* in the dissemination of this disease among communities, in spite of the fact that the typhoid bacillus cannot survive desiccation, which I think is an argument against its being carried by dust. The danger incident to the carriage of this non-spore-bearing bacillus in the digestive tract of the fly probably exceeds that resulting from direct carriage on the appendages and body of the insect.

The possibility of flies becoming infected with *Bacillus typhosus* is increased with the frequent accessibility of flies in numbers to infected material. This fact is most clearly demonstrated in military and construction camps, and during typhoid epidemics in the slum and insanitary portions of towns and cities.

A new danger is also added by the discovery of typhoid
"carriers." Although cases of walking typhoid or ambulatory
enteric fever have been known for a number of years and the
occurrence of chronic carriers was recognised in Germany, it is
only within the last few years that the attention of medical men
generally, in Europe and America, has been drawn to this important fact. The occurrence of these chronic carriers, who are
not ill but continue to give out the typhoid bacillus in their
excreta and urine, is discussed at some length by Howard (1911).
He quotes from an important article in the *Boston Medical
and Surgical Journal* and gives the following quotations and
cases :

"It is asserted by Kutscher that, in south-western Germany,
direct contact is a more important factor in the spread of typhoid
fever than polluted water, and that about four per cent. of typhoid
patients become chronic carriers of the specific bacilli which they
excrete in both urine and faeces, sometimes for long periods.
Doerr, for example, cites cases reported by Drober and Hunner, in
which the bacilli were isolated from the gall-bladder seventeen
and twenty years after recovery, and Lentz asserts that if after
ten weeks convalescence the excretion of the bacilli has not ceased,
it will most likely continue permanently and uninterruptedly, in
spite of medication.   He cites a number of cases in which, after
ten, thirty and even forty years after recovery, the excretion
continued.   Levy and Kayser report that in the autumn of 1905
a number of cases of typhoid fever occurred in an insane asylum,
in which two years previously an inmate had had the disease and
had recovered.   On the appearance of these later cases this person
was examined and was found to be excreting the bacilli in her
faeces.   Further examinations were made at intervals of several
weeks and the bacilli were found ten times.   In October 1906,
she died of a typhoid bacillery septicemia due to auto-infection
from the gall-bladder, and on autopsy the bacilli were isolated
from the spleen, liver, bile, wall of the gall-bladder and from the
interior of a large gall stone."

Nieter and Liefmann report a similar case occurring in an
insane asylum containing 250 inmates among which seven chronic
carriers were found.

To continue the account of carrier cases:

"Klinger found, among 1700 persons, twenty-three typhoid carriers, ranging in age from eighteen months to sixty years, eleven of whom had no typhoid history. Of 842 convalescents from the disease, sixty-three, or thirteen and one-tenth per cent., were found to be excreting the bacilli, and eight were still doing so six weeks after recovery.

"Kayser, tracing outbreaks to their sources, found a boy of twelve years, a member of a milkman's family, to be a chronic carrier and the probable source of infection in a number of cases. Another outbreak in which seventeen persons were seized (two deaths) was traced to a woman who had no typhoid history but was excreting the specific bacilli. She was employed in the dairy from which the persons seized had obtained their milk. Of 260 cases of typhoid fever investigated, 60 were traced to infected milk. Among the sixty victims were thirty maids and kitchen girls, twelve bakers and forty-four persons engaged more or less in kitchen work. In all, twenty-eight cases were traced directly to apparently healthy typhoid carriers."

The final case recorded is probably that of the now celebrated "Typhoid Mary." Six years previous to the institution of the enquiry this woman appeared to have had a mild attack of typhoid fever. "Since that time there have been undoubtedly twenty-eight cases of typhoid fever in the families in which she worked. The number of cases in a family within a few weeks of her advent varied from one or two up to six out of seven members. The evidence seemed so strong that she was at once removed to Reception Hospital by force. Examinations of her faeces and urine were made and the typhoid bacilli found in her faeces confirmed positively our suspicions (says the writer of the account) with regard to the possibility of her conveying typhoid fever."

The significance of the occurrence of so large a number of chronic carriers as these selected observations indicate is very great, especially when considered in relation to the question of the possibility of flies having access to infected matter under what one might call normal conditions. Flies would have as equal access to the infected faeces of these unrecognised "carriers" as to the faeces of a healthy person. Further investigations are certain to disclose a considerable percentage of "carriers" who are, unknowingly, distributors of the typhoid bacillus.

THE DISSEMINATION OF TYPHOID FEVER, EPIDEMIOLOGICAL AND CIRCUMSTANTIAL EVIDENCE.

In discussing the possible methods of the dissemination of the typhoid bacillus in the report on the prevalence of typhoid fever in the District of Columbia, Kober (1895) states: "The agency of flies and other insects in carrying the germs from box-privies and other receptacles from typhoid stools to the food cannot be ignored"; and in discussing certain special cases he says: "There is abundant evidence of unlawful surface pollution...and as the germs find a suitable soil in such surroundings it is possible that the flies, which abound wherever surface pollution exists, may carry the germs into the houses and contaminate the food." Later he states: "A large percentage of the cases occurred in houses supplied with box-privies which, apart from being an important cause in soil pollution, are believed to be otherwise instrumental in the dissemination of germs chiefly through the agency of flies."

There is a very large amount of testimony given as to the rôle played by flies in the spread of enteric in military stations and camps, and especially during the two wars—the Spanish-American and the Boer War. All the conditions most favourable for the dissemination of the bacilli by flies were, and in many military stations are still, present; open latrines or filth-trenches accessible to flies on the one hand and on the other the men's food within a short distance from the latrines. I cannot do better than quote the evidence in the words of the witnesses and allow it to speak for itself.

Vaughan, a member of the United States Army Typhoid Commission of 1898, states:

"My reasons for believing that flies were active in the dissemination of typhoid fever may be stated as follows:

"(a) Flies swarmed over infected faecal matter in the pits and then visited and fed upon the food prepared for the soldiers in the mess-tents. In some instances where lime had recently been sprinkled over the contents of the pits, flies with their feet whitened with lime were seen walking over the food.

"(b) Officers whose mess-tents were protected by screens suffered proportionately less from typhoid fever than did those whose tents were not so protected.

"(c) Typhoid fever gradually disappeared in the fall of 1898 with the approach of cold weather and the consequent disabling of the fly.

"It is possible for the fly to carry the typhoid bacillus in two ways. In the first place faecal matter containing the typhoid germs may adhere to the fly and be mechanically transported. In the second place, it is possible that the typhoid bacillus may be carried in the digestive organs of the fly and may be deposited with its excrement."

One of his conclusions was that infected water was not an important factor in the dissemination of typhoid in the national encampments of 1898, since only about one-fifth of the soldiers in the national encampments during the summer of that year developed typhoid fever, whereas about 80 per cent. of the total deaths were due to this disease. In the latter connection Sternberg (1899) refers to a report of Reed upon an epidemic in the Cuban War, in which it was stated that the epidemic was clearly not due to water infection but was transferred from the infected stools of the patients to the food by means of flies, the conditions being especially favourable for this means of dissemination.

Sternberg, as Surgeon-General of the U.S. Army, issued the following instructions[1]: "Sinks should be dug before a camp is occupied or as soon as practicable. The surface of the faecal matter should be covered with fresh earth or quicklime or ashes three times a day."

I think that the instructions of that ancient leader of men, Moses, who probably had experienced the effects of flies, were even better than these. He said (Deut., ch. xxiii, vv. 12, 13): "Thou shalt have a place also without the camp whither thou shalt go forth abroad; and thou shalt have a paddle (or shovel) among thy weapons; and it shall be, when thou sittest down abroad, thou

---

[1] *Circular No. 1 of the Surgeon-General of the U.S. Army*, April, 1898.

shalt dig therewith, and shalt turn back and cover that which cometh from thee."

Sternberg is of the opinion that typhoid fever and camp diarrhoea are frequently communicated to soldiers through the agency of flies, "which swarm about faecal matter and filth of all kinds deposited upon the ground or in shallow pits, and directly convey infectious material attached to their feet or contained in their excreta to the food which is exposed while being prepared in the common kitchen, or while being served in the mess-tent."

Veeder (1898), in referring to the conditions existing in the camps of the Spanish-American War, says that in the latrine trenches he saw "faecal matter fresh from the bowel and in its most dangerous condition, covered with myriads of flies, and at a short distance there was a tent, equally open to the air, for dining and cooking. To say that the flies were busy travelling back and from between these two places is putting it mildly." Further, he says, "There is no doubt that air and sunlight kill infection, if given time, but their very access gives opportunity for the flies to do serious mischief as conveyers of fresh infection wherever they put their feet. In a very few minutes they may load themselves with the dejections from a typhoid or dysenteric patient, not as yet sick enough to be in hospital or under observation, and carry the poison so taken up into the very midst of the food and water ready for use at the next meal. There is no long and round-about process involved. It is very plain and direct. Yet when the thousands of lives are at stake in this way the danger passes unnoticed, and the consequences are disastrous and seem mysterious until attention is directed to the point; then it becomes simple enough in all conscience."

The following statements which I have noted in reading the voluminous report of the Commission which was charged with the investigation of the origin and spread of typhoid fever in the United States military camps during the Spanish War in 1898 (see Reed, Vaughan and Shakespeare, 1904), give the opinions of observers in different camps:—p. 62, the epidemic was due to flies, not to the water; the careless disposition of filth; p. 74, flies were swarming over sinks, kitchens and mess tables; p. 88, the epidemic was due to the infection of food by flies; p. 104, thoroughly convinced that the chief agent in the distribution of typhoid fever was the fly; p. 107, the cause was not the water-supply but fly contamination; flies "had inflicted greater loss upon the American soldiers than all the arms of Spain"; p. 206, the cook observed flies with feet covered with disinfecting lime, on the food; p. 273, where mosquito netting was used little or no typhoid was contracted; p. 279, flies observed carrying lime from privy vaults; repeated mention of sinks and flies in millions, of sinks near to mess tents; p. 535, the flies and the heat made a visit to the sinks (latrines) like a visit to purgatory; p. 665, flies swarmed so numerously that the first faecal droppings were covered before defaecation was complete.

The above statements are all the more significant, when it is remembered that at that time (1898) the idea of the house-fly as a disease carrier had hardly been conceived except in the minds of a limited number of investigators. The evidence which was afforded, however, was convincing enough and it was only natural that the Commission came to the conclusion that: "flies undoubtedly served as carriers of infection." From this onward, the belief in the disease-carrying powers of the house-fly increased in strength and became firmly established in people's minds.

Chmelicek (1899) in recording his experience of the conditions of camp life at Tampa, Florida, during the Spanish-American war, states: "The pits were only about forty feet from the entrance of the kitchen tent and the number of flies around these holes was countless." He calls attention to the fact, to which Aldridge (1907) referred later, that the greatest incidence of typhoid was among the mounted troops, and he attributed it to the larger number of flies present. In reference to the conditions in the kitchens he mentions the fact that the flies travelled from the latrines to the kitchen tent where sugar, which was exposed for hours, was almost black with them and looked more like a bag or box full of raisins than of sugar.

Dutton (1909) gives an interesting figure to demonstrate the manner in which flies would be carried from sources of typhoid infection (division hospitals and latrines) in the camps of the United States Army at Fernandina and Tampa to different parts of these camps. He states that Sergeant Brady, who was stricken with typhoid fever at Fernandina, mentioned to him that the lime used about the latrines and garbage dumps was carried by flies to the food which was being used in the camps.

In the South African War, a year or two later, the same conditions existed, and there was a very heavy loss of life from enteric fever. Writing on the subject, Dunne (1902) says: "The plague of flies which was present during the epidemic of enteric at Bloemfontein in 1900 left a deep impression on my mind, and, as far as I can ascertain from published reports, on all who had experience on that occasion. Nothing was more noticeable than the fall in the admissions from enteric fever coincident with the killing off of the flies on the advent of the cold nights of May and June. In July, when I had occasion to visit Bloemfontein, the hospitals there were half empty, and had practically become convalescent camps."

A similar experience is related by Tooth (1901). Referring to the *rôle* of flies he says: "As may be expected, the conditions in these large camps were particularly favourable to the growth and multiplication of flies, which soon became terrible pests. I was told by a resident in Bloemfontein that these insects were by no means a serious plague in ordinary times, but that they came with the army. It would be more correct to say that the normal number of flies was increased owing to the large quantities of refuse upon which they could feed and multiply. They were all over our food, and the roofs of our tents were at times black with them. It is not unreasonable to look upon flies as a very possible agency in the spreading of the disease, not only abroad but at home. It is a well-known fact that with the first appearance of the frost enteric fever almost rapidly disappears....It seems hardly credible that the almost sudden cessation of an epidemic can be due to the effect of cold upon the enteric bacilli only. But there can be no doubt in the mind of anybody who has been living on the open veldt, as we have for three or four months, that flies are extremely sensitive to the change of temperature, and that the cold nights kill them off rapidly." In the discussion on this paper Church stated that "many nurses told me that if one went into a tent or ward in which the patients were suffering from a variety of diseases, one could tell at once which were the typhoid patients by the way in which the flies clustered about their mouths and eyes while in bed." It was further stated in the discussion that where the Americans used quicklime in their latrines the cooks in the neighbouring kitchens found that the food became covered with quicklime from the flies which came from the latrines to the kitchens.

Dr Tooth, in a letter to me, says: "I am afraid my written remarks hardly express strongly enough the importance that I attach to flies as a medium of spreading infection. Of course, I do not wish to under-rate the water side of the question, but once get, by that means, enteric into a camp the flies, in my opinion, are quite capable of converting a sporadic incidence into an epidemic. A pure water supply is an obvious necessity, but the prompt destruction of refuse of every description is every bit as important."

Smith (1903), in speaking of his experience in South Africa, says: "On visiting a deserted camp during the recent campaign it was common to find half a dozen or so open latrines containing a foetid mass of excreta and maggots." Similar observations were made by Austen (1904), who, describing a latrine that had been left a short time undisturbed, says: "A buzzing swarm of flies would suddenly arise from it with a noise faintly suggestive of the bursting of a percussion shrapnel shell. The latrine was certainly not more than one hundred yards from the nearest tents, if so much, and at meal times men's mess-tins, etc., were always invaded by flies. A tin of jam incautiously left open for a few minutes became a seething mass of flies (chiefly *Pycnosoma chloropyga* Wied.), completely covering the contents."

Howard (1900), referring to an American camp where no effort was made to cover the faeces in the latrines, says: "the camp contained about 1200 men, and flies were extremely numerous in and around the sinks. Eggs of *Musca domestica* were seen in large clusters on the faeces, and in some instances the patches were two inches wide and half an inch in depth, resembling little patches of lime. Some of the sinks were in a very dirty condition and had a very disagreeable odour."

A few examples of the prevalence of conditions favouring the dissemination of enteric by flies in permanent camps may be noted. Cockerill (1905), in describing camp conditions in

Bermuda, mentions kitchens within one hundred yards of the latrines; the shallow privy, seldom or never cleaned out, and middens are found which contain masses of filth swarming with flies. He states that in more recent years the period of greatest incidence is in the summer, being chiefly due to flies and contaminated dust. Wanhill (1909) has furnished an interesting report on the typhoid conditions in Bermuda. From 1893 to 1902 Bermuda had the highest enteric fever rate among the troops of any command occupied by British troops. Wanhill was placed in charge in 1904 and in two years the disease was almost eradicated. He considered that flies were the most important agents in the dissemination of the bacilli.

Quill (1900), reporting on an outbreak of enteric in the Boer camp in Ceylon, states: "During the whole period that enteric fever was rife in the Boer camp flies in that camp amounted to almost a plague, the military camp being almost similarly infested, though to a less extent. The outbreak in the Boer camp preceded that among the troops; the two camps were adjacent, and the migration of the flies from the oṇe to the other easy." Weir, reporting on an outbreak of enteric fever in the barracks at Umbala, India[1], says that most of the pans in the latrines were half or quite full, and flies were very numerous in them and on the seats, which latter were soiled by the excreta conveyed by the flies' legs. The men stated that the plague of flies was so great that in the morning they could hardly go to the latrines. He found that the flies were carried from the latrines to the barrack-rooms on the clothes of the men. This state of affairs suggests another mode of infection, namely, *per rectum*. As Smith has pointed out (*l.c.*) it is not improbable that flies under these conditions may be inoculators of dysentery.

Aldridge (1907) gives some highly suggestive statistics showing the influence of the presence of breeding-places of flies. Flies are found in greater numbers in mounted regiments than in infantry, and he shows how this affects the incidence of enteric fever. In the British army in India, 1902-05, the ratios per 1000 per annum of cases admitted were: cavalry, 5·74, and infantry 4·75. He states that: "A study of the incidence of enteric fever shows that

---

[1] *Army Medical Department Report*, 1902, p. 207.

stations where there are no filth trenches, or where they are a considerable distance from the barracks, all have an admission-rate below the average, and all but one less than half the average."

Jones (1907) gives an interesting account of his experience of typhoid fever in the army at Nasirabad, India. He states: "We have been led in this station to regard fly infection as the principal cause of the unenviable prevalence of enteric fever"— and "Believing as we do that flies are the chief carriers of enteric fever in India, any plan which gets rid of them is worthy of consideration." Howard (1911) calls attention to the significance of the method adopted by Jones to persuade the high-caste natives to adopt and carry out his views. Making use of the word "kakophagy" (excrement-eating) Jones writes: "I presume no one wishes to be a kakophagist; yet we are so in spite of ourselves, if flies bred in filth pits alight on our food just before we eat it." This kind of argument would surely appeal to the most indifferent high-caste native or other person who was inclined to regard flies as "wholesome and appetising." Stratton (1907) discussing the seasonal prevalence of enteric at Meerut, India, states that the fever recurs with the reappearance of the dust and flies after the rainy season; during the monsoon there is a great diminution of enteric.

Ainsworth (1908) has studied the relation between enteric fever and climatic conditions at the military stations at Poona and Kirkee, India. His conclusion is that it is highly significant and at least suggests that a *prima facie* case has been established for further investigation. By means of curves he illustrates the rise and fall of the number of flies, the typhoid evidence and the rainfall. As Nuttall, in referring to this paper, says: "The curves given on pages 497 and 498 are certainly very striking, the fly curve reaching its acme about two weeks before the maximum number of cases of typhoid occurred."

All these facts are equally applicable to the conditions in our own towns and cities. Where the old conservancy methods are used, such as pails and privy middens, the incidence of typhoid fever is greater than in those places where the system of water disposal has been adopted. I have examined the annual reports of the medical officers of health of several large towns where such conversions are being made, and they show a falling-off of the typhoid fever-rate coincident with this change. In Nottingham, for example[1], in the ten years 1887–1896, there was one case of typhoid fever for every 120 houses that had pail-closets, one case for every 37 houses with privy middens, and one case for every 558 houses with water-closets. The last were scattered, and not confined to the prosperous districts of the town.

Klein (1908) in discussing an outbreak of typhoid fever at Wilshaw (Eng.) describes what is a common occurrence in very many outbreaks in towns and cities. He says " After the occurrence of a case of enteric fever in a house forming one of a row, a number of typhoid cases making their appearance in the neighbouring houses. All known channels of transmission, for example, personal contact, defective drainage, polluted water or milk, could be excluded. The only condition common to all the houses of the row was this—that they were swarming with flies." Klein made a bacteriological examination of some of the " flies " (it is presumed they were *Musca domestica*) after crushing them and found *Bacillus coli* and *B. typhosus*.

The insanitary conditions found in many of our towns and cities are admirably suited in every detail for the breeding of flies, the disease-carriers and for their carriage of the germs. In a previous paper (1908) I have called attention to these facts. It was pointed out that : " wherever there are collections of either excremental products or decaying and waste vegetable substances and food stuffs, house-flies are able to breed. Consequently, where such conditions as the following exist we shall almost certainly find house-flies : heaps of stable-manure and other excremental products and house refuse which have stood for more than eight or nine days, the time occupied by the development of the fly ; such systems of excremental disposal as middens, dry ashpits and pails which are not regularly removed within the same time. I find that public tips frequently form permanent breeding places for flies on account of the variety of substances tipped thereon. House refuse and excremental substances should not be deposited on public tips in the vicinity of houses but should be removed

[1] "Typhoid Fever and the Pail System at Nottingham," *Lancet*, Nov. 29th, 1902, p. 1489.

immediately, as their mere presence for a short time attracts the
flies and their continued presence serves to multiply the nuisance."
In this connection I called attention to the following evidence
indicating the relation of the presence of flies to insanitary conditions and the presence of nuisances: "In his report on the
sanitary conditions of Wigan ('Reports of Medical Inspectors of
the Local Government Board on the General Sanitary Circumstances and Administration of the County Borough of Wigan, with
especial reference to Infantile Mortality and to Endemic Prevalence
of Enteric Fever and Diarrhoea,' 22 pp., 1906) Dr Copeman says
(p. 18): 'At the Miry Lane Depot, as previously mentioned, there
is always stored (awaiting removal by farmers) an enormous amount
of nightsoil mixed with ashes which in hot weather especially, is
not only exceedingly offensive, but is beset by myriads of house-
flies. As the result of personal enquiry at the various houses in
the neighbourhood in which during the year 1905 deaths from
diarrhoea had occurred, I learnt that considerable nuisance from
the foul odour was apt to be experienced during the prevalence of
hot weather, especially with the wind in the south or south-west,
i.e. blowing from the Depot to the special area; so much on
occasion as to render it necessary to shut all the windows, while
the inhabitants of certain of the houses nearest the Corporation
Depot stated that at certain times of the year their rooms were
apt to be invaded by a veritable plague of flies *which swarmed
over everything of an edible nature on the premises.* (The italics
are mine, C. G. H.)   This being so, it would not appear improbable
that these flies, some of which have doubtless had an opportunity
of feeding on and becoming contaminated with excremental
material of human origin, may have been the means of carrying
infectious material to certain foodstuffs, such, more particularly, as
milk and sugar and so, indirectly, of bringing about infection of
the human subject.' "

The county of Durham has the highest death-rate from enteric
fever of any county in England or Wales.   In discussing a preliminary
report of an investigation into the subject Dr Newsholme, Medical
Officer of the Local Government Board, as reported in *The Times*
(weekly edition) of August 5th, 1910, states: "We come finally
to the conditions with which the excess of enteric fever in the

county of Durham can in a large measure be causally associated. These are the extremely filthy domestic arrangements, by which excremental matters are retained in the immediate vicinity of dwellings." The account goes on to state "The general position as regards the relation between conservancy methods and enteric fever may be stated thus : throughout England and Wales counties persisting in the use of conservancy methods of dealing with excremental matters and not having adopted the water-carriage system have excessive enteric fever, in all instances in which industrial conditions imply considerable aggregations of population."

One of the most important investigations on the relation of flies to intestinal disease was that of Jackson (1907). He investigated the sanitary condition of New York Harbour and found that in many places sewer outfalls had not been carried below low-water mark, consequently solid matter from the sewers was exposed on the shores, and that during the summer months on and near the majority of the docks in the city a large amount of human excreta was deposited. This was found to be covered with flies. The report, considered as a mere catalogue, is a most severe indictment against the insanitary condition of this great water front. By means of spot-maps he shows that the cases of typhoid are thickest near the points found to be most insanitary. He shows, as English investigators have also shown, how the curves of fatal cases correspond with the temperature curves and with the curves of the activity and prevalence of flies which were obtained by actual counts. He also adduced bacteriological evidence, and it is stated that one fly was found to be carrying over one hundred thousand faecal bacteria.

An instructive example of the part which flies may play not only in the carriage of typhoid bacilli to persons near the infected matter, but also, through the medium of milk, to a larger number of people, is communicated by Taylor (Colorado State Board of Health, U.S.A.) to the New York Merchants' Association. He says : "In the city of Denver we had a very sad as well as a plain demonstration of the transmission of typhoid fever by flies and milk. Early in August of this year the wife of a dairyman was taken with typhoid fever, remaining at home about three weeks before her removal to the hospital, August 28th. During the first two weeks in September we received reports of numerous cases of typhoid fever in the northern portion of Denver, and upon investigation found that all these cases had been securing their milk from this dairy. An inspection of this dairy was then made, and in addition to learning of the illness of the dairyman's wife, we also found the dairyman himself suffering with a mild case of typhoid fever, but still up and delivering milk. The water supply of this dairy was fairly good. However, we found that the stools of both the wife and husband had been deposited in an open privy vault located thirty-five feet from the milk-house, which was unscreened and open to flies. The gelatine culture exposed for thirty minutes in the rear of the privy vault and in the milk-house among the milk-cans gave numerous colonies of typhoid bacilli, as well as colon bacilli and the ordinary germ-life. The source of infection in the dairyman's wife's case is unknown, but I am positive that in all the cases that occurred on this milk route the infection was due to bacilli carried from this vault by flies and deposited upon the milk-cans, separator and utensils in the milk-house, thereby contaminating the milk. The dairyman supplied milk to 143 customers. Fifty-five cases of typhoid fever occurred, and six deaths resulted therefrom[1]."

Washburn (1910) has shown how insanitary conditions in localities where little attention is paid to the prevention of health render easy the infection of food by flies and the dissemination of the typhoid bacillus. In the report of an investigation into the prevalence of typhoid fever at Charlestown, West Virginia, U.S.A., Ridlon (1911), in discussing the epidemiological factors, states: " The most probable source of infection in five cases was from flies. These cases were located within two hundred feet or less of other cases where the disinfection of the stools was inefficient, where there were no screens and where the abundant flies had free access to both dejecta of patients and the food." He continues: " That flies under the proper conditions can be a prominent factor in the spread of infection is an undisputed fact, as is also the fact that their prevalence can be greatly diminished by proper care of

[1] *The House-fly at the Bar, Indictment Guilty or not Guilty?* The Merchants' Association of New York, April, 1909, p. 48.

their breeding places, including stable-manure, household refuse and garbage."

An outbreak of typhoid fever, which was recently reported upon (October 1912) by Drs Amyot and McClenahan, occurred in the Insane Asylum at Hamilton (Ontario, Canada). Since July 1912 there had been fourteen undoubted cases with two deaths. The investigation showed that the flies had been the responsible agents, carrying the infection from unscreened lavatories to the patients. The manure piles in the stables had been left exposed.

Terry (1913) describes the very great reduction in the typhoid death-rate in Jacksonville, U.S.A., which followed the control of the fly-borne typhoid by compulsory rendering of privies fly-proof.

The dissemination of Typhoid Fever, Bacterio-logical Evidence.

The bacteriological evidence, the result of exact experiments, indicates conclusively the ability of flies to carry the typhoid bacillus in a viable condition, not only internally but also externally.

A non-spore-bearing bacillus such as *B. typhosus* is less adapted to external transference than a spore-bearing bacillus. It is never-theless interesting to note the length of time which elapses before *B. typhosus* is moribund. Howard (1911) states that Dr Mohles informs him that *B. typhosus* will live in butter under common market conditions for 151 days and still be able to grow when transferred to suitable conditions. In milk kept under market conditions they retain active mobility for twenty days; after this they lessen in numbers and finally disappear in the forty-third day. An important fact is that observed by Delepine, namely, that *B. typhosus* can remain in a viable condition on the walls of a privy for twelve months. The bearing which this fact has on the possibility of flies becoming infected subsequent to an epidemic of typhoid fever will be readily understood.

In order to form a correct estimate of the extent to which and the manner in which flies distribute pathogenic and other bacteria, a study of the natural bacterial flora of the digestive tract of the house-fly would appear to be desirable, if not essential. Until recently little attention has been paid to this matter. Nicoll (1911) has begun such a study with interesting results. He finds that the house-fly may carry at least twenty-seven varieties of *Bacillus coli*, by far the most frequent of which are *B. coli commune* and MacConkey's bacillus No. 71. From the character of these colon bacilli it would appear that the house-fly derives its bacterial flora from excremental matter and other sources. The presence of colon bacilli in the digestive tract of the fly is only to be expected from the filthy feeding-habits of the insect; in fact, the absence of these bacilli would be more than remarkable. Nicoll also finds that certain non-lactose fermenting bacilli appear to be capable of multiplying in the intestine of the fly. Of these Morgan's bacillus No. 1 is not an infrequent inhabitant of the fly's intestine and *B. paratyphosus* B. has been found on two occasions.

Graham-Smith (1909) has also examined flies captured in various places with a view to ascertaining what percentage were infected with bacilli of the colon group. The results of his examinations of flies from the different sources was as follows: from the neighbourhood of decaying animal matter, Cambridge, 28 flies were examined and 54·5 per cent. were infected; from a railway-siding, Islington, 48 flies were examined and 25 per cent. were infected; from a room used by men at a gas-works, 24 flies were examined and 16·6 per cent. were infected; from a house near a glue works, 26 flies were examined and 15 per cent. were infected; from the kitchen of a London County School, 18 flies were examined and 22·2 per cent. were infected; from a house about fifty yards distant from a jam factory, Bermondsey, 40 flies were examined and 10 per cent. were infected. The flies which were examined were *M. domestica*, *Fannia canicularis*, *Calliphora erythrocephala* and *C. vomitoria*, *Stomoxys calcitrans* and a few other small flies. Cultures were made of both the intestines and of the surfaces of the flies' bodies. Altogether 35 lactose-fermenting organisms of the colon group were isolated, 22 from surface cultures and 13 from the intestines. Although the numbers were comparatively small, the experiments indicated that the flies were infected with bacilli of the colon group in proportion to the opportunities offered by the locality they were frequenting. The highest degree of infection was found on those flies frequenting decaying animal matter and the next highest on those caught near manure.

Torrey (1912), in the United States, has also made a study of the numbers and types of bacteria carried by flies under city conditions. Flies examined up to the latter part of June were found to be free from faecal bacteria and carried a homogeneous flora of coccal forms. During July and August there occurred periods in which the flies examined carried several millions of bacteria, alternating with periods in which the number of bacteria was reduced to hundreds. The scanty flora, he considered, probably indicated the advent of large numbers of recently-emerged flies. Faecal bacteria of the colon type were first encountered in abundance during the early part of July. The bacteria in the intestines of the fly were 8·6 times as numerous as those occurring on the external surface of the insects; an important fact to note. On the surface of the flies bacteria of the colon group constituted 13·1 per cent. of the total and within the intestine they constituted 37·5 per cent. of the total. Of the lactose-fermenting bacteria which were isolated and identified 79·5 per cent. belonged to the colon-aerogenes group and 20·5 per cent. belonged to the acidilactici group. Fifteen cultures of *Streptococci*, isolated and identified, were distributed among the *equinus*, *faecalis* and *salivarius* groups. He found none of the *pyogenes* type. The most important isolations were three cultures of *Bacillus paratyphosus*, Type A. Bacteria of the paracolon type, causing a final intense alkaline reaction in litmus milk and fermenting only certain monosaccharides, were frequently found during August.

Further references to studies of this nature are given in a later section (p. 288). The foregoing experiments, apart from the indications which they give as to the nature of the flora of what one might call a normal fly, demonstrate how necessary it is to exercise great caution in making deductions from the results of isolated examinations and cultures of flies both externally and internally.

Reference has already been made to the fact that Jackson gave bacteriological evidence as to the ability of house-flies to carry *B. typhosus*. There are, however, a number of workers who have carried out experiments along bacteriological lines and their results will be given.

Celli (1888) recovered *Bacillus typhi abdominalis* from the dejections of flies which had been fed on cultures of the same, and he was able to prove that they passed through the alimentary tract in a virulent state by subsequent inoculation experiments.

Firth and Horrocks (1902), in their experiments, took a small dish containing a rich emulsion in sugar made from a twenty-four-

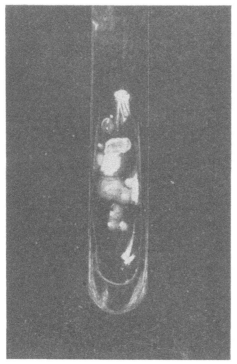

FIG. 97.   Agar-agar slope culture of bacteria and moulds deposited by *M. domestica* caught in the author's laboratory (Jan. 1910) and allowed to make a single journey over the culture medium.

hour agar slope of *Bacillus typhosus* recently obtained from an enteric stool and rubbed up with fine soil.   This was introduced with some infected honey into a cage of flies together with sterile litmus agar plates and dishes containing sterile broth, which were placed at a short distance from the infected soil and honey.   Flies were seen to settle on the infected matter and on the agar and broth.   The agar plates and broth were removed after a few days, and after incubation at 37° C. for twenty-four hours colonies of *Bacillus typhosus* were found on the agar plates, and the bacillus was recovered from the broth.   In a further experiment the infected material was dusted over with fine earth to represent superficially buried dejecta, and the bacillus was isolated from agar

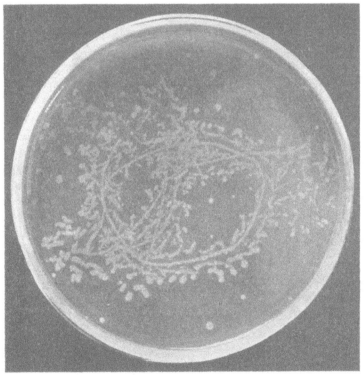

FIG. 98.   Agar plate culture of tracks of *M. domestica* caught in a room and allowed to walk across and around the medium.   Natural size.   (Prepared by H. T. Güssow.)

plates upon which the flies had subsequently walked, as in the former experiment.   They also found the bacillus on the heads, wings, legs and bodies of flies which had been allowed to have access to infected material.

Hamilton (1903) recovered *Bacillus typhosus* five times in eighteen experiments from flies caught in two undrained privies, on the fences of two yards, on the walls of two houses, and in the room of an enteric fever patient.

Ficker (1903) found that when flies were fed upon typhoid cultures they could contaminate objects upon which they rested. The typhoid bacilli were present in the head and on the wings and legs of the fly five days after feeding. He also recovered *B. typhosus* from the flies 23 days after they had been infected. The bacillus was isolated from flies caught in a house in Leipzig, where eight cases of typhoid had occurred. He calls attention to the fact that the value of Celli's conclusions is diminished owing to the fact that at the time the experiments were carried out the differentiation between *B. typhosus* and organisms of a similar character was hardly possible.

Buchanan (1907) was unable to recover the bacilli from flies taken from the enteric ward of the Glasgow Fever Hospital. Flies were allowed to walk over a film of typhoid stool and then transferred to the medium (Grünbaum and Hume's modification of MacConkey's medium), and subsequently allowed to walk over a second and a third film of medium. Few typhoid bacilli were recovered and none from the second and third films.

Sangree (1899) performed somewhat similar experiments to those of Buchanan and recovered various bacilli in the tracks of the flies. This method of transferring the flies immediately from the infected material to the culture plate is not very satisfactory, as I have already pointed out (1908), as it would be necessary for the flies to be very peculiarly constructed not to carry the bacilli. The fly should be allowed some freedom before it has access to the medium to simulate natural conditions. Experiments of this kind were carried out in the summer of 1907 by Dr M. B. Arnold (Superintendent of the Manchester Fever Hospital) and myself. Flies were allowed to walk over a film of typhoid stool and then were transferred to a wire cage, where they remained for twenty-four hours with the opportunity of cleaning themselves, after which they were allowed to walk over the films of media. Although we were unable to recover *B. typhosus* the presence of *B. coli* was demonstrated. *B. coli* was also obtained from flies caught on a public tip upon which the contents of pail-closets had been emptied; the presence of *B. coli*, however, may not necessarily indicate recent contamination with human excrement.

Aldridge (1907) isolated a bacillus, apparently belonging to the paratyphoid group, from flies caught in a barrack latrine in India during an outbreak of enteric fever. In appearance and behaviour to tests it was very similar to *B. typhosus.*

A series of careful experiments were made by Sellars[1] in connection with Niven's investigations on the relation of flies to infantile diarrhoea. Out of thirty-one batches of house-flies carefully collected in sterilised traps in several thickly populated districts in Manchester he found, as a result of cultural and inoculatory experiments, that bacteria having microscopical and cultural characters resembling those of the *Bacillus coli* group were present in four instances, but they did not belong to the same kind or variety.

In a report of investigations carried out at the Central Research Institute of India by Thomson (1912), the following conclusions are given: " The ingestion of typhoid germs in large numbers has no bad effect on the health of the flies; they can retain living typhoid bacilli within their bodies and transmit infection thereby for a period of twenty-four hours after ingestion; they can carry the living germs on the exterior of their feet or bodies for a period of six hours, and so transmit infection."

Graham-Smith (1910) carried on infection experiments with *B. typhosus.* Flies were fed on syrup infected with this bacillus and afterwards on plain syrup. Sixteen hours after the removal of the infected syrup cultures on Drigalski-Conradi medium were made of the intestinal contents of five flies; eight flies were allowed to walk on culture plates and plates were also sown with emulsified faeces. *B. typhosus* was recovered in all cases. This experiment was repeated with flies, two days, three days, and daily to six days after infection. The results showed that *B. typhosus* may remain alive in the intestinal canal for at least six days and that flies may infect plates upon which they walk for at least forty-eight hours after infection.

A typical instance of a small epidemic of typhoid fever is

[1] Recorded in the *Report on the Health of the City of Manchester*, 1906, by James Niven, pp. 86—96.

reported by Cochrane (1912). This occurred at St George's, Bermuda, and flies were considered to be the probable carriers. There were eight cases of fever; three of these were in soldiers and the five civilian cases were in three families. Two of the three soldiers were employed in two of the families. A fatal case of typhoid had occurred at a house 300 yards from the house where the first case occurred in September of the previous year; three cases occurred in April and May. A dry earth latrine was used where the fatal cases of typhoid occurred. A fly was caught near this latrine. It was put into 5 c.c. of sterile salt solution for a minute and cultures made. The fly was emulsified and other cultures made. *Bacillus typhosus* was isolated from the washings. The house and the latrine were cleaned and disinfected and no further cases have occurred in the vicinity since.

CHAPTER XXII

THE RELATION OF FLIES TO SUMMER DIARRHOEA
OF INFANTS

THE conclusive nature of the evidence that the house-fly is an important factor in the spread of typhoid fever naturally directed the attention of medical men and investigators to the possibility of their being concerned in the dissemination of this serious intestinal complaint of infants. Niven (1910) defines summer or epidemic diarrhoea as being "a term applied to an affection marked by a somewhat definite group of symptoms, in which vomiting sickness, copious diarrhoea, rice-watery and green stools, and finally convulsions play a conspicuous part. This condition is not rarely somewhat prolonged, and is often attended with some degree of fever. On the one hand it shades into typhoid and paratyphoid fevers, and on the other it is not rarely the termination of a tuberculous enteritis or some wasting affection." This disease is responsible for an enormous mortality among infants under two years of age, and in fact accounts for more deaths than any other disease.

In referring to epidemic diarrhoea in Portsmouth, Fraser (1902), cited by Nuttall and Jepson, states that "on visiting the houses in question I find that in all, almost without exception, the occupants have suffered from a perfect plague of flies. They told me every article of food is covered at once with flies....I repeat that to this, and this alone, I attribute the diarrhoea in the Goldsmith Avenue district."

Nash was one of the first medical observers to call attention (in 1902) to the remarkable coincidence between the abundance of flies and the prevalence of this serious infantile disease. In the years 1902 and 1903 the summers were wet and therefore un-favourable to the breeding and activity of *M. domestica*, and in these years the diarrhoeal diseases were less prevalent and the infantile mortality rate was considerably below the average. He suggested (1903), in a paper read before the Epidemiological Society of London in January, 1903, that flies carried the infective material from all kinds of filth to the food supplies and were re-sponsible for the spread of this disease and supported his contention with a further instance, namely, that "in the early part of Sep-tember, 1902, flies became prevalent, and co-incidentally diarrhoea, which had hitherto been conspicuous by its absence, caused thirteen deaths in Southend. Then came a spell of cold weather; the flies rapidly diminished in number, and no further deaths from diarrhoea were recorded" (1905). In 1904, by means of a "spot map" he found that the great majority of deaths from diarrhoea occurred in the proximity of brick fields in which were daily deposited some thirty tons of house refuse, an admirable breeding place for this insect. He has shown the actual danger which exists in flies carrying bacterial organisms to milk as many other in-vestigators have shown, and the danger resulting from the co-incident occurrence of uncovered milk and infected flies is too obvious to need emphasis.

Newsholme (1903) discussed the possibility of food infection by flies in the houses of the poor. He states: "The sugar used in sweetening the milk is often black with flies which may have come from a neighbouring dustbin or manure heap, or from the liquid stools of a diarrhoeal patient in a neighbouring house. Flies have to be picked out of the half-empty can of condensed milk before its remaining contents can be used for the next meal." The observations of Copeman (p. 241) have already been mentioned and similar instances of the relation between flies and the incidence of summer diarrhoea have been referred to by Snell (1906) and other medical officers of health in their reports. Sandilands (1906) states that there are "good grounds for the supposition that in this disease, which in some respects is analogous to typhoid fever and cholera, flies may be carrying agents of the first im-portance." He observes that the meteorological conditions which influence the prevalence of diarrhoea exercise a precisely similar effect upon the prevalence of flies. He states: "The immunity of well-to-do infants may be explained, partly by the distance that separates the sick from the healthy, and partly by the small number of flies in their neighbourhood. In poor districts six or seven babies may occupy the tenements of one house with a common yard where flies congregate and flit in and out of the open windows, themselves conveying infected excrement to the milk of healthy infants, or depositing the excrement in the dustbin, whence it may again be conveyed into the house by other flies. Calm weather promotes diarrhoea and high winds are unfavourable to the spread of diarrhoea and to the active migration of flies alike. Loose soil and fissured rock, containing organic filth in its crevices, favour the spread of diarrhoea and the breeding of ·flies, whilst solid rock is unfavourable to both." Ainsworth (1909) studied the relation of flies to infantile diarrhoea in Poona and Kirkee, India, and by means of a yearly curve illustrates the relation, which, as in the case of similar curves constructed from English statistics, affords, or would appear to afford according to the critics, evidence of a close relationship between flies and summer diarrhoea.

The most exhaustive epidemiological study of the relation of flies to this disease has been made by Niven (1910) in Manchester. He commenced to make observations in 1903, and from 1904 systematic captures of flies were made at selected stations. After a consideration of the various factors which have been studied in relation to summer diarrhoea such as soil, temperature, etc., he states: "What we require for the explanation of the facts of summer diarrhoea is the presence of some transmitting agent rising and falling with the rise and fall of diarrhoea, the features pertaining to which must correspond and explain the features of the annual wave of diarrhoea. None of the facts of which we have cognizance do afford such an explanation, and we come by exclusion to consider the house-fly. The process of conveyance is not striking and arresting as it is in military camps abroad; nor does the number of flies usually approach that observed in tropical and sub-tropical countries. We are, therefore, obliged to attack the question *de novo*, and examine such evidence as we possess to see whether we may rest reasonably confident that in flies we have found the transmitting agent sought for. If the house-fly is the transmitting agent in summer diarrhoea, the following conditions should be fulfilled:—

"(1) (a)   There should be evidence that the house-fly carries bacteria under the ordinary summer conditions; (b) house-flies should be present in sufficient numbers in houses invaded by fatal diarrhoea.

"(2)   There should be a close correspondence between the aggregate number of house-flies in houses and the aggregate number of deaths from diarrhoea week by week.

"(3)   The life-history of the house-fly should explain any discrepancy between the observed number of flies and the observed number of deaths.

"(4)   The minority of breast-fed children not apparently accessible to infection should receive explanation.

"(5)   There should be a closer correspondence of diarrhoeal fatality with the number of flies than with any other varying seasonal fact.

"(6)   Any other closely-corresponding seasonal fact should be capable of interpretation in terms of the number of house-flies.

"(7)   Any variation from district to district in the annual curve of deaths should be accompanied by a similar variation in the curve of flies.

"(8)   It will be at once manifest when we come to enteric fever that the house-fly plays but a minor direct part in the production of the annual wave. Such part, however, should have reference to the number of flies and of pre-existing centres of infection. If it can be shown that that portion of the enteric wave which is connected with flies changes from one period of time to another in such a manner as to be explainable in terms of flies, but not of meteorological conditions, the evidence in favour of flies will be greatly strengthened.

"(9)   No other available hypothesis must be capable of explaining the course of summer diarrhoea."

For the purposes of this study daily counts of the flies captured at the different observation stations which were selected

were made from July (in one year from the end of May) until
November during the years 1904, 1905, 1906, 1908 and 1909.
The complete data obtained are given; these include for each
week the number of flies captured, the number of deaths from
diarrhoea, the number of fatal cases commencing, mean temperature in the shade, rainfall in inches, and the underground

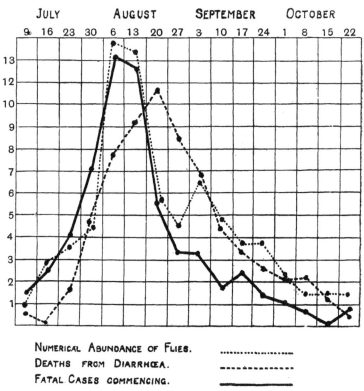

NUMERICAL ABUNDANCE OF FLIES.
DEATHS FROM DIARRHŒA.
FATAL CASES COMMENCING.

FIG. 99.   Chart illustrating the relation of the numerical abundance of house-flies
to summer diarrhoea in the city of Manchester in 1904.   Prepared from
statistics and chart given by Niven.

temperatures at 1 foot and 4 feet respectively.   From these data
charts are constructed.   I have prepared Figs. 99 and 100 from the
statistics and chart given for the years 1904 and 1909.   Although
in Fig. 99 the temperature curve has been omitted, it may be
pointed out that in Niven's charts there is no direct relation

between the curves of temperature or rainfall and those of flies or deaths from diarrhoea. In addition he points out the manner in which the curve of fatal cases or deaths falls away from the curve of flies in the middle of the decline.

In pointing out the intimate correspondence of the curves of the cases and deaths to the number of flies captured week by

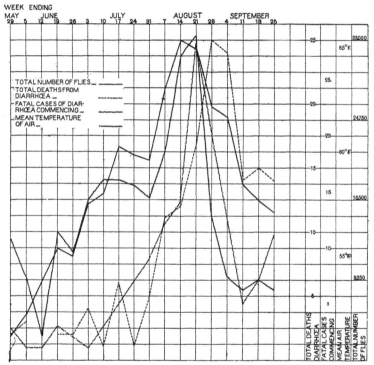

Fig. 100. Chart illustrating the relation of the numerical abundance of house-flies to summer diarrhoea in the city of Manchester in 1909. Prepared from the statistics given by Niven.

week, Niven states: "The number of cases begins to increase until the flies captured attain a maximum. The maximum number of cases commencing in the years 1904 and 1905 is in the same week as the maximum number of flies. The figures for the deaths are still more striking. The shape of the curves at an interval of a week to a fortnight near the maximum point is practically identical in the two. The errors of the curves will be subjected to examination afterwards. But, even with their manifest and

H. H.-F.   17 necessary defects, they show a degree of correspondence which creates a high degree of probability that flies are the transmitting agents in summer diarrhoea. In all the curves it will be seen that deaths diminish more rapidly than do flies in the middle part of the decline. For this there are two causes. In years of high diarrhoea incidence the more susceptible and exposed infants have been killed off or rendered immune. In every year towards the close of the fly season the flies are attacked by *Empusa muscae*, and are hindered by cold from leaving the house, so that they cease to act as transmitting agents."

For a full discussion of the careful observations made by Niven in his exhaustive study of the epidemiology of this disease in Manchester, the reader is referred to the original paper. In summarising the results of his analysis Niven states, after declaring that summer diarrhoea is an infectious illness: "The health of infants prior to attack—in other words, the social condition—has much to do with the fatality. The summer wave is not due to dust, nor is it conditioned by any growth of bacteria in or on the soil. There is nothing to support the view that the infective organisms are of animal origin, and the connection between privy middens and diarrhoea goes far to prove the contrary. The disease becomes more fatal only after house-flies have been prevalent for some time, and its fatality rises as their numbers increase and falls as they fall. The correspondence of diarrhoeal fatality is closer with the number of flies in circulation than with any other fact. The next closest connection is with the readings of the four-foot thermometer, with which, however, diarrhoeal fatality can have no direct relation. Flies and the readings of the four-foot thermometer are both functions of air and surface temperatures and of rainfall. Certain facts in the life-history of the fly throw light on discrepancies arising in the decline of flies and cases. The close correspondence between flies and cases of fatal diarrhoea receives a general support from the diarrhoea history of sanitary sub-divisions of the Manchester district. The few facts available for the study of the correspondence of flies and fatal cases in different sub-divisions, in the course of the same year, also lend support to this view. No other explanation even approximately fits the case."

In the investigation which Hamer (1908, 1909) carried out in London with a view to determining the relationship which the presence of accumulations of refuse and offensive matter bears to the fly nuisance, opportunity was afforded, and wisely taken advantage of, to study the question of the possible relationship of flies to summer diarrhoea. Hamer indicates what appears to him to be a difficulty in the way of accepting this theory. He states: " It should be pointed out that there are certain difficulties in the way of accepting the thesis that the correspondence exhibited in the curves (he refers to the fly curve and diarrhoea curve) affords reason for concluding that flies and summer diarrhoea stand to one another in relation of cause and effect. At the commencement of the hot summer weeks, when the number of flies has begun to show marked increase, the diarrhoea curve is rapidly rising. After some weeks the number of flies reaches the maximum, and then diminishes, and so, in almost precise correspondence, does the amount of diarrhoea. A period is later reached, towards the close of the hot weeks, at which the number of flies is still as markedly excessive as at the earlier period when the amount of diarrhoea was increasing, but at the later period the amount of diarrhoea is declining; it even anticipates decline in the number of flies. If the fly is to be regarded as the carrier of the organism which causes diarrhoea, it might perhaps have been anticipated that at the later period—the number of flies still being excessive and infective material being then presumably more widely distributed than ever before—the amount of diarrhoea, instead of showing early and rapid decline, would still be increasing. It would almost appear that the advocate of the ' fly-borne diarrhoea hypothesis ' must necessarily fall back in support of his theory upon the hypothetical organism, conveyed by the fly, which he may claim is affected by temperature in such a way as to bring about correspondence between the diarrhoea curve and the fly curve. The very closeness of the correspondence between these two curves may indeed from this point of view be thought of as constituting a difficulty rather than a point in favour of the hypothesis that summer diarrhoea is caused by flies."

Against Niven's suggestion that one of the explanations of the decline of the diarrhoeal curve while the number of flies still remains excessive Hamer brings forward two considerations: first, that the comparatively early fall in the amount of diarrhoea is observed in years of very low mortality as well as in years of excessively high mortality in which such an explanation might hold good. Secondly, the hypothesis does not adequately account for the almost identically similar behaviour of the two curves both in their ascent and descent. The close correspondence between the curves, Hamer believes, accords better with the view that both are dependent upon variations of temperature than with the hypothesis that diarrhoea stands in direct causal relation to fly prevalence. In his last report and in the light of three years' records (1907—1909) Hamer is still unconvinced that the evidence available can be considered to support a causal relation between summer diarrhoea and flies and a critical attitude is still maintained.

In regard to Hamer's chief criticism, namely, that while the number of flies is still excessive, the diarrhoeal curve has begun to decline, I have previously pointed out (1910) that a consideration of the habits of the house-fly will probably afford an explanation of this difficulty. Flies are very susceptible to changes of temperature. When the temperature falls flies become less active and retire into the shelter of houses and other buildings, although their numbers, as indicated by captures in traps set indoors, may still be considerable. In so far then as the activity of the flies is associated with the temperature, the temperature curve should be studied in addition to the fly and diarrhoea curves. If this is done, it is usually found that a fall in the number of flies is preceded by a fall in the temperature and that these two curves are associated somewhat closely, that is, the numerical activity of the flies—since the numbers caught are more indicative of their numerical activity than of their numerical abundance—is dependent upon the temperature and also, I have found, upon the state of the weather and sky. Therefore, if the flies become less active, they will be less liable to transmit the organisms causing summer diarrhoea, and although the numbers caught in the houses may exceed in numbers those caught earlier in the season when the diarrhoea curve was rising, those which are very active will be less in number and consequently instead of increasing, the diarrhoea curve begins to fall. The dissemination of summer diarrhoea is brought about chiefly owing to the activity of the flies outside the houses as well as inside. A fall in temperature or a spell of dull weather decreases considerably this outside activity and will, therefore, cause a decline in the number of diarrhoea cases. The number of cases of diarrhoea is dependent on the *activity* of the flies and this is dependent on climatic conditions, chief of which is temperature. Considered in the light of these facts this seeming difficulty is not an argument against the idea that we hold on the relation of flies to summer diarrhoea, but rather one in support of it.

The relation of temperature and the activity of the flies has also been commented upon by Nash (1909).

The great difficulty with which we are faced in discussing the question of the relation of flies to the prevalence of summer diarrhoea is that it has not been proved to the satisfaction of most investigators what the specific pathogenic organism is; or perhaps there are associated organisms. It is possible that more than one etiological factor exists. Various organisms have been found in the diarrhoeal stools. Some epidemics in the United States have been associated with the dysentery (Flexner's) bacillus in the stools; in other epidemics dysentery bacilli were not found. Metchnikoff (1909) believes that *Bacillus vulgare* may be the causative organism. Morgan (1906–7) isolated a bacillus which he designated " No. 1," and which may be an important factor in the causation of the disease. In a further paper Morgan and Ledingham (1909) give a more complete account of their researches on Morgan's bacillus which belongs to the non-lactose fermenting group, to which group all the pathogenic bacteria inducing affections of the intestinal tract belong, namely, the typhoid and paratyphoid bacilli, the dysentery and food-poisoning organisms. In 1905, 58 cases of infantile diarrhoea were examined and Morgan's bacillus was found in 48·2 per cent.; in 1906, in 54 cases it was found in 55·8 per cent.; in 1907, 191 cases were examined and it occurred in 16·2 per cent., and in 1908 it occurred in 53 per cent. of the cases, numbering 166 that were examined. It was found that rats and monkeys were susceptible to infection by feeding and that they succumbed after a period of diarrhoea.

One of the most interesting and highly suggestive results of the research was the discovery of Morgan's bacillus in flies. " Batches of flies came for examination from infected and uninfected houses in Paddington and from a country house situated many miles from London, where no cases of diarrhoea had occurred, at any rate, within a radius of two miles. The flies were killed with ether vapour and crushed with a sterile rod in peptone broth. The result was that Morgan's bacillus was isolated from nine of the thirty-two batches from infected houses and from one of the thirty-two batches from uninfected houses[1]. It was also got in five out of twenty-four batches from the country house." Dr Morgan in the course of a letter to me says: "I certainly think they are carriers of summer diarrhoea, and the variety I especially suspect of doing this is the *Musca domestica.*"

In a study of the micro-organisms occurring in flies caught in normal surroundings and in diarrhoea-infected locations Graham-Smith (1912) found that a greater proportion of flies is infected with the non-lactose-fermenting bacteria during August and the early part of September than at other times. Of the groups of bacilli into which these organisms can be divided the Morgan or Ga group is the only one which occurs frequently in flies from diarrhoea-infected houses and rarely in non-diarrhoea-infected houses. He considers it certain that flies infected with Morgan's bacillus can contaminate materials on which they feed or over which they walk.

The epidemiological evidence for and against the hypothesis that the house-fly is an agent in the dissemination of summer diarrhoea has been subjected to a most careful and detailed examination and analysis by Martin (1913), whose paper is in my opinion the most judicious criticism of the problem as it now stands.

After pointing out that the observations of Niven and Hamer indicate the dependence of both the number of flies and the epidemic upon the cumulative effect of previous warm weather, Martin calls attention to the fact that a notable feature of the

---

[1] Morgan's bacillus has also been isolated from naturally infected flies by Nicoll (1911) and Cox, Lewis and Glynn (1912). See pp. 245 and 296.

curve is that the same fly population is accompanied by a rise in the number of cases in early August and by a fall in early September. If the time relations of fly prevalence and diarrhoea cases is to be regarded as something more than an interesting coincidence a satisfactory explanation must be found for these facts. He rightly states that the number of flies is dependent upon the accumulated effect of temperature because a considerable period of warm weather, weeks or months, is required to produce an abundance of flies from the few which come out of hibernation. The dependence of the epidemic upon this factor is not so obvious in the absence of exact knowledge as to the etiology of diarrhoea. But assuming an infective agent of a bacterial nature with which food might become contaminated he shows that the dose of infection would not be dependent upon the accumulated effect of temperature during the previous three months. It is just conceivable that the virulence of the organism might be enhanced by the continued influence of warm weather. Describing the excellent *a priori* reasons for supposing that flies could transmit the infective agent of diarrhoea he says : " Anyone familiar with the domestic *ménage* of the average working man on a hot summer day, with the baby sick with diarrhoea and other small children to care for, must realize that the opportunities afforded for fly transmission are adequate enough."

After discussing the data relating to the question to which I have previously referred, namely, the decline of the epidemic while the number of flies is still considerable, he groups the two factors which may contribute in a varying degree to the decline of the epidemic. First, a fall in temperature, diminishing (*a*) the activity and number of the supposed transmitters, and (*b*) the dose of infection which the child ingests owing to the effect upon the rate of multiplication of the infective agent. Second, the exhaustion of the more susceptible individuals, as pointed out by Niven (*l.c.*), and also by Peters (1910) in describing an epidemic in Mansfield, where, under similar conditions, in one section of the town the epidemic was nearly finished whilst in another the aforementioned factors appears to predominate.

In commenting upon the evidence of Nash and Niven, which I have already given, upon the relation between the number of flies and the distribution of cases of infantile diarrhoea, Martin states, and I think justly: "I doubt very much whether any evidence of great value could be obtained upon this point. Even supposing it to be true that fly carriage is of first importance, I should expect that, if all the facts were known, a much higher correlation would be discovered between diarrhoea and carelessness with regard to disposal of excreta and protection of food from the visitation of flies than between diarrhoea and fly prevalence."

As a conclusion to this *résumé* of the evidence for and against the hypothesis that the house-fly is an important agent in the dissemination of summer diarrhoea, in regard to which idea a critical attitude is, in my opinion, not only justified but conducive to a more satisfactory elucidation of the problem, I do not think I can do better than to quote Martin's concluding paragraph. He says: "Many of the facts which I have brought forward merely indicate some form of infective agent and do not necessitate recourse to the hypothesis that carriage of flies dominates the situation. I would point out, however, that:

"1. The fly-carrier hypothesis is the only one which offers a satisfactory interpretation of the extraordinary dependence of the epidemic upon the accumulated effect of temperature.

"2. That it offers a ready explanation of the spread of infection to neighbouring children who have no direct personal contact with the patient.

"3. That the peculiarities of the relation in time between fly prevalence and the epidemic in different localities are not inconsistent with the view that fly carriage is essential to epidemicity. No other interpretation which is at present forthcoming is nearly so satisfactory, and it is at least worthy to guide in the meantime our efforts at prevention[1]."

[1] In a recent Bulletin Armstrong (1914) gives an account of an effort made in the summer of 1913 by the Department of Social Welfare of New York to determine the importance of the house-fly in the transmission of diarrhoeal disease among infants. Two congested areas in the Italian quarter were selected. In one every step was taken to make the houses sanitary and flyless by education, screening, cleaning up, etc.; in the other nothing was done. A careful census and weekly inspections were made. The statistical findings, if not entirely conclusive, were exceedingly interesting and demonstrated an apparent marked reduction in the

amount of diarrhoeal disease in the protected area. In the protected area there were, in children under five years of age, twenty cases of diarrhoeal disease, and fifty-seven cases in the unprotected or filthy area. In the former the duration of the disease was 273 days, in the latter the total days of sickness were 984. Of the other suggestive statistics contained in this report the following may be mentioned. A bacterial count of the flies was made. The average number of bacteria on an agar culture from flies in the clean area was 13,986 and in the dirty area 1,106,017. The average number per fly of bacteria on Conradi plates (indicating intestinal organisms) was 4489 on flies from the clean area and 292,117 from flies from the dirty area.

CHAPTER XXIII

THE DISSEMINATION OF OTHER DISEASES BY FLIES

ANTHRAX

IN considering the relation of flies to anthrax several facts should be borne in mind. As early as the eighteenth century it was believed that anthrax might result from the bite of a fly, and the idea has been used by Murger in his romance *Le Sabot Rouge*. A very complete historical account of these earlier ideas is given by Nuttall (1899). Most of the instances in support of this belief, however, that flies may carry the infection of anthrax, refer to biting flies. As I have already pointed out, *M. domestica* and such of its allies as *F. canicularis*, *C. erythrocephala*, *C. vomitoria* and *Lucilia caesar* are not biting or blood-sucking flies. The nearest allies of *M. domestica* which suck blood in England are *S. calcitrans*, *Haematobia stimulans* Meigen and *Lyperosia irritans* L.; the rest of the blood-sucking flies which may be considered in this connection belong to the family Tabanidae, including the common genera Haematopota, Tabanus, and Chrysops. These biting and blood-sucking flies live upon the blood of living rather than dead animals. But it is from the carcases and skins of animals which have died of anthrax that infection is more likely to be obtained, and I believe that such flies as the blow-flies (*Calliphora* spp.) and sometimes *M. domestica* and *Lucilia caesar* which frequent flesh and the bodies of dead animals for the purpose of depositing their eggs and for the sake of the juices, are more likely to be concerned in the carriage of the anthrax bacillus and the causation of malignant pustule than are the blood-sucking flies. Consequently, as *M. domestica* and its allies only are under consideration, and for the sake of brevity, the relation to anthrax of the non-biting flies only will be considered here.

The earliest bacteriological evidence in support of this belief was Raimbert (1869). He experimentally proved that the house-fly and the meat-fly were able to carry the anthrax bacillus, which he found on their proboscides and legs. In one experiment two meat-flies were placed from twelve to twenty-four hours in a bell-jar with a dish of dried anthrax blood. One guinea-pig was inoculated with a proboscis, two wings and four legs of a fly, and another with a wing and two legs. Both were dead at the end of sixty hours, anthrax bacilli being found in the blood, spleen and heart. He concludes: "Les mouches qui se posent sur les cadavres des animaux morts du Charbon sur les dépouilles, et s'en nourris-sent, ont la faculté de transporter les virus charbonneux déposé sur la peau peut en traverser les differentes couches." Davaine (1870) also carried out similar experiments with *C. vomitoria* which was able to carry the anthrax bacillus. Bollinger (1874) found the bacilli in the alimentary tract of flies that he had caught on the carcase of a cow dead of anthrax. Sangree (1899) allowed a fly to walk over a plate culture of anthrax and then transferred it to a sterile plate; colonies of anthrax naturally developed in its tracks. Buchanan (1907) placed *C. vomitoria* under a bell-jar with the carcase of a guinea-pig (deprived of skin and viscera) which had died of anthrax. He then transferred them to agar medium and a second agar capsule, both of which sub-sequently showed a profuse growth of *B. anthracis*, as one might expect. Specimens of *M. domestica* were also given access to the carcase of an ox which had died of anthrax; they all subsequently caused growths of the anthrax bacillus on agar. I entirely agree with Nuttall, who says: "It does seem high time, though, after nearly a century and a half of discussion, to see what would be the result of properly carried out experiments. That ordinary flies (*M. domestica* and the like) may carry about and deposit the bacillus of anthrax in their excrements, or cause infection through their soiled exterior coming in contact with wounded surfaces or food, may be accepted as proven in view of the ex-perimental evidence already presented."

These experiments only prove the ability of flies to mechanically transfer the anthrax bacilli from infected to uninfected matter. Graham-Smith (1910 and 1911), however, has carried these experiments further with a view to discovering, among other points, the length of time that flies may carry the bacilli or its spores, *B. anthracis* being a spore-bearing bacillus and consequently more adapted for transference. Flies were placed for one hour in a' cage containing the body of a mouse just dead of anthrax, its body having been opened to enable the flies to feed upon its blood. The flies were afterwards transferred to a clean cage which on the following morning was found to contain red spots of vomit and yellowish faeces. *B. anthracis* was found in the former both microscopically and by cultures. The flies were transferred daily to fresh cages and fed on syrup; at intervals specimens were removed and dissected and cultures were made on agar from their legs, wings, heads, crops and intestinal contents. Cultures were also made from the faeces. As a result of these careful examinations, it was found that the non-spore-bearing anthrax bacilli did not remain alive on the external parts of the fly for more than twenty-four hours. They remained alive in the intestine for three days and in the crop for five days, especially when this organ contained partially coagulated blood. The bacilli were present in the faeces deposited forty-eight hours after infection. No spore-bearing forms were obtained from film preparations made at various times from the contents of the crop and the intestine. Experiments were also carried on with the spores of *B. anthracis*. An emulsion of an old anthrax culture was made and heated to 70° C. for fifteen minutes, after which a number of flies were allowed to feed on it. These flies were then transferred to fresh cages daily and fed on syrup and as in the previous experiment specimens were caught and dissected at intervals, agar cultures being made from their legs, wings, heads, crops and intestinal contents and faecal deposits. Smears were made from the crop and intestinal contents at various times but the absence of anthrax bacilli on microscopic examination demonstrated that the spores do not develop in the fly. This experiment showed that flies infected with anthrax spores may carry the spores upon their legs and wings for at least twelve days, and that the spores are present in considerable numbers in the crop and intestinal contents for at least seven days. The spores remained in a living condition in the vomit and faecal deposits for six days or longer. Cultures were also made from drops of sugar after the flies had been allowed to feed upon them and anthrax bacilli were obtained on the tenth day after the flies had fed. It was also shown in another experiment that the anthrax spores may remain alive for at least twenty days upon the legs and wings and in the intestinal contents of the fly and that faeces passed fourteen days after infection contained living spores. Dried faeces and vomit were shown to contain the spores in a vital condition for twenty days. Cultures were obtained from the bodies of dead flies four hundred and twenty-eight days after death and proved to be virulent by animal inoculations.

In connection with the experiments of Faichnie and others on flies bred from larvae infected with *B. typhosus* the results of Graham-Smith's experiments with *B. anthracis* are of interest. The larvae of *C. erythrocephala* were fed on meat infected with anthrax spores. The flies bred from these larvae were heavily infected for at least two days after emerging. In a single series of experiments *B. anthracis* could not be cultivated either from the limbs or intestinal contents of flies more than fifteen or nineteen days old. It was found that flies were able to infect, during the first two days after emerging, materials over which they walk, and to deposit infected faeces.

In a further report Graham-Smith (1912) states that he finds that a large proportion of *Musca domestica* which develop from larvae infected with the spores of *B. anthracis* are infected.

These results confirm the suggestion made by Joseph (1887) and later by Nuttall (*l.c.*) that the non-biting flies, when infected, may spread anthrax by depositing bacilli upon wounds or food and they have a significant bearing upon the spread of the disease among domestic animals and the production of malignant pustule in man.

TUBERCULOSIS.

With the proven existence of so many factors contributing to the dissemination of the tubercle bacillus, the significance of experiments of a positive nature indicating the ability of flies to transfer the virulent germs may appear to be lessened. A careful consideration of the facts, nevertheless, will show that this is far from being the case. My experiments and observations on the feeding habits of the fly have shown that this insect is especially fond of and attracted to sputum. This is a matter of common observation where spittoons or cuspidors are used and are not kept in a clean condition. There is no lack of opportunity, under natural conditions, for flies to infect themselves externally and internally with the tubercle bacilli. To what extent they may prove the means of infection depends upon their access to food. In my opinion their greatest danger lies in the possibility of their coming into contact with the mouths or food of helpless infants.

Spillman and Haushalter were the first to carry out bacteriological investigations on the dissemination of *Bacillus tuberculosis* by flies. As early as 1887 they found this bacillus in large numbers in the intestines of flies from a hospital ward, and also in the dejections which occurred on the windows and walls of the ward. Hoffmann (1886) also found tubercle bacilli in the excreta of flies in the room where a patient had died of tuberculosis, and he also found the bacilli in the flies' intestinal contents. One out of three guinea-pigs which were inoculated with the flies' intestines died; two inoculations with the excreta had no effect, which led him to believe that the bacilli became less virulent in passing through the fly's alimentary tract. But Celli (*l.c.*) records experiments in which two rabbits inoculated with the excreta of flies fed with tubercular sputum developed the disease.

Hayward (1904) obtained tubercle bacilli in ten out of sixteen cultures made from flies which had been caught feeding on bottles containing tuberculous sputum. Tubercle bacilli were also recovered from cultures made from faeces of flies which had fed in the same manner, which apparently caused a kind of diarrhoea in the flies, and they died from two to three days afterwards. Faeces of flies fed on tubercular sputum were rubbed up in sterile water and injected into the peritoneal cavity of guinea-pigs, which developed tuberculosis. Buchanan (1907) allowed flies to walk over a film of tubercular sputum and then over agar; the agar was then washed with water and a guinea-pig died of tuberculosis in thirty-six days by inoculating it with the resulting solution.

Cobb (1905) is of the opinion that by the infection of human food after feeding upon tubercular sputum flies may be an important factor in the dissemination of tuberculosis, and the force of his remarks is not mitigated by the acrid criticisms of Mays (1905).

Lord (1904) made a careful series of experiments as a result of which he reached the following conclusions: Flies may ingest tubercular sputum and excrete tubercle bacilli, the virulence of which may last for at least fifteen days. The danger of human infection from tubercular fly specks is by the ingestion of the specks on food. Spontaneous liberation of tubercle bacilli from fly specks is unlikely. If mechanically disturbed, infection of the surrounding air may occur. He suggests that tubercular material (sputum, pus from discharging sinuses, faecal matter from patients with intestinal tuberculosis, etc.) should be carefully protected from flies lest they act as disseminators of the tubercle bacilli. During the fly season greater attention should be paid to the screening of rooms and hospital wards containing patients with tuberculosis and laboratories where tubercular material is examined. As these precautions would not eliminate fly infection by patients at large, food stuffs should be protected from flies which may already have ingested tubercular material. The importance of these conclusions will be realised by those who are acquainted with recent work on tubercular infection by way of the alimentary tract.

Graham-Smith (1910) carried out a series of experiments with flies artificially infected with *B. tuberculosis*. A large number of flies freshly caught were allowed to feed upon an emulsion of a culture of human tubercle bacilli in syrup. After feeding, the flies were transferred to a clean cage and fed daily on syrup. Smear preparations were made from the crop and intestinal contents of these flies which were caught at intervals. Smears were also made from vomit and faecal material. As a result it was found that under experimental conditions the tubercle bacilli were present in the crop for three days. In the intestine they were found in considerable numbers up to six days and were still present after twelve days and possibly longer. In the faeces they were numerous up to the fifth day and occasionally found up to the fourteenth day after infection. In a further experiment flies were fed upon tuberculous sputum and afterwards on non-tuberculous sputum. The tubercle bacilli were found in the intestinal contents for at least four days, during which time the faeces were also infected. These careful experiments confirm the conclusions of the previous investigators as to the ability of flies to carry *B. tuberculosis* in a virulent condition.

CHOLERA.

The necessity of guarding food against flies in the belief that they might disseminate cholera was called attention to by Moore in 1853, according to Nuttall and Jepson (1909). He stated that "flies in the East have not far to pass from diseased evacuations or from articles stained with such excreta, to food cooked and uncooked."

One of the first to suggest that flies may disseminate the cholera spirillum was Nicholas (1873), who, in an interesting and prophetic letter, said : " In 1849, on an occasion of going through the wards of the Malta Hospital where a large amount of Asiatic cholera was under treatment, my first impression of the possibility of the transfer of the disease by flies was derived from the observation of the manner in which these voracious creatures, present in great numbers, and having equal access to the dejections and food of the patients, gorged themselves indiscriminately, and then disgorged themselves on the food and drinking utensils. In 1850 the *Superb*, in common with the rest of the Mediterranean squadron, was at sea for nearly six months ; during the greater part of the time she had cholera on board. On putting to sea the flies were in great force, but after a time the flies gradually disappeared and the epidemic slowly subsided. On going into Malta harbour, but without communicating with the shore, the flies returned in greater force, and the cholera also with increased violence. After more cruising at sea the flies disappeared gradually, with the subsidence of the disease. In the years of 1854 and 1866 in this country the periods of occurrence and disappearance of the epidemics were co-incident with the fly season." In 1886, Flugge, according to Nuttall and Jepson (*l.c.*), observed that flies may infect food during cholera times and that they must play an important part in the dissemination of the disease when they are numerous. He also draws attention to the fact that the worst cholera months are those in which insects abound.

Buchanan (1897), in a description of a gaol epidemic of cholera which occurred at Burdwan in June, 1896, states that swarms of flies occurred about the prison, outside which there were a number of huts containing cholera cases. Numbers of flies were blown from the sides where the huts lay into the prison enclosure, where they settled on the food of prisoners. Only those prisoners which were fed in the gaol enclosure nearest the huts acquired cholera, the others remaining healthy.

Tsuzuki (1904), reporting upon the cholera outbreak in Northern China in 1902, states that "flies in China are a terrible infliction to the stranger," and remarks that if they are capable of carrying about the cholera germ they must play an important part in the spread of the disease. His experiments, mentioned later, demonstrated that flies are able under natural conditions to carry the cholera spirillum.

Bacteriological evidence.

Maddox (1885) appears to have been the first to conduct experiments with a view to demonstrating the ability of the flies to carry cholera spirillum, or as it was then called, the "comma bacillus." He fed the flies *Calliphora vomitoria* and *Eristalis tenax* (the "drone-fly") on pure and impure cultures of the spirillum, and appears to have found the motile spirillum in the faeces of the flies. He concludes that these insects may act as disseminators of cholera. During a cholera epidemic Tizzoni and Cattani (1886), working in Bologna, showed experimentally that flies were able to carry the "comma bacillus" on their feet. They also obtained, in two of these experiments, the spirillum from cultures made with flies from one of the cholera wards. Sawtchenko (1892) made a number of careful experiments. Flies were fed on bouillon culture of the cholera spirillum, and to be certain that the subsequent results should not be vitiated by the presence of the spirillum on the exterior of the flies, he disinfected them externally and then dissected out the alimentary canal, with which he made cultures. In the case of flies which had lived for forty-eight hours after feeding, the second and third cultures represented pure cultures of the cholera spirillum.

Simmonds (1892) in Hamburg placed flies on a fresh cholera intestine, and afterwards confined them from five to forty-five minutes in a vessel in which they could fly about. Roll cultures were then made and colonies of the cholera spirillum were obtained after forty-eight hours. Colonies were also obtained from a fly one and a half hours after having access to a cholera intestine, and also from flies caught in a cholera post-mortem room. Uffelmann (1892) fed two flies on liquefied cultures of the cholera spirillum, and after keeping one of them for an hour in a glass he obtained 10,500 colonies from it by means of a roll culture; from the other, which was kept two hours under the glass, he obtained twenty-five colonies. In a further experiment he placed one of the two flies, similarly infected with the spirillum, in a glass of sterilised milk, which it was allowed to drink. The milk was then kept for sixteen hours at a temperature of 20—21° C., after which it was shaken, and cultures were made from it; one drop of milk yielded over one hundred colonies of the spirillum. The other fly was allowed to touch with its proboscis and feed upon a juicy piece of meat that was subsequently scraped. From one half of the surface twenty colonies, and from the other half one hundred colonies, of the spirillum were obtained. These experiments show the danger which may result if flies having access to a cholera patient, and bearing the spirillum, have access also to food. Macrae (1894) records experiments in which boiled milk was exposed in different parts of the gaol at Gaya in India, where cholera and flies were prevalent. Not only did this milk become infected, but the milk placed in the cowsheds also became infected. The flies had access both to the cholera stools and to such food as rice and milk.

Tsuzuki (l.c.) caught flies in a cholera house in Tientsin and isolated the cholera vibrios from them by incubating the flies in bouillon and making plate cultures from the bouillon. Flies confined in a cage were also shown to transfer the cholera vibrios from a cholera culture to a culture plate of sterile agar.

Chantemesse (1905) isolated cholera vibrios from the feet of flies seventeen hours after they had been contaminated.   Ganon (1908) found that flies could transmit infection for at least twenty-four hours after feeding upon infected matter and that during such a period they may be carried long distances in railway carriages. He was unable to show that flies could retain the power of infecting for more than four days, as the experimental flies did not live longer than that time.

Graham-Smith (1910) has carried on a few experiments on the distribution of the cholera vibrios by flies.   Flies were fed for an hour on a cholera culture emulsified in broth, afterwards they were transferred to a fresh cage.   At intervals specimens were caught and cultures were made of head, leg, wing, crop, and intestinal contents, the cultures being subsequently plated out and also examined microscopically.   The cholera vibrios were found on the legs up to thirty hours after infection but not later.   They persisted in the intestine and crop for forty-eight hours but could not be found after that time.   The faeces passed thirty hours after feeding were infected.

The foregoing experiments prove beyond doubt the ability of flies to carry the cholera organism both internally and externally, in a virulent condition and to infect food for a significant length of time after feeding upon or coming in contact with infected matter. Among those authors who have expressed their belief in the possibility of flies acting as agents in the dissemination of cholera, the names of Marpmann (1897) and Geddings (1903) may be mentioned.

OPHTHALMIA.

Flies are now generally recognised as active and important agents in the spread of ophthalmia, and although, so far as I have been able to discover, we have little bacteriological evidence at present to support this belief, the circumstantial evidence is sufficiently strong to warrant it.   Nuttall and Jepson (l.c.) point out that Budd as early as 1862 considered it was fully proven that flies serve as carriers of Egyptian ophthalmia.

In speaking of its occurrence at Biskra, Laveran (1880) says that in the hot season the eyelids of the indigenous children are covered with flies, to the attentions of which they submit; in this way the infectious discharge is carried on the legs and proboscides of the flies to the healthy children.

Abel (1899) quotes the statements of Howe (1888) to the effect that the number of cases increases rapidly from the moment when flies are present in large numbers. Eye trouble occurs in the same places when flies are numerous, *e.g.* the delta of the Nile; in the desert where there are few flies there are also few cases of illness. Natives and especially children are remarkably indifferent to the attacks of flies, they allow the flies to settle in crowds about their eyes, sucking the secretions, and never think of driving them away. It is of interest to note that Howe states that an examination of the flies captured on diseased eyes revealed bacteria on their feet which were similar to those found in the conjunctival secretion. Howard (1911), whose attention was called by Howe to these facts, sent to Egypt for specimens of the flies commonly swarming about the eyes of ophthalmic patients which on examination proved to be *M. domestica.* I have also received specimens from Egypt.

Dr Andrew Balfour, formerly of the Gordon College, Khartoum, in a letter to me, says that the Koch-Weeks bacillus is generally recognised as being the exciting cause of Egyptian ophthalmia. He says, " Ophthalmia is not nearly so common in the Sudan as in Egypt, nor are flies so numerous; doubtless the two facts are associated." Dr MacCallan, of the Egyptian Department of Public Health, in answer to my inquiries, says that acute ophthalmias are more liable to transmission by flies than trachoma. In his opinion the spread of the latter is, to a comparatively small extent, through the agency of flies, but it is mainly effected by direct contact of the fingers, clothes, etc.

The Koch-Weeks bacillus was first seen by Koch (1883) in Egypt in cases of acute catarrhal ophthalmia. He found that two distinct diseases were referred to under that name; in the severe purulent form he found diplococci, which he identified as very probably *Gonococci*; in the more catarrhal form he found small bacilli in the pus corpuscles. He ascribed the propagation of the disease to flies, which were often seen covering the faces of children. Axenfeld (1908) states that " almost the only organisms occurring in acute epidemics of catarrhal conjunctivitis are the

Koch-Weeks bacillus (perhaps also influenza bacillus), and the pneumococcus (in Egypt the gonococcus also, rarely *subtilis*). Other pathogenic conjunctival organisms[1] only exceptionally occur." And, further, "Gonococci and Koch-Weeks bacilli evidently lose their power of causing a conjunctivitis very slowly indeed, and are very independent of any disposition." His statement that "on account of their great virulence and the marked susceptibility to them, a very small number suffices," is important in considering the relation of flies to the spread of the disease, although, as he remarks, every infection does not produce the disease. The fact that the Koch-Weeks bacillus cannot resist dryness cannot be urged as an argument against the spread of the infection by flies, or the same would apply to the typhoid bacillus, whose carriage by flies is proven. Axenfeld mentions L. Müller and Lakah and Khouri as advocating the view that flies may spread the infection more readily. In view of the fact that, as the same author states, "Koch-Weeks conjunctivitis is to be classed with the most contagious infectious disease which we know of," it is important that the *rôle* of flies should be recognised.

Notwithstanding the occurrence in temperate climates of flies in less numbers than in such countries as Egypt, it would be well to bear in mind the probable influence of flies in cases of acute conjunctivitis, such as those described by Stephenson (1897) in England. The sole difference between the disease in Egypt and in England is, as Dr Bishop Harman points out to me in a letter, that "the symptoms produced (in Egypt) are, from climate and dirtiness of the subjects, more severe, and that there is found a greater number of cases of gonorrhoeal disease than in England"; and, I would add, a far greater number of flies. This disease is eminently suited for dissemination by flies, both on account of the accessibility of the infectious matter in the form of a purulent discharge from the eyes and on account of the flies' habit of frequenting the eyes.

A number of writers, among whom are Braum (1882), Demetriades (1894) and German (1896), refer to the agency of flies in communicating gonorrhoeal and similar infections of the eye.

---

[1] In this connection he states (p. 236): "We can make the general statement that the *Staphylococcus* in the conjunctiva is not contagious."

Abel cites Welander (1896) who describes the infection of a woman in a hospital. This patient's bed was next to that of another patient suffering from blennorrhoea, but a screen which did not reach to the ceiling separated the two beds. All means of infection, except through the agency of flies, appeared to be excluded. Welander found that flies bore living gonococci upon their feet three hours after they had been contaminated with secretion.

PLAGUE.

Although fleas are considered to be the chief agents in the dissemination of the plague bacillus, in spite of the fact that the proof is not as yet considered by all to be absolutely convincing, it is nevertheless interesting from an historical point of view to refer to the ideas that have prevailed and the experiments which have been carried out in reference to the relation of flies to plague. Nuttall and Jepson (1909) refer to the earlier writings on this subject. The prevalence of large numbers of flies during outbreaks of plague has been referred to by Knud as early as 1498 and by Varwich in 1577. Mercurialis (1577) referred to the contamination of food by flies which had been frequenting plague patients. The infection of healthy persons by contaminated flies was also suggested by Lange (1791). Haesar (1882) is cited by the above authors as referring to Bengasi, Tripolis, where an epidemic of plague occurred in 1858, being known to the Turks by the name of the " Kingdom of Flies."

Yersin (1894) observed the presence of dead flies in the laboratory in which autopsies on plague animals were made. He demonstrated the presence of virulent plague bacilli in such dead flies by inoculation experiments. Nuttall (1897) conclusively proved that flies were able to carry the plague bacillus and that they subsequently died of the disease. The flies were fed on organs of animals which had died of plague. He found that such flies might survive eight days at 12—14° C. and that they still contained the virulent bacilli for forty-eight hours or more after they were transferred to clean vessels. At temperatures of 14° C. and higher, the infected flies died more quickly than did the control flies which had been fed on the organs of healthy animals.

In two of the experiments the infected flies were all dead on the seventh and eighth days respectively at temperatures of 14° C. These facts indicate that flies should not be allowed to have access to the bodies or excreta of cases of plague or to the food.

SMALL-POX.

Nuttall and Jepson (*l.c.*) give one reference only to flies in relation to small-pox. Hervieux (1904) states that Laforgue at Tamorna-Djedida, Province of Constantine, observed that during an epidemic of small-pox at that place all the children who were attacked lived in the south-west of the village and there was no small-pox in the northern part of the village. This distribution of the disease was attributed to the direction of the prevailing winds, and observations indicated that flies and mosquitoes were distributed with the wind. Laforgue believed that flies played an important part in spreading the virus of small-pox.

DIPHTHERIA.

While it is hardly likely, as Nuttall and Jepson have pointed out, that under natural conditions flies would play any part in the dissemination of diphtheria, it is conceivable that, if the necessary conditions of infection occurred they would carry the infection. Dickenson (1907) cites Smith (1898) who carried out the usual and hardly valuable experiment of allowing flies to walk over infected matter and afterwards over culture media, with the natural positive results. The unreliable character of such experiments is indicated by the results of the experiments which Graham-Smith (1910) carried out with *Bacillus diphtheriae*. Two series of experiments were made. In the first flies were allowed to feed for thirty minutes on an emulsion of *B. diphtheriae* in saliva and then transferred to a fresh cage. At intervals from one hour up to seventy-two hours after feeding flies were killed and cultures were made on transparent serum medium from their legs, wings, heads, crops and intestinal contents. In the second series of experiments the flies were allowed to feed for one hour on an emulsion of *B. diphtheriae* in broth and afterwards they were treated similarly to the flies in the first series. From the tabulated results it would appear that *B. diphtheriae* seldom remains alive on the legs and wings for more than a few hours. In the crop and intestine they may live for twenty-four hours or occasionally longer. The faeces passed during the first few days after infection are frequently infected. These experiments would indicate, therefore, the ability of the house-fly to carry infection if suitable conditions occur.

YAWS (*Framboesia tropica*).

This disease which is widely distributed throughout the tropics and is especially common on the west coast of Africa is extremely contagious. It is characterised by ulcerous papules which develop into fungus-like incrustations of a spreading and intensely disagreeable nature. Gudger (1910) has called attention to a very early suggestion that flies carry the infection of this disease. This is contained in Bancroft's *Essay on the Natural History of Guiana in South America* which volume was published in 1769. The author states: " The yaws are spungey, fungous, yellowish, circular protuberances, not rising very high, but of different magnitudes, usually between one and three inches in circumference. These infest the whole surface of the body, and are commonly so contiguous that the end of the fingers cannot be inserted between them; and a small quantity of yellowish pus is usually seen adhering to their surface, which is commonly covered with flies through the indolence of the Negroes. This is a most troublesome, disagreeable disorder, though it is seldom fatal. Almost all the negroes once only in their lives, are infected with it, and sometimes the whites also, on whom its effects are much more violent. It is usually believed that this disorder is communicated by the flies that have been feasting on a diseased object, to those persons who have sores, or scratches, which are uncovered, and from many observations, I think that this is not improbable, as none ever receive this disorder whose skins are whole; for which reason the Whites are rarely infected, but the backs of the Negroes being often raw with whipping and suffered to remain naked, they scarce ever escape it."

Gudger (*l.c.*) also calls attention to Kosters' *Travels in Brazil in the years from* 1809–1815, published in Philadelphia in 1911 in which this author says, in reference to yaws: "This horrible disorder is contracted by inhabiting the same room with a patient and by inoculation; this is effected by means of a small fly, from which every precaution is often of no avail. Great numbers of the insects of this species appear in the morning, but they are not so much seen when the sun is powerful. If one of them chances to settle upon the corner of the eye or mouth, or upon the most trifling scratch, it is enough to inoculate the *bobas*, if the insect comes from a person who labours with the disease." In reference to these two statements, as in many others where the word "flies" is used, it may not be *M. domestica* to which the authors refer, in fact the statement that the flies "are not so much seen when the sun is powerful" militates against the idea, although it does not in the least diminish the possibility of their being active agents in the dissemination of infected matter.

Other observers adduce similar evidence. Wilson (1868) states that in the West Indies there is a prevalent belief that flies convey the disease from one person to another. Two cases are reported by Hirsch (1896) for which he believed flies were responsible. Both patients were living among Fijian children who were affected with the disease. The necessary raw places for the reception of the infection were present, one patient having an uncovered ulcer and the other sores on his feet, both of which exposed surfaces would attract flies. Cadet (1897) also points out that skin lesions such as ulcers, the bites of insects or other animals, scratches, etc., are necessary for infection which may take place and through direct contact of infected clothes or by means of flies carrying the diseased secretions on their legs.

Experimental evidence is brought forward by Castellani (1907). He allowed *M. domestica* to feed upon infected matter obtained by scraping slightly ulcerated papules. They were also fed upon the semi-ulcerated papules on the skin of three yaw patients. In both cases *Spirochaeta pertenuis*, the causative organism of this disease, was found in microscopic preparations made from the mouth parts and legs of the flies. Monkeys were also infected with the disease by allowing specimens of *M. domestica* which had been fed as in the previous cases to come in contact with lesions made on the eyebrows of the monkeys. Castellani is of the opinion that yaws is generally transmitted from person to person by direct contact but under certain circumstances it may be conveyed by flies and possibly by other insects.

Robertson (1908) took about 200 flies which had been captured on yaws lesions and shook them up in sterile water. After standing for twenty-four hours the water was centrifugalized and smears were made from the precipitate. In four slides the organism *Spirochaeta pertenuis* were found.

In St Lucia, Windward Islands, Nicholls (1912) after studying the disease concluded that the majority of cases of yaws in the West Indies were caused by the inoculation of surface injuries by the fly *Oscinis pallipes*. This insect feeds on the skin discharges of man and other animals. The flies are very persistent and engorge themselves with pus, blood, serum or sebaceous secretion.

LEPROSY.

Experiments with flies and human and rat lepra are recorded by Wherry (1908). It was found that such flies as *M. domestica*, *C. vomitoria* and *Lucilia caesar* take up enormous quantities of lepra bacilli from the carcase of a leper rat and deposit them with their faeces; but the bacilli apparently do not multiply in the intestinal tracts of the flies, as the latter are clear of bacilli in less than forty-eight hours. Larvae which have hatched out in the carcase of a leper rat become heavily infected with lepra bacilli. When they are removed and fed upon uninfected meat they pass out most of the lepra bacilli and the flies hatching out from the pupae of these larvae are generally uninfected. If the larvae of *C. vomitoria* be fed almost continuously on the carcase of a leper rat they remain heavily infested with lepra bacilli and on pupating such heavily infested specimens appear to be incapable of further development.

A house-fly, *M. domestica*, caught on the face of a human leper was found to be infected with lepra bacilli. At the beginning of the observation these were few in number but on the third day more than 1115 lepra-like bacilli were present in each speck deposited. However, only one bacillus was found in the specks deposited between the third and sixth days. The acid-proof bacilli in the fly were not infective when injected into the subcutaneous tissue of the guinea-pig.

Lebœuf (1912) has studied the dissemination of the leprous bacillus by the house-fly. Flies (*M. domestica*) frequently settle on leprous ulcers left exposed. Of twenty-three flies caught on such ulcers known to contain many bacilli, nineteen were found to contain the leprous bacillus in their intestines. The bacilli were sometimes present in very large numbers and were excreted in the faeces of the flies and showed no signs of degeneration even after a day or more in the intestine of the fly. Bacilli were found in flies caught in the room and not directly on the ulcers but they occurred less frequently in such cases. The author examined twenty-three flies caught in his own house which was situated about one hundred and fifty yards from the hospital reserved for very advanced cases of leprosy, but in no case did he find the leprous bacillus, which he takes as indicating that the range of dissemination is not very great. While the foregoing experiments would indicate that *M. domestica* is capable of disseminating the bacillus, it apparently can be infected only from patients who present open lesions or from infected discharges. *B. leprae* was not found in twenty-nine flies caught in rooms of patients with only nervous symptoms or with unbroken skin lesions. Lebœuf concludes: (1) that *M. domestica* is capable of absorbing enormous quantities of the leprous bacillus; (2) that the bacillus is found in large quantities, apparently in perfect condition, in the excrement of infected flies; (3) that there does not appear to be any multiplication of the bacillus within the fly, but the organism does not seem to be degenerate; (4) that it is possible that flies passing from a leprous patient and depositing their excrement on the nasal orifices, perhaps during sleep, or upon raw cutaneous wounds of otherwise healthy persons living in the immediate neighbourhood of leper patients, may disseminate the bacilli in this way.

Minett (1911) has also discussed the question as to the dissemination of leprous bacilli by flies.

These experiments and observations, although conducted with a comparatively small number of flies are, I think, sufficiently conclusive to warrant the conclusion that, under the conditions indicated by Lebœuf, namely, abundance of flies, proximity of leprous ulcers or infected discharges and exposed places for infection on healthy persons, house-flies are able to transmit the leprous bacillus.

DYSENTERY.

It would appear to be not improbable that flies, in view of their relation to typhoid fever, should sometimes be agents in the spread of dysentery. The etiology of this disease has been a matter of considerable controversy but it is generally accepted now that in the bacillary type of the disease one of the causative organisms is *B. dysenteriae*, these organisms being regarded as the cause of epidemic dysentery, while in the sporadic or endemic dysentery, usually known as amoebic dysentery, the causative organism belongs to the *Amoeba* group of protozoa, of which there are probably several species. The possibility of flies carrying the infection of both types of disease appears to me to be highly probable, especially in the amoebic type of the disease. The amoebae multiply on the intestine by fission and are passed out with the faeces. When the faeces become hard the amoebae encyst. It is not unreasonable to suggest in view of positive evidence in other cases of a similar nature, that flies bred in or feeding upon infected faeces, might ingest the amoebic cysts and with these infect food.

Unfortunately we have no exact evidence as to flies carrying dysenteric infection. Smith (1903) states: "An old idea of some Anglo-Indian surgeons was that dysentery could be caught by using the same latrine as a dysentery patient. There may be something in this...the ubiquitous fly may, therefore, be a dysentery inoculator in open camp latrines." While such inoculation is not unlikely, the probability of flies infected in the latrine visiting food or patients in the hospital is greater. The possibility of flies playing in dysentery a similar part to that which they do in typhoid and cholera is referred to by Bergey (1907). An epidemic of one hundred and thirty-six cases of dysentery which occurred in an insane asylum at Worcester,

Massachusetts, U.S.A., is described by Orton (1910) who considered that house-flies were responsible for the epidemic. In this paper which contains interesting observations on the breeding of the flies in spent hops and malt barley, the author describes experiments which were carried out with a view to testing his hypothesis. The possibility of infection being distributed in such an institution was appreciated to a greater extent owing to the difficulty of confining the intestinal discharges to the proper place. The clothing and bedding of the patients were brought to the laundry in which flies were abundant. In the experiments *B. prodigiosus*, being easily recognisable, was exposed in the laundry. This bacillus was recovered from flies subsequently at intervals in other rooms of the hospital. Such results would appear to indicate a strong probability of the carriage of the dysenteric organism under similar conditions which were known to exist.

I feel confident that further investigations into the relation of house-flies to the distribution of the causative organisms of dysentery of both types, bacillary and amoebic, will give positive results[1].

ORIENTAL SORE.

This skin disease which is characterised by a slowly spreading, ulcerating condition of the skin is endemic in certain tropical and sub-tropical regions such as Northern Africa, Sahara (Biskra), Egypt, Asia Minor, Mesopotamia (Bagdad) and India. Wright in 1903 discovered the so-called Leishman parasites in the granulation cells. Manson also discovered the same organism in cases of oriental sore and considered them, or a parasite morphologically identical, as the causative organism of the sore and the latter investigator (1907) suggested that flies, bugs or other insects might be responsible for the indirect method of infection.

---

[1] Krontowski (1913) infected the larvae of various flies including *M. domestica* with *B. dysenteriae* and allowed them to pupate. An examination of the faeces of the resulting flies gave negative results. The author elaborated Auché's experiments with adult flies and came to the conclusion that the bacteria can retain their virulence when on the feet or proboscis, or in the alimentary tract of the fly, whence they are eventually voided with the excreta. Wherever flies have free access to food and to infected human faeces the danger of dysentery and, according to Krontowski, epidemic typhus is great.

Wenyon (1911) has recently made a very thorough study of oriental sore in Bagdad in which he investigated the possible relation of house-flies to the disease. He found that house-flies appeared to diminish in numbers to some extent during the hottest part of the summer when the maximum shade temperature reached 110° F. Flies swarm about the faces of the children, especially those having the sore. Such flies collected from the face of a child suffering from an ulcerating type of sore are found to have the intestine filled with the exudation of the sore in which the parasites (*Leishmania tropica*) were readily found. Such a fly feeding immediately afterwards upon some fresh abrasion of the skin must certainly in a number of instances inoculate the sore parasite. The parasite could not be found in the intestines of *Stomoxys* feeding on the faces of children with sores. In infection experiments with house-flies no evidence of the development of the sore parasite could be found and they seemed to disappear quickly. Wenyon concludes that the limited distribution of the disease and the widespread prevalence of the house-fly would not appear to confirm the view held by some authorities that the house-fly is the normal carrier of the disease organism.

In discussing tropical sore Nuttall and Jepson (1910) record Seriziat's assertion that flies convey "Bouton de Biskra." In his study of the disease at Biskra, Laveran (1880) found that from September to October the slightest wound tends to become a sore. He has seen it develop from small pimples and pustules and from wounds caused by burns or blisters. He believed that flies carried the virus on their feet and proboscides and thus distributed infection.

There is no doubt that in tropical countries the organism of. the various types of sore can be mechanically transferred by flies from the granulated areas to new wounds and thereby inoculate the same. Its spread by flies would be governed by their abundance, by the viability of the organism of the fly and the opportunities for obtaining and distributing infection. In this connection the observations of Patten[1] (1912) are of interest. Referring to an idea entertained by some workers at one time

[1] The Etiology of Kala-Azar, *Nature*, Vol. LXXXIX. pp. 306–308, 1912.

that flies might act as one of the carriers of the tropical disease Kala-Azar, a disease caused by a species of *Leishmania*, he stated that two years ago he had fed a large number of bred house-flies (*Musca nebula*, an Indian species) on fresh splenic juice and had found that the parasites disappeared from the alimentary tract in a few hours. It was difficult, therefore, to understand how the parasite could be transmitted in this way. Such an observation was against the theory that flies feeding upon the discharge from an ulcerated sore might carry the infection. In another paper Patton (1912) gives further details and states that flies fed upon the discharge from the sores and afterwards on abrasions or scratches did not produce the sore although the experiment was carried out daily for about a month. The same investigator has since brought forward very conclusive evidence as to the organism being carried by the bed-bug.

SYPHILIS.

I have been able to discover one reference only to the possibility of flies acting as agents in the spread of this disease by the mechanical transfer of the *Spirochaeta*, the syphilitic organism, from a patient to a healthy person. Dr Kerr of Morocco, in a paper on "Some Prevalent Diseases in Morocco," read before the Glasgow Medico-chirurgical Society (Dec. 7th 1906), described epidemics of syphilis, where, according to the author, the disease was spread by flies which had been feeding upon the open sores of a syphilitic patient.

CHAPTER XXIV

MISCELLANEOUS EXPERIMENTS ON THE CARRIAGE OF MICRO-ORGANISMS BY FLIES, BY BOTH NATURAL AND ARTIFICIAL INFECTION

IN addition to the experimental evidence which has accumulated on the relation of flies to specific diseases, there are also on record a considerable number of experiments on the carriage of various micro-organisms, not in every case pathogenic, by flies. As these experiments have usually some particular interest I have brought together such as seemed worthy of record.

I have previously recorded (1910) a very interesting series of experiments carried out in 1908 by Güssow. Their particular interest lies in the fact that they clearly demonstrate the varied bacterial and fungal flora which the house-fly normally carries.

Güssow's experiments were as follows:

Experiment No. 1.

A fly was caught in his living room (Norwood, London) at 10 a.m. on May 4th and allowed to walk over nutrient agar-agar in a Petri dish; the necessary precautions being taken to prevent extraneous infection of the medium. The Petri dish was placed in an incubator and kept at 28—30° C. At 6 p.m. on the same day there were plain indications of colonies forming, but they were too small to allow a separation count.

May 5th, 10 a.m.   7 colonies of bacteria and 4 of fungi showing.
  „   6th, 10 a.m.   16   „   „   5   „
  „   7th, 10 a.m.   23   „   „   6   „
  „   8th, 10 a.m.   30   „   „   6   „

That is, in 96 hours, 30 colonies of bacteria and 6 colonies of fungi were observed.

The fungi were examined and identified as follows:

    2 colonies of *Saccharomyces* sp.
    2   „   *Penicillium glaucum*
    1 colony of *Aspergillus niger*
    1   „   *Cladosporium herbarum*

The bacteria were removed in the ordinary manner and were sub-cultured, plated out and identified as follows:

Micrococcus ureae	2	colonies
Bacillus subtilis	7	„
Bacillus coli commune	11	„
Sarcina lutea	2	„
Bacilli stained by Gram ...	3	„
Bacilli not stained by Gram ...	5	„

Experiment No. 2.

A fly was caught at 11.30 a.m. on May 4th out of doors on Central Hill, Norwood, London, and was allowed to walk over nutrient agar-agar at 12 o'clock noon.

May 4th, 6 p.m.   Colonies were plainly forming.

„   5th, 10 a.m.   13 colonies of bacteria and 6 colonies of fungi.

„   6th, 10 a.m.   21   „   „   7   „

„   7th, 10 a.m.   39   „   „   7   „

„   8th, 10 a.m.   46   „   „   7   „

That is, in 94 hours, 46 colonies of bacteria and 7 colonies of fungi were obtained from this fly No. 2.   The fungi were identified as follows:

2 colonies of *Macrosporium* sp.

3   „   *Penicillium glaucum*

1 colony of *Cladosporium herbarum*

1   „   *Fusarium roseum*

The bacteria after being sub-cultured and plated out were identified as:

Bacillus tumescens	18	colonies
Micrococcus pyogenes aureus ...	9	„
Sarcina lutea	2	„
Sarcina ventriculi	1	colony
Bacillus amylobacter	4	colonies
Acid-fast bacillus	1	colony
Bacilli stained by Gram ...	4	colonies
Bacilli not stained by Gram ...	7	„

Experiment No. 3.

This experiment was perhaps the most interesting of the three as the fly was captured at 10.30 a.m. on May 4th on a dust-bin (Norwood, London), a situation in which flies are frequently found.   It was allowed to walk over the surface of nutrient agar-agar.

May 4th, 6 p.m.   Signs of colonies observed.

„   5th, 10 a.m.   18 colonies of bacteria and 7 colonies of fungi.

„   6th, 10 a.m.   58   „   „   9   „

„   7th, 10 a.m.   113   „   „   10   „

„   8th, 10 a.m.   116   „   „   10   „

That is, after 95¾ hours, 116 colonies of bacteria and 10 colonies of fungi were obtained from this single fly. The fungi were identified as:

Penicillium glaucum	... ...	4 colonies
Eurotium sp.	... ...	1 colony
Saccharomyces sp.	... ...	2 colonies
Fusarium roseum	... ...	1 colony
Aspergillus niger	... ...	1 ,,
Mucor racemosa ...	... ...	1 ,,

The bacteria after having been sub-cultured and plated out were identified as :

Bacillus coli commune ...	...	34 colonies
Bacillus subtilis...	... ...	16 ,,
Bacillus tumescens	... ...	8 ,,
Bacillus lactis acidi	... ...	4 ,,
Sarcina lutea ...	... ...	12 ,,
Sarcina ventriculi	... ...	2 ,,
Micrococcus pyogenes aureus	...	21 ,,
Micrococcus ureae	... ...	11 ,,
Acid-fast bacilli...	... ...	2 ,,
Bacilli stained by Gram	...	4 ,,
Bacilli not stained by Gram...		2 ,,

The extremely large number and preponderance of bacilli carried by this fly No. 3 shows very strikingly the infection which a fly frequenting such miscellaneous household refuse as is contained in the average household dust-bin or garbage may carry, and the results of such careful experiments as those which are recorded above demonstrate clearly not only that flies normally carry about the spores of fungi and bacteria and the extra infection which they obtain by frequenting refuse, but also their liability to carry and disseminate such bacteria, pathogenic and non-pathogenic, with which they may come into contact in their wanderings.

Manning (1902) obtained cultures of the following bacteria from infected flies: B. pyocyaneus, Staphylococcus pyogenes aureus, B. typhi-abdominalis, and B. coli commune.

If flies have access to wounds of an inflammatory and suppurative nature they are liable to transport the Staphylococci to other spots. Buchanan (1907) allowed M. domestica to walk over a film of Staphylococcus pyogenes aureus from an abscess, and afterwards over agar; a mixed growth resulted, in which S. pyogenes aureus predominated. Buchanan (l.c.) also experimented with the bacillus of swine fever. Nine blow-flies (C. vomitoria) were caught on the carcases of pigs which had died from swine fever during an epidemic in 1905. Each fly was allowed to walk over an agar culture plate for about one minute. From one of the plates the bacillus of swine fever was isolated. The flies were swarming on the carcases of the dead animals and were frequently seen to pass directly from each source of infection to the feeding troughs of the pigs.

Staphylococci were found by Joly (1898) on a house-fly caught in the laboratory. Celli (*l.c.*) also records experiments which indicated that *S. pyogenes aureus* retains its virulence after passing through the intestine of the fly.

Reference has been previously made to experiments with *Bacillus prodigiosus* which was used on account of its suitability in the case of experiments with flies, being non-pathogenic, non-spore-bearing and easily recognisable. There are a number of records of other experiments on the carriage of the bacillus by flies.

Abel (1899, cited by Nuttall and Jepson) refers to an experiment of Otto Helm in 1875 who stated that the slimy masses containing the bacillus "*Monas prodigiosus*" are easily conveyed from one food substance to another. Abel placed cultures of *B. prodigiosus* and clean potatoes in different parts of a room. To the potatoes he added putrid meat so that the odour would attract the flies. After 2—3 days colonies of *B. prodigiosus* appeared in all the potatoes. Negative results were obtained when the room was rid of flies. Similar experiments by Burgess are mentioned by Hart and Smith (1898). Flies were fed on material containing *B. prodigiosus* and then allowed to fly in a large room. After a few hours the flies were recaptured and allowed to walk over slices of sterilised potato on which colonies of the bacillus subsequently developed. Abel records two instances in which he observed spontaneous infection of food by flies in houses where they were abundant. Flies were captured in these houses and placed singly in tubes containing sterilised potato. Seven of twenty-eight flies so captured in one house gave *B. prodigiosus* and a similar positive result was given by three out of thirty-three flies caught in the other house. Abel accordingly points out the possibility of flies carrying the bacilli of typhoid and cholera in a similar manner.

19—2

TABLE SHOWING SOURCES OF BACTERIA FROM FLIES

Date	Source	Total number of bacteria	Total acid bacteria	Rapid liquefying bacteria	Slow liquefying bacteria	Bacterium lactis acidi. Group A, Class 1	Coli aerogenes Group A, Class 2
1907							
July 27	(a) 1 fly, bacteriological laboratory	3,150	250	600	100	—	—
,, 27	(b) 1 fly, bacteriological laboratory	550	100	—	—	—	—
Aug. 6	(c) 19 cow-stable flies	7,980,000	220,000	—	20,000	—	—
	Average per fly	420,000	11,600	—	1,000	—	—
,, 14	(d) 94 swill-barrel flies	155,000,000	8,950,000	—	—	4,320,000	4,630,000
	Average per fly	1,660,000	95,300	—	—	46,000	49,300
,, 14	(e) 144 pig-pen flies	133,000,000	2,110,000	100,000	266,000	933,000	1,176,000
	Average per fly	923,000	18,700	700	1,150	6,500	12,000
Sept. 4	(f) 18 swill-barrel flies	118,800,000	40,480,000	—	14,500,000	10,480,000	30,000,000
	Average per fly	6,600,000	2,182,000	—	804,000	582,000	1,600,000
,, 21	(g) 30 dwelling-house flies	1,425,000	125,000	—	12,500	—	—
	Average per fly	47,580	4,167	—	417	—	—
,, 21	(h) 26 dwelling-house flies	22,880,000	22,596,000	120,000	34,000	—	—
	Average per fly	880,000	869,000	4,600	1,300	—	—
,, 27	(i) 110 dwelling-house flies	35,500,000	13,670,000	8,840,000	125,000	—	—
	Average per fly	322,000	124,200	80,300	1,100	—	—
Aug. 30	(j) 1 large blue-bottle blow-fly	308,700	(a)	—	—	—	—
	Total average of 414 flies	1,222,570	367,300	7,830	73,500	211,500	553,800
	Average per cent. of 414 flies	—	30	6	6	7	18
	Average per fly of 256 flies, experiments (d), (e), and (f)	3,061,000	765,000	230	268,700	211,500	553,800
	Average per cent. of 256 flies, experiments (d), (e), and (f)	—	25	—	8	7	18

(a) 2200 mould spores

In 1907 experiments were carried out by Dr M. B. Arnold and myself with *B. prodigiosus*. Flies which had just emerged from the pupae, and therefore not already contaminated with an extensive bacterial flora, were allowed to walk over a film of the bacillus, after which they were confined to sterile glass tubes. At varying periods they were taken out and allowed to walk over the culture plates. Those contained for over twelve hours retained the bacillus on their appendages and transferred them subsequently to the culture media, but they were not recovered from those flies which were kept in confinement for twenty-four hours; a large number of flies, however, were not used.

In discussing the relation of flies to typhoid fever reference was made to the infection of milk. In this connection an interesting investigation was made by Esten and Mason (1908) on the *rôle* which flies play in the carriage of bacteria to milk.

The flies were caught by means of a sterile net; they were then introduced into a sterile bottle and shaken up in a known quantity of sterilised water to wash the bacteria from their bodies and to simulate the number of organisms that would come from a fly falling into a quantity of milk. They summarised their results in the table given on the opposite page.

While the counts of the bacteria can only be considered as comparative the results indicate clearly the nature of the source of infection. Commenting on these results the authors state that "early in the fly season the numbers of bacteria are comparatively large. The place where flies live also determine largely the numbers they carry."

Poliomyelitis.

Experiments have been carried on and are being continued by Flexner and Clarke (1911) on the contamination of the house-fly with the virus of poliomyelitis, more generally known as infantile paralysis or spinal meningitis. It was found that flies contaminated with the virus of poliomyelitis harboured the virus in a living and infectious state for at least forty-eight hours. It was not shown that this is the limit of the period of survival and the experiments threw no light on the question as to whether the virus is retained merely as a superficial contamination or whether it could survive in the fly's gut. Further experiments recorded by Howard and Clarke (1912) demonstrated that *M. domestica* retained the virus either in or on their bodies for at least twenty-four and forty-eight hours respectively. They also showed that the virus may remain in a viable condition in the alimentary tract of the fly for at least six hours. The possibility of flies obtaining infection from the infected discharges from the nose and throat or intestine is indicated. Further reference to the transmission of this disease is made under *Stomoxys calcitrans*.

Trypanosomes.

On account of its non-blood-sucking habits, few experiments have been carried out with a view to demonstrating the possibility of *M. domestica* carrying Trypanosomes. Reference has already been made to investigations on the relation of flies to the allied organisms causing Tropical Sore. Experiments are recorded[1] in which *M. domestica* was fed for 3—4 minutes on blood from a guinea-pig infected with *Trypanosoma hippicum*; after an interval of about 30 seconds the flies were placed over the scratched skins of mules for about five minutes and it was demonstrated that this trypanosome may be transmitted by the flies[2].

Surra.

Mitzmain (1913) states that in experiments with *Stomoxys calcitrans* and live stock, Surra organisms have been demonstrated in the mouth parts and stomachs of house-flies.

Danysz virus.

An interesting experiment on the carriage of Danysz rat virus was carried out by Graham-Smith (1910). Flies artificially infected with the virus by feeding were allowed to settle and feed on a piece of bread soaked in milk. After one hour the bread was given to a mouse. A mouse which had fed upon bread given to flies forty-eight hours after infection died in two days and the virus was isolated from the spleen; another mouse fed on bread given to the flies four days after infection died in two days. In

---

[1] *Report Dept. Sanit. Isthmian Canal Comm.*, Dec. 1911, pp. 42, 43.

[2] The possibility of *M. domestica* acting as a vector of *T. hippicum* is convincingly treated by Darling (1913) who suggests that more attention be paid to the possible relation of *M. domestica* to trypanosome diseases.

another experiment a mouse was fed on bread soaked in an emulsion of the flies' faeces passed about forty-eight hours after infection and scraped from the walls of the cage. The mouse died in two days and the organism was isolated from it. These experiments showed that flies which have fed on the virus are capable of infecting to so great an extent food on which they settle and feed that mice fed on it became infected.

Rabies.

Experiments with a view to discovering whether flies would carry the virus of rabies obtained during the larval state have been carried out by Fermi (1911). In the first series of experiments the author fed fly larvae on the brains of rabies cases and then tested their virulence by emulsifying and injecting subcutaneously. In a second series a fixed virus and fly larvae were rubbed into an emulsion and likewise injected subcutaneously. The results indicate that rabies virus cannot be transmitted through fly larvae. It appears that the fly emulsion has an attenuating effect upon a fixed virus, either through its direct action upon the virus or through its indirect action upon the organism. It possesses no absolute lyssicidal power since a virus mixed with fly larvae emulsion is found to be virulent when administered subdurally.

Fungal spores.

In the experiments of Güssow, already mentioned, it was shown that flies normally carry the spores of moulds such as *Penicillium, Eurotium, Mucor,* yeasts, etc. Consequently their frequent infection of food materials, which may be observed if attention is given to the matter, is readily understood. Gayon (1903) also cultivated several species of moulds from flies which he caught and dropped into nutrient gelatin. Experimenting with yeasts, Graham-Smith (1910) found that the yeast organisms did not appear to survive for more than a few hours on the legs and wings, but that they could be found in cultures of the crop and intestine for at least three days, and were present in the faeces 48 hours after the fly had been infected.

Cobb (1906) made studies of the extent to which flies transported the spores of a fungus attacking sugar cane. The feet of a fly which had been feeding upon the spores of this fungus left tracks of fungal spores on the sides of the glass vessel in which it had been contained. The spores from five of the tracks were calculated and the number of spores per track was estimated to be 860,000. The possibility of the spores being carried by flies in this way was indicated by the fact that such spores germinated under suitable conditions.

Bacteria on flies captured under natural conditions.

In addition to the investigation of Güssow, which I have recorded at the beginning of this chapter, a series of carefully conducted experiments has been carried out by Cox, Lewis and Glynn (1912) with a view to ascertaining the number and varieties of bacteria carried by flies infected under natural conditions in sanitary and unsanitary city districts in Liverpool. The flies were captured in sterilised wire traps which were exposed for twenty-four hours. Flies from various districts were allowed to swim in measured quantities of sterile water to simulate the pollution of liquids when flies fall into them and to estimate the rate at which the bacteria are given off. This experiment also served to indicate the comparative number of bacteria set free from the bodies of flies from dirty or more cleanly areas. The same flies were afterwards ground up in a sterile mortar with a sterile pestle to find the gross number of bacteria carried on and in a fly and to ascertain whether the number carried inside a fly is always greater or less than the number set free even after struggling in a liquid for 30 minutes. Over 450 naturally infected flies were caught during September and the early part of October 1911 in different parts of the city of Liverpool and the number and kinds of bacteria carried and contained by them were investigated. Their experiments showed that:

1. The number of bacteria derived from flies while struggling in a liquid may be very large and increase with the time they remain in the liquid. The number of bacteria varies from 2000, the lowest figure for five minutes, to 350,000, the highest figure for 30 minutes. This number may be taken as a measure of their capacity to pollute liquid with their vomit, excreta or bodies. The number of

bacteria carried inside the fly is very much greater than those carried externally.

2. Flies caught in insanitary or congested areas of the city carried and contained far more bacteria (aerobic), including those of the intestinal group, than flies from the more sanitary, that is, cleaner, less congested or suburban districts. The number of aerobic bacteria obtained from flies caught in insanitary districts varied from 800,000 to 500,000,000 per fly; flies from the cleaner or less congested areas gave from 21,000 to 100,000 bacteria per fly.

3. Flies caught in the dwelling rooms of different corporation houses forming two sides of a street about 400 yards long which constituted a sanitary " oasis " in the middle of a slum district carried and contained less bacteria of all kinds than those from the dwelling rooms of a street with insanitary property on each side.

4. The number of intestinal bacteria as indicated by glucose bile salt fermenters is greater in insanitary or congested areas, where they vary in number from 10,000 to 333,000,000 than in the more sanitary areas where from 100 to 10,000 are carried per fly.

5. Pathogenic bacteria and those allied to the food-poisoning group were only obtained from the congested or moderately congested areas and never from the suburban districts.

6. Flies caught in milk shops apparently carry and contain more bacteria than those from shops with exposed food in a similar neighbourhood. The authors attribute this fact to the milk being a suitable culture medium for bacteria after having been inoculated by the flies, later they re-inoculate themselves.

7. A comparison of the number of bacteria carried by flies and blue-bottles caught in an eating-house opposite to slaughter-houses showed that the latter carried a much greater number.

The morphological characters and cultural reactions of 123 strains of bacteria were examined by the authors. Among those identified were two *Streptococci* and several *Staphylococci* and *Sarcinae*. One hundred and six were small gram-negative non-spore-bearing bacilli; these were grouped as follows:

Chromogenic group: two strains of *B. pyocyaneus* were isolated from flies from a knackers' yard.

Colon group: 41 colonies of this group were picked off haphazard and classified according to McConkey as follows:

B. acidi lactici type	...	19·5 per cent.
B. coli communis type	...	12·2 ,, ,,
B. neapolitanus type	...	19·5 ,, ,,
B. lactis aerogenes type	...	46·4 ,, ,,

Salmonella group: one bacillus gave identical reactions to *B. enteritidis* Gaertner, except that the serological tests were negative.

Morgan's Infantile Diarrhoea group: one identical to Morgan's No. 1, and many others closely resembling it and Morgan's Nos. 2 and 3 were obtained.

Others were included in the proteolytic, acid lactose-sucrose (saccharose), and miscellaneous groups.

The authors conclude: "It is clear that flies from the suburbs where infantile diarrhœa is rare carry far less bacteria than those in the city where it is common. It was, nevertheless, impossible in the time at our disposal to correlate exactly the number or varieties of bacteria carried by flies in the city with the number of cases and deaths from infantile diarrhœa in individual streets.

"As the amount of dirt carried by flies in any particular locality, measured in terms of bacteria, bears a definite relation to the habits of the people and the state of the streets it demonstrates the necessity of efficient municipal and domestic cleanliness if the food of the inhabitants is to escape pollution, not only with harmless but also with occasional pathogenic bacteria."

In his study of the micro-organisms carried by flies under normal conditions to which reference has already been made (p. 245), Graham-Smith found that more than one-third of all the flies examined were infected with lactose-fermenting bacilli of the colon type.

TABLE SHOWING PERIOD DURING WHICH ORGANISMS WERE USUALLY RECOVERED AND THE LAST OCCASION AFTER INFECTION ON WHICH THEY WERE RECOVERED IN THE EXPERIMENTS OF GRAHAM-SMITH (1910), FROM WHOM THE TABLE IS TAKEN.

Organism	Period during which organisms were frequently recovered						Latest occasion on which organism was recovered					
	Legs	Wings	Head	Crop	Gut	Faeces	Legs	Wings	Head	Crop	Gut	Faeces
Bacillus typhosus ...	—	—	—	—	2 days	2 days	—	—	—	—	6 days	2 days
B. enteritidis ...	—	—	—	—	1 day	0	7 days	—	7 days	8 days	7 ,,	0
B. tuberculosis (culture)	—	—	—	3 days	12 days	8 days	—	—	—	3 ,,	16 ,,	13 days
B. tuberculosis (sputum)	—	—	—	—	3 ,,	3 ,,	—	—	—	—	7 ,,	5 ,,
Yeast	0	0	0	0	3 ,,	2 ,,	2½ hrs.	2½ hrs.	2½ hrs.	2 days	3 ,,	2 ,,
B. diphtheriae	0	0	2 hrs.	2 hrs.	2 hrs.	6 hrs.	5 ,,	5 ,,	5 days	7 ,,	5 ,,	2 ,,
B. anthracis (no spores)	0	0	2 days	3 days	3 days	2 days	2 days	0 ,,	4 ,,	5 ,,	3 ,,	2 ,,
V. cholerae	5 hrs.	5 hrs.	5 hrs.	5 hrs.	2 ,,	30 hrs.	30 hrs.	5 ,,	5 hrs.	2 ,,	2 ,,	30 hrs.
B. prodigiosus	1 day	12 ,,	2 days	5 days	15 ,,	2 days	8 days	12 ,,	11 days	5 ,,	17 ,,	6 ,,
Anthrax spores ...	10 days	4 days	4 ,,	8 ,,	7 ,,	5 ,,	20 ,,	20 days	20 ,,	13 ,,	20 ,,	13 ,,

Intestinal Protozoa.

Stiles (1913) in an interesting note refers to the possibility of flies transferring intestinal protozoa such as *Entamoeba coli, Lamblia duodenalis* or *Trichomonas intestinalis* from faecal material to food supplies. He suggests that the presence of such protozoa in food supplies might be taken as an indication of contamination[1]. The possibility of flies becoming infected, owing to their habits, with intestinal amoebae has also been discussed by Converse (1910) and others.

In considering experiments on artificially infected flies it should be remembered that the flies are enabled to obtain, in most cases, a much grosser infection than they might be able to obtain under natural conditions. Further, many factors which might possibly affect the degree of infectivity under natural conditions have not exercised possible adverse influences. Pending the results of further investigations therefore, experiments which have been carried out under unusually favourable artificial conditions must be considered in conjunction with those performed under natural conditions, which are chiefly recorded in the accounts given of the various specific diseases. The experiments of Güssow and of Cox, Lewis and Glynn which have been described in this section admirably demonstrate the nature and extent of natural infection.

---

[1] Stiles and Keister (1913) have investigated this matter further with the result that they have arrived at the conclusion that the evidence that flies commonly act as carriers of the spores of intestinal protozoa is not very conclusive.

CHAPTER XXV

THE RELATION OF FLIES TO MYIASIS AND TO THE SPREAD OF INTESTINAL WORMS

MYIASIS.

THE occurrence of the larvae or maggots of flies of different species in the human body, where they most commonly are found in the intestinal tract and less frequently in the urinary passages, usually leads to a diseased condition to which the term Myiasis is applied. For many years cases of myiasis have been recorded and references to such cases are widely scattered through medical and scientific literature. In a large number of instances, especially in the cases of the earlier records, the identity of the species of larva was not determined and in fact, until recently, the determination of dipterous larvae was a matter of considerable difficulty owing to our lack of knowledge of the developmental histories of even the commoner species.

In a recent brief review of our knowledge (1912) I have collected a number of the more important and typical cases of myiasis of the intestinal and urinary tracts. Austen (1912) has also recorded in a very complete and excellent account instances of British flies which have been found in cases of myiasis in man, to which account I am indebted for many of the cases to which reference will be made.

The species whose larvae have been recorded as causing myiasis in man are as follows:

The House-fly, *Musca domestica.*
The Lesser House-fly, *Fannia canicularis.*
The Latrine fly, *Fannia scalaris.*

The Blow-flies and Blue- or Green-bottle flies, *Calliphora* spp., *Lucilia* spp. and *Sarcophaga* spp.

Muscina stabulans.

The Root Maggot fly, *Anthomyia radicum.*

The Cheese Maggot fly, *Piophila casei.*

The Drone fly, *Eristalis tenax.*

Thereva sp.

The House-fly, Musca domestica.

In view of the abundance of this species and its habits it is somewhat remarkable that it has not been more frequently recorded in cases of intestinal myiasis. Austen states that only two cases have been brought to his notice; in both cases the larvae occurred in infants. In one case, the larvae of *Musca domestica* were "voided from the alimentary canal of a male infant aged seven months" together with the larvae of *Fannia canicularis.* The larvae were of different ages. Larvae of *M. domestica* have been found by me (1909) in the stools of a child. Cohn (1898) also records the occurrence of the eggs in like material. Nicholson (1910) records three cases of intestinal rectal myiasis[1].

The Lesser House-fly, Fannia canicularis, *and*
the Latrine fly, F. scalaris.

These two species would appear to be most common in cases of myiasis of the intestinal and urinary tracts. As long ago as 1839 Jenyns recorded the case of a clergyman about 70 years of age, who complained of general feebleness, loss of appetite and a disagreeable epigastric feeling of a tremulous character. These symptoms began in the spring of 1836 and it was not until the autumn that the larvae were observed. They were expelled repeatedly in large numbers and their expulsion in this manner continued for several months. The larvae were about equal in size and extremely active on their appearance. The malady did

[1] Felt (1913) records a case of myiasis caused by the larvae of *M. domestica*. The infestation presumably arose from canned sardines which had probably been left exposed, as eggs and larvae were found in the fish of four out of six boxes examined.

Jones (1913) describes the occurrence of twenty to thirty living larvae of *M. domestica* in the stomach of a fatal case of hepatic abscess. The disease and the occurrence of the larvae was no doubt a coincidence.

not recur and the evacuation of the larvae ceased shortly; the patient's health gradually improved but not completely. The author calls attention to the fact that the symptoms made their appearance in the spring, but the larvae were not expelled until the summer and autumn following. It would appear, therefore, that they entered the stomach in the egg state and after hatching passed into the intestine where they completed their growth. From the description and figures which the author gives of these larvae they would appear to be *F. scalaris* and not *F. canicularis* as was supposed.

In 1876 Judd described the discharge of the larvae of *F. scalaris* from the intestine by a boy in Kentucky, U.S.A.

Stephens (1905) records the passage of two larvae *per rectum*, one of which was described as *F. canicularis* and the other as *Musca corvina*, but from the author's description I am inclined to believe that the latter was *M. domestica*.

The occurrence in the intestine of a ·youth of what would appear to be the larvae of one of these species of *Fannia* is recorded by Cattle (1906). This patient consulted the author in September 1905 and stated that he had passed the larvae a basinful at a time *per anum*. For some weeks he had not been feeling well and now complained of abdominal discomfort. The chief trouble was apparently an imaginative one, induced no doubt by the sight of the living larvae in his faeces. He had no vomiting or other gastric or intestinal symptoms. The larvae gradually left the patient although so late as March 1906 one or two at a time were occasionally seen.

Tulpius (1672) records the passage of 21 small larvae from the urethra. From the figure which is given it would appear that these are *F. canicularis*. In 1792 Veau de Launay recorded the occurrence of and figured a larva which resembles *F. canicularis*.

Chrevil (1909), in an admirable and complete summary of previously recorded cases of myiasis of the urinary tract of which a critical examination is made, gives an additional case of the occurrence of the larva of *F. canicularis* in a woman of fifty-five who suffered from albuminuria and urinated with much difficulty. On May 26th thirty or forty larvae of *F. canicularis* of different sizes were passed.

In a case of the occurrence of *F. canicularis* in the United States reported by Blankmeyer (1907) it is stated that "the symptoms consisted of abdominal pains and distention and bloody diarrhoea followed by constipation." A saline purgative resulted in the passing of a bulky stool which was found alive with the larvae numbering from 1000 to 1500. A few larvae continued to be passed for a few weeks.

Austen refers to the only case of urethral myiasis known to him in England. In this case which was reported by Dr J. F. Palmer to the Chelsea Clinical Society in 1901 a single larva identified by Austen as *F. scalaris* was passed *per urethram* by a male patient.

A case of vaginal myiasis in an old beggar woman is described by Pieter (1912).

Laboulbene (1856) records the rearing of the larvae belonging to this genus from the intestine of a woman who had suffered for some time from stomachic pains with loss of sleep and appetite. On October 12th she took castor oil and after violent efforts and a further dose of an emetic she vomited altogether about seventy larvae. The expulsion of the larvae was followed by a regaining of the appetite and sleep. In 1909 I recorded the occurrence of the larvae of *F. canicularis* in the stools of patients suffering from intestinal disorders. Soltau (1910) has recorded the occurrence at Plymouth on May 28th of the larva of *F. canicularis* in the stools of a man who had not previously had intestinal pains. The occurrence in September 1909 in the faeces of a boy aged 12 of the larvae of a species of *Fannia* has been described by Garrood (1910).

The occurrence of dipterous larvae in the intestine is recorded by Hope in 1840, but in these earlier records it is frequently impossible to determine the species.

The Blow-flies and Blue- or Green-bottle flies, Calliphora *spp.*, Lucilia *spp. and* Sarcophaga *spp.*

The larvae of these species of flies are chiefly sarcophagous feeding upon wounded and ulcerated surfaces in various mammals including man. Cleland (1912) records the occurrence of the larvae of *Calliphora* in an ulcer from a leper. In 1905 some eggs were taken from the stool of a patient suffering from diarrhoea in the Manchester General Infirmary and were sent to me for examination; they proved to be the eggs of *C. erythrocephala.* I am unaware of any other records of the occurrence of the larvae of these flies in the intestinal or urinary tracts.

Muscina stabulans.

Portchinsky (1913) mentions the case of a Russian peasant which was brought to his notice in which, according to the physician's report, the larvae were in the man's body from about November 1909 to March 1910, causing during the whole period great pain and sickness with vomiting. From January 1910 onwards his faeces contained blood. Finally, during the two days following the injection of tannin, about 50 larvae of *M. stabulans* were passed. From Portchinsky's account it would appear that this species is not infrequently the cause of intestinal myiasis.

The Root Maggot Fly, Anthomyia radicum.

The breeding habits of this fly have already been discussed (p. 214). Austen (*l.c.*) mentions a single record of the occurrence of the larvae of this species in the faeces of a child which did not display any symptoms of ill health. The larvae which were passed in the faeces for some days soon disappeared after the administration of castor oil.

The Cheese Maggot Fly, Piophila casei.

Austen (*l.c.*) records three cases of the occurrence in the human body of this fly. These larvae are the well known maggots or "skippers" which are found in cheese. In 1896 a case was reported in London of a woman of forty-nine years of age who had been attending the throat hospital for eighteen months for chronic pharyngitis, etc. For three weeks she noticed a profuse watery discharge from the nose and experienced sharp pains in the left frontal region. The discharge was never purulent. Various lotions were used without success until dilute Mandl solution was employed when four larvae of *P. casei* were discharged. Austen has examined adult flies bred from a case of intestinal myiasis in London, and Rondani records the expectoration of the larvae of this species in Italy by a patient suffering from an affection of the chest.

Paris (1913) records the discharge of larvae of *Piophila* together with those of *Anthomyia* from a young man (20) of Dijon suffering from intestinal haemorrhage and complaining of pains in the anal region. A decoction of absinthe leaves successfully caused the discharge of the larvae, numerous larvae appearing subsequently at regular intervals. A cure was effected by a few weeks' treatment with chloroform enemata and strong doses of thymol.

Mode of Infection.

In the majority of cases of intestinal and urinary myiasis the larvae of the two species of *Fannia* have been found. Less frequently the larvae of *M. domestica* have occurred. The larvae of these species and also of *A. radicum* breed in excrement and decaying vegetable products, and the female flies guided by their sense of smell and impelled by their natural instincts seek such substances. Owing to the fact that these flies are attracted to excrement, decaying, putrefying or purulent substances or matter several methods of infection are rendered possible.

In the case of intestinal myiasis, the flies may have deposited their eggs in or upon rotting or decaying fruit, vegetables or other food which may be eaten in a raw state, and thus the eggs or young larvae will be taken into the digestive tract. Or, the flies which are generally to be found depositing their eggs in the old style privies, may deposit their eggs in or near the anus, especially if the person is somewhat costive. The larvae on hatching, make their way into the rectum and thence into the intestine. This latter mode of infection is probably the common one in the case of infants belonging to careless mothers. Such infants are sometimes left about in an exposed and not very clean condition, in consequence of which flies are readily attracted to them and deposit their eggs.

The infection of the urinary tract is more difficult to understand. The flies are no doubt attracted to the genital apertures by the different albuminous secretions, spermatic, menstrual, gonorrheal or leucorrhoeal. The larvae would feed upon the muco-purulent secretions. It is easier to understand the infection of the urinary tract of a woman rather than that of a man. The case recorded by Chevril indicates fairly clearly how the female urinary tract may be infected by the continued or prolonged exposure of the organ. As the flies are frequently found in bedrooms the infection of both sexes during hot weather is sometimes rendered possible. Infection is chiefly facilitated by uncleanliness and carelessness.

The whole subject of the relation of these flies to myiasis of the intestinal and urinary tracts is one which has received comparatively little attention. Certainly not the attention it deserves on account of the complications incident to such infections that may arise.

Miscellaneous Cases of Myiasis.

In addition to the aforementioned cases of myiasis, which have been ascribed in most cases to definite species of dipterous larvae, medical and other literature contains numerous references to the occurrence of living dipterous larvae, or maggots in human beings[1].

The remarkable occurrence of dipterous larvae in the anterior chamber of the human eye has been recorded by Ewetzky and Kennel (1904) and also by Thomas and Parsons (1908). I was enabled, through the kindness of my friend Dr Shipley, to examine sections of the latter case which showed the larvae in situ. Portchinsky (1913) has described and figured the occurrence of the larva of Hypoderma bovis, the ox warble, in the anterior chamber of the eye which it would appear to have reached by way of the nasal sinus. The occurrence of dipterous larvae in the orbit is also recorded by Keyt (1900) and Gann (1902).

Cleland (1912) has given a number of Australian records. Two cases are given of the occurrence of dipterous larvae in the ear; in 1865 twenty maggots were removed from the ear of a small boy at Castlemaine, Victoria (Austr. Med. Journ. Vol. x. p. 95). Dr G. H. Salter of Ballan, Victoria, described (Austr.

---

[1] Since the above was written Graham-Smith (1913) has given an excellent review of Myiasis caused by non-blood-sucking flies. In addition, the following authors have cited cases: Francaviglia (1912), Balzer, Dantin and Landesmann (1913), Surcouff (1913), Zepeda (1913), Neiva and Gomes de Faria (1913), de Moura (1913), Heckenroth and Blanchard (1913), Sergent (1913), Edgar (1913), Field (1913), Hall and Muir (1913), Candido (1913), Miller (1910).

Med. Gaz. Mar. 1895) four cases of myiasis. An alcoholic woman complained of pain in the eye and a feeling as though pebbles were rolling over the eye-ball. The eye was red and much swollen and the lids more or less glued together with pus and blood. On separating them the space between the lower eyelid and the globe was teeming with maggots, about twenty or thirty in number, from their size apparently two or three days old. The cornea was cloudy and ulcerated in two or three spots and vision was materially impaired two months later. In Gisborne, a few months previously, Salter removed a number of full-grown maggots from the nose of a child aged seven months, the subject of hereditary syphilis. On another occasion the presence of the maggots accounted for a discharge from a child's navel. In a woman with epithelioma of the left temporal region a large number of full-grown maggots were removed which had almost demolished the growth as well as the left lower eye-lid and had found their way into the left orbit. In the same publication Russell of Adelaide, as cited by Cleland, described a case in which twenty-three maggots were removed from between the eye-lids, and a case in which five maggots were extracted from a boy's ear, from which there was a foetid discharge. In describing the case of the maggots from the eye-lids he says : "There was a rounded open ulcer the size of a sixpenny piece at the inner corner of the eye, which had not ulcerated through the whole thickness of the lid. There was much conjunctivitis but no ulceration."

These cases may readily be explained in the light of our knowledge of the feeding and breeding habits of *M. domestica* and its allies responsible for myiasis in man and other animals.

The Rôle played by Flies in the Spread of Intestinal Worms.

It appears to have been an early surmise that house-flies might, owing to their habits of seeking and frequenting excrement, serve as a means of disseminating parasitic worms by either infecting by contact the exterior of their bodies with the eggs of the worms or by ingestion of the eggs in feeding. The human excreta when moist are attractive to flies and the comparatively large size of the eggs of the parasitic worms does not preclude their dispersal by flies as experiments have indicated.

Grassi (1883) appears to have been the first to demonstrate the ability of flies to ingest the eggs. He broke up segments of the common tape-worm (*Taenia solium*) in water; they had previously been preserved in alcohol for some time. Flies sucked up the eggs in the water and he found them unaltered in the faeces of the flies. The eggs of *T. solium* measure, according to Nicoll (1911), ·035 mm. in length and ·025 mm. in breadth. Eggs of *Oxyuris* were also passed unaltered. In another experiment flies were allowed to feed on the eggs of *Trichocephalus* and he found the eggs some hours afterwards in the flies' faeces which had been deposited in the room beneath the laboratory; he also caught flies in the kitchen with their intestines full of eggs.

Nuttall (1899) records a personal communication from Stiles who placed the larvae of *Musca* with female *Ascaris lumbricoides* which they devoured together with the eggs which these large nematodes contained. The larvae and adult flies contained the eggs of the *Ascaris* and as the weather at the time the experiment was carried out was very hot the *Ascaris* eggs developed rapidly and were found in different stages of development in the insect, thus proving, as Nuttall points out, that *M. domestica* may serve as a disseminator of this parasite.

Callandruccio (1906) examined flies which had settled upon faeces containing the ova of the tape-worm *Hymenolepis nana* and the ova were found in the flies' intestines. Flies which had fed upon material containing the eggs of *H. nana* deposited excrement containing these eggs on sugar. Twenty-seven days later the eggs of this tape-worm were found in the stools of a girl who had eaten some of this sugar ; as other possible sources of infection were carefully excluded this experiment clearly demonstrates a method of infection by flies.

Galli-Valerio (1905) found that flies could carry not only the eggs but also the larvae of the American hook-worm *Necator americanus*. He was unable to find either eggs or larvae in the intestines of the flies. Leon (1908) recovered the eggs of the tape-worm *Dibothriocephalus latus* from the excrement of flies which had been fed upon honey with which the eggs of the tape-worm had been mixed.

The most complete and valuable series of experiments on the dispersal of the eggs of parasitic worms of flies has been carried out by Nicoll (1911) in connection with the Local Government Board enquiry. He recognised the isolated nature of the work of previous investigators, and while the account of his experiments is described as being of a preliminary character it contains many results and observations of considerable value. Ten species of parasitic worms were experimented with, namely, the tape-worms, *Taenia serrata*, *T. marginata* and *Dipylidium caninum* from the dog, *Hymenolepis diminuta* from the rat, *Moniezia expansa* from the ox, and the nematodes *Trichuris trichiuris* from man, *Toxascaris limbata* and *Ankylostoma caninum* from the dog, and *Sclerostomum equinum* and *Ascaris megalocephala* from the horse. Of these the following only are human parasites : *T. trichiuris*, *D. caninum*, *H. diminuta*, and *T. limbata*. The main object of Nicoll's experiments was to ascertain, first, to what extent, and second, for how long a period flies could carry eggs by actually ingesting them and retaining them within their intestines, and, third, how great was the partiality they displayed towards feeding on infective material. He points out that the only parasites with which it is possible that flies can directly affect man are *Taenia echinococcus* and *T. solium* together with the following species which do not require an intermediate host:

Class a. Those in which the larval worm remains within the egg-shell.

Ascaris lumbricoides,	size of egg	·060 mm. long,	·045 mm. broad
Toxascaris limbata	„	„	·080 „	„	·070 „	„
Belascaris mystax	„	„	·075 „	„	·070 „	„
Oxyuris vermicularis	„	„	·050 „	„	·020 „	„
Trichuris trichiuris	„	„	·050 „	„	·025 „	„
Hymenolepis nana	„	„	·040 „	„	·040 „	„

Class b. Those in which the larva is liberated from the egg-shell and spends its life in water.

Ankylostoma duodenale,	size of egg	·060 mm. long,	·040 mm. broad
Necator americanus	„	„	·065 „	„	·040 „	„
Schistosomum haematobium „	„	·115 „	„	·045 „	„
Schistosomum japonicum	„	„	·075 „	„	·040 „	„

The feeding habits of the house-fly have already been described in an earlier chapter, but on account of its particular interest in connection with the possibility of the dissemination of parasitic worms by flies frequenting excrement, I have reserved a quotation of Nicoll's own observations on this point for the present section ; these observations have been repeatedly confirmed by myself and doubtlessly by other investigators. He says : "It is a matter of common observation that fresh and moist faeces attract flies much more readily than old dried faeces. Flies feed on warm fresh faeces with considerable avidity, and they will do so even although they have been previously feeding on other material. To flies which have not fed for some time the presence of fresh human faeces acts as an immediate source of attraction, and in some of my experiments the eagerness with which they attacked it was most striking. When the portion of faeces was so small that the flies could not find standing room upon it or around it, they struggled together and pushed each other aside, and more than once I have seen them so closely packed together that each fly could find room only for the tip of its proboscis, the flies on the top practically standing on their heads, supported by the bodies of those around. Their behaviour towards older faeces, however, is very different. When the material has become cold it does not attract flies nearly so readily. So long as it remains moist it continues to attract and does so quite as much as moist bread, although very much less so than moist sugar. When it has become dry it possesses little or no attraction, but this is increased when it is moistened again. It is evident, therefore, that the presence of moisture plays an important part in a fly's attitude towards faeces as an article of food.

"When the alternatives of fresh faeces, sugar and bread were offered, the flies did not confine their attention to any one of these articles but made repeated excursions from one to the other."

He proceeds to describe some interesting observations in regard to flies feeding on segments of tape-worms. Such tape-worm segments may be deposited together with the faeces or independently and in the case of some species of parasitic worms the eggs are conveyed to the exterior in the detached segments instead of being shed singly into the gut. It was found that such detached tape-worm segments possessed a great attraction for flies. When an intact segment of a tape-worm such as *Taenia serrata*, *T. marginata* or *Dipylidium caninum* mixed with moderately fresh faeces was presented to some flies they appeared to select the tape-worm and feed upon it in preference to the faeces. This observation was repeated on several occasions. Further, when an isolated tape-worm segment, some faeces and some sugar were separately introduced into the fly cage, the flies showed a decided preference for the tape-worm, which they attacked with much assiduity. This preference was shown not only when the tape-worm segment was fresh but even when it had lain a day or two.

Flies, it was found, are able gradually to pierce the fairly tough external covering of the tape-worm segments and to extract the internal contents containing the eggs. When flies had been feeding on a tape-worm segment for 5—10 hours their crops were found greatly distended with the white milky juice of the tape-worm and tape-worm eggs were found in the flies' intestines.

Special observations were made on the feeding of the fly larvae on parasitic worms. While it was found that even after a lapse of three or four days, the adult flies were unable to penetrate the thick cuticular investment of round worms such as *Ascaris megalocephala* and *Toxascaris limbata*, such difficulties did not exist in the case of the larvae. Fresh round worms when offered to the larvae were at once attacked, but although they swarmed over the worms, unless the cuticle of the worms was torn or ruptured in some way, they were unable to penetrate them and in the absence of other food they died. On the other hand, if the round worms were cut up or broken before being given to the larvae, the latter devoured the internal parts with great rapidity. Commencing at one end of a broken piece they would eat their way through the soft tissues leaving nothing but the cuticular tube. It was found that within two or three days half a dozen larvae would devour a large worm 20 or 30 times their own bulk. When larvae had fed upon female egg-bearing worms large numbers of eggs were found surrounding them but not actually adhering to them and in the intestines of such larvae no intact eggs were found, but fragments of shells were always visible. No embryonic worms in any stage were found. Nicoll is of the opinion that even full-grown larvae are unable to swallow un-ruptured eggs as large as those of the worms used, namely ·07 mm.

Details are given by Nicoll of the most important of his experiments on the feeding of infective material to flies. The material was offered to the flies in four different ways. 1. Faeces containing ova. 2. Complete worms, or intact parts of them. 3. Broken or damaged segments of worms. 4. Suspensions of ova in water. After feeding flies were examined at varying intervals; the bodies and appendages were first examined; after-wards the alimentary tract was dissected out and examined. The results of his experiments were as follows:

(a)   *Hymenolepis diminuta* (ova ·07 by ·065 mm.).

1.   11 flies were fed on contents of caecum of rat containing numerous eggs. Negative results were obtained.

2.   12 flies fed on rat faeces containing numerous ova. Negative results obtained.

3.   6 flies fed on ripe segments of *Hymenolepis diminuta* containing numerous ova. Negative results.

4.   6 flies fed on emulsion of tape-worm in water, containing numerous ova. Negative results.

These experiments demonstrated the inability of *Musca domestica* to ingest eggs as large as those of *Hymenolepis diminuta*.

(b)   *Toxascaris limbata* (ova ·08 by ·07 mm.).

1.   8 flies fed on dog faeces containing numerous eggs. Negative results.

2 and 3.   Two lots of 6 flies each fed on intact and broken female worms respectively. Negative results in each case.

(c)   *Ankylostoma caninum* (ova ·06 by ·04 mm.).

Negative results were obtained after feeding flies on dog faeces containing ova.

(d)   *Trichuris trichiuris* (ova ·05 by ·025 mm.).

From 12 flies which had been fed on human faeces containing a few ova, the ova of this tape-worm were recovered from one fly and from the faeces of another fly.

(e) *Dipylidium caninum* (ova ·04 by ·04 mm.).

1. 4 *Musca* and 1 *Fannia* were fed on unbroken ripe segments of the worm. The eggs were recovered from 2 flies (*Musca*) 26 hours after the infected faeces had been removed, demonstrating the ability of the flies to suck the eggs of the tape-worm and to carry them for a considerable time.

2. 6 flies were fed on dog faeces containing ova. The ova were found in 4 of these flies; in the case of 2 of the flies 43 hours after the infected faeces had been removed, showing that the flies can carry the eggs in their intestines for at least that length of time.

(f) *Taenia marginata* (ova ·035 by ·035 mm.).

1. The eggs of this species were removed from a blow-fly (*Calliphora*) on the second day after it had been fed on mashed up segments containing ova.

2. Eggs were recovered from *Musca* and *Fannia* nearly three days after they had fed on ruptured segments containing numerous ova.

3. Eggs were recovered from *Musca* the day after it had fed on an emulsion of ripe segments.

(g) *Taenia serrata* (ova ·035 by ·035 mm.).

1 and 2. Eggs were found in flies in very large numbers, 500 in 3 flies and 400 in 2 flies, 2 hours after feeding on ripe segments in water, and on ruptured segments.

3. Eggs were found in a fly 21 hours after feeding on intact segments containing numerous ova.

4. Negative results were obtained from 4 flies fed on dried segments.

5. A single egg was recovered from one of 7 flies 8 hours after they were fed on intact segments containing numerous ova and one egg was found in the flies' faeces about 20 hours after the segments had been removed.

6. From 3 flies which had fed on an emulsion of segments containing numerous ova, 1, 22 and 312 eggs respectively were recorded up to 7 hours after the emulsion had been removed.

200 eggs were recovered from faeces of other flies. 171 eggs were recovered from a fly 24 hours after feeding; these eggs were fed to a rabbit in which 23 *Cysticerci* were afterwards found, showing that the eggs remained capable of infecting for at least one day. Eggs were also recovered from a fly and from faeces 48 hours after feeding.

7. 10 flies were fed on faeces containing ruptured mature segments; a piece of sugar was also introduced and the faeces were kept moist. The results demonstrated the important fact that faeces containing tape-worm segments may continue to be a source of infection from which food, such as sugar, may be contaminated for as long as a fortnight.

A series of experiments was carried out in which fly larvae were fed upon faeces containing tape-worms or ripe segments.

Taenia serrata. 5 eggs were recovered from three larvae two days after they had been placed on dog faeces containing ripe segments but no eggs were recovered from pupae or flies developing from larvae which had been fed on infected faeces.

Toxascaris limbata. Larvae were placed on dog faeces containing mature female worms but no embryos or larval tape-worms were recovered from the fly larvae, pupae or adult flies.

Ascaris megalocephala. Several larvae were put in horse faeces containing female worms with numerous eggs. Negative results were obtained from an examination of the fly larvae, pupae and adult flies. Nicoll points out that these results are at variance with the result of Stiles already quoted, but, as he shows, the eggs of *Ascaris lumbricoides,* which Stiles used, are smaller than those of *A. megalocephala.*

Nicoll also experimentally shows that the well-known habit which flies have of cleaning their proboscides, appendages and bodies after feeding militates very materially against their carrying eggs on the exterior of their bodies. The longest period after which eggs of *Hymenolepis diminuta,* for example, were found adhering to flies was about three hours. During this time, however, they may travel some distance from the source of infection. It was demonstrated that the eggs of this species of tape-worm, which are too large to be ingested by the flies, could be carried externally by the flies and that food (sugar) could be infected.

The careful experiments of this investigator show that, under experimental conditions, the eggs of the following parasitic worms may be carried by *Musca domestica*: *Taenia solium, T. serrata, T. marginata, Dipylidium caninum, Dibothriocephalus latus* (?), *Oxyuris vermicularis, Trichuris* (*Trichocephalus*) *trichiuris*, both internally and externally; *Necator americanus, Ankylostoma caninum, Sclerostomum equinum, Ascaris megalocephala, Toxascaris limbata* (= *Ascaris canis* e.p.), *Hymenolepis diminuta* externally only.

The practical significance of the results of the aforementioned studies of the relation of *Musca domestica* and such of its allies as have similar coprophagous or coprophilous habits is great. The necessity of preventing flies from gaining access to faeces or tape-worm segments is most clearly demonstrated and the possibility of flies infecting food with tape-worm eggs will undoubtedly furnish an explanation in many otherwise obscure cases of infection.

PART VI

CONTROL MEASURES

CHAPTER XXVI

PREVENTIVE AND REMEDIAL MEASURES

THE significance of the house-fly as a carrier of the causative organisms of certain of our most common diseases renders its control fundamentally necessary in any effort towards sanitary reform or in any system of preventive medicine. The disgust which its filthy habits call forth should be a sufficient impulse in the direction of its control; the fact that it can be no less dangerous than the mosquito or the tse-tse fly, should opportunity occur, should merit and demand the attention of all charged with or interested in the care of the people's health. The evidence which has been adduced is more than sufficient to demonstrate that the prevention of many diseases cannot be undertaken with any hope of success so long as this factor in their dissemination is ignored.

PREVENTION OF BREEDING.

Of all control measures this is by far the most important. It is the key of the whole situation. The fly and mosquito problems are essentially similar. Malarial and yellow fever are eradicated by the abolition and protection of the breeding places of the mosquito. Similarly, by the abolition, and protection or treatment of the breeding places of the house-fly its control could be effected and its significance as a disease-carrier nullified.

The study of the breeding habits of the house-fly has indicated the places and materials in which it breeds. The chief breeding places are collections of horse-manure or stable refuse. Flies must not be allowed to have access to such stable refuse. In order to attain this end it must be kept in covered fly-proof receptacles and regularly removed or it must be treated with some insecticidal substance. An increasing number of cities and towns are recognising the importance of the former of these requirements and are passing and, what is more important, enforcing bye-laws regarding the erection of fly-proof pits or chambers for the temporary storage of manure. It is further necessary to have the floor and interior of the stable well constructed; the floor should be of solid masonry or concrete to permit good drainage and thorough cleansing.

Many municipalities have the necessary bye-laws on their statute books but lack the incentive or courage to enforce them. The plain facts already set forth should convince such municipalities of the grave danger to the health of the people they are elected to serve that the presence of breeding places of house-flies constitutes.

A necessary adjunct to the construction of sanitary stables and the storage of the horse-manure in fly-proof receptacles is the regular removal of the manure. It was observed that flies prefer to oviposit in the warm excreta and on this account eggs are often deposited in the manure before it is thrown into the storage receptacle. From such infested manure flies would emerge if it were not removed well within the shortest time that is occupied in the completion of the life-cycle of the fly. This time is shorter in the summer than in the winter. Therefore, during the summer and autumn months, from June to October, the manure should be removed regularly at intervals not exceeding seven days. A large number of towns and cities require its removal twice a week during this period and this is a wise precaution. During the remainder of the year the period may be extended to nine days or three times each month[1].

[1] Levy and Tuck (1913) and Hutchison (1914) propose that advantage should be taken of the habit which the larvae of *M. domestica* have of migrating from the manure to pupate in the soil or in a drier situation. By placing the manure in receptacles from which the larvae can escape through wire gauze sides and bottoms, the larvae can be caught and killed in pans placed beneath such receptacles. By the use of such "maggot-traps" Hutchinson was able to show that 98 or 99 per cent. of the total number of larvae can be made to leave the manure, provided it is kept moist.

While the advent of the automobile has undoubtedly decreased the breeding places of the house-fly very materially and will continue to do so, there remains still much to be done in the way of segregating or localising livery stables in addition to the enforcement of the aforementioned sanitary regulations. This principle is being adopted in certain places and its more general practice would have an appreciable effect on the problem of fly control. One great advantage of such segregation would be the increased opportunities afforded to sanitary inspectors of supervising the proper care and treatment of the stables and of the stable refuse.

Too much stress cannot be laid upon the necessity of prohibiting the storage of stable refuse and other breeding materials of the house-fly (see p. 94) in such places as railway depots, canal wharves and similar places pending its removal. In the majority of cases outbreaks of flies have been traced to the adoption of such practices and such an abundance of flies may have dangerous consequences.

As an alternative to, but preferably in conjunction with the storage of horse-manure in fly-proof receptacles, the treatment of the manure with an insecticidal substance with a view to destroying any house-fly larvae can be adopted with very marked success.

In 1897 Howard conducted a series of experiments with a view to discovering an insecticidal substance which could be used for the destruction of the larvae in the heaps of manure in which they were breeding. He found that both lime and gas lime were not efficacious. In an experiment in which 8 lbs. of horse-manure containing larvae were treated with a pint of kerosene, which was washed down into the manure with water, it was found that all the larvae were killed. He also found that by treating 8 lbs. of well-infested horse-manure with one pound of chloride of lime all the larvae were killed, but the results were not satisfactory when a quarter of the quantity of chloride of lime was used. On experimenting with the kerosene treatment on a large scale he found that it was not only laborious but also not entirely successful, as is sometimes the case in the practical application on a large scale of successful experimental methods. He, therefore, devised another method of treating the horse-manure of stables. A chamber six feet by eight feet was built in the corner of the stable with which it communicated by means of a door; it was provided also with a window furnished with a large screen. The manure was thrown into the chamber every morning and a small shovelful of chloride of lime scattered over it. At the end of ten days or a fortnight the manure was removed through an open door and carted away. The experiment was carried out in the stable of the U.S. Department of Agriculture and a marked decrease in the number of flies was observed.

That chloride of lime can be used with beneficial results and on a large scale has been -demonstrated by one of my correspondents, Messrs McLaughlin, Bros., lumber manufacturers of Arnprior, Ontario, Canada. In response to my request for a brief statement of their experience they write (April, 1913) as follows:

"Four years ago we began using chloride of lime in our stables here to keep down house-flies. There are about 120 horses in the stables at night and most of them are brought in to feed at noon. The manure is removed every morning and again in the afternoon when the mid-day feed is over. After each cleaning-up, chloride of lime is taken in in a shovel and scattered lightly just outside each stall. It must not be put too close in, for if the horse lies on it it will burn the hair off him. The risk of this accident is, however, very slight, as during the four years we have used the lime, it occurred only once to one horse and a very little care will obviate it entirely. The smell of the chlorine given off by the lime while it is scattered in the stables doubtless tends to drive away the flies, and as the manure and lime are swept up together, they are well mixed, and the chlorine having thus a good chance to exercise its germicidal properties, a large proportion of the flies' eggs and larvae are no doubt destroyed by it.

"It is, of course, impossible to determine with accuracy the decrease in the number of flies due to the use of the chloride of lime, but our men are satisfied that the reduction must be about 75 per cent. One striking evidence of the diminution is the comparative quietness of the horses now at night. A few years ago, when passing the stables on a warm summer night, one was astonished at the noise caused by the never-ceasing tramping of the fly-pestered animals. Since we began to use chloride of lime, this noise has practically ceased.

"We use about eight or ten pounds of the lime a day, or about 1000 lbs. during the fly season. Bought by the barrel of some 300 lbs. it costs 2 cents per lb. in Montreal."

The above statement has been given in full as it affords striking evidence of the practical use of chloride of lime and its effect, not only from an insecticidal standpoint but as a means of securing for the horses the rest they need and usually deserve, the value of which rest is evident.

I am informed by chemists that it is doubtful whether the admixture of a small quantity of chloride of lime would seriously affect the manurial properties of the stable-manure. Exact information on this point, however, has not yet been secured.

Howard (1911) records the results of experiments with kerosene as an insecticide. While it was found that on a small scale if eight quarts of horse-manure were sprayed with one pint of kerosene which was afterwards washed down with a quart of water all the larvae were killed; on a larger scale it was not wholly successful; "a considerable proportion of the larvae escaped injury." Even had all the larvae been destroyed the cost of the treatment would be prohibitive.

Herms (1911) suggests that "when the manure pile can be spread out to a depth of about half a foot it may be drenched with a distillate petroleum, which possesses a high flash point, i.e. does not ignite easily, and has the necessary insecticidal property. The petroleum oils, sold as proprietary compounds on the market as 'miscible oils,' 'spray emulsions,' and the like, should be applied at the rate of one part of the oil to ten parts of water. If kerosene oils of a low flash point are used about stables and out-buildings the danger from fire must be considered."

On account of the danger incident to its use and the cost of the treatment, apart from the fact that its employment is not supported by our experimental results, one is compelled to hesitate in recommending the general use of kerosene or paraffin oil for the destruction of fly larvae.

Forbes had a series of experiments carried out in Illinois, U.S.A., on the destruction of the larvae in manure heaps. These are recorded by Howard (1911). In one series hydrated high calcium lime was used. When three pounds of this lime were mixed with fifteen pounds of horse-manure, ninety-four per cent. of the larvae were killed; two pounds of lime mixed with twelve pounds of manure killed sixty-nine and one-tenth per cent. of the larvae: four pounds mixed with twelve pounds of the manure killed sixty-one and three-tenths per cent. The diminished percentage in the last two experiments was accounted for by the fact that the larvae were nearly full-grown. The most successful results were obtained by the use of iron sulphate. It was found that the breeding of the house-fly in horse-manure could be effectively controlled by spraying the manure with a solution of iron sulphate. A solution of two pounds of iron sulphate in one gallon of water for each horse per day was used. It was calculated that the average city horse produces about fifteen pounds of manure per day and the heavier draught horses produce twenty to thirty pounds per day. The amount to be treated is, of course, much less than this as the horses are out of the stables a large proportion of the day. The average cost of treatment in Illinois would work out at less than one penny (one and one-half to two cents) per horse per day. Not only is the iron sulphate stated to deodorize the manure but it does not injure its manurial properties. In fact, it is extremely probable that it increases the fertilising properties of the manure. Unfortunately, at the time of writing we have no experimental evidence as to the effect of this and other insecticides on the fertilising value of the manure. It is anticipated that investigations on this aspect of the question will be carried on[1].

It is claimed that the treatment of the manure with equal parts of acid phosphate and kainit will repel the flies and prevent their oviposition. Such treatment would certainly increase the fertilising value of the manure.

One is frequently asked how the farmer is to undertake the control of the house-fly about his premises when he is compelled to

[1] Since the above was written I have carried out a series of experiments in August and September, 1913 (see *Journ. Econ. Ent.*, vol. 7, p. 281), with a view to obtaining information on this aspect of the problem. It was found that the greatest mortality was produced by chloride of lime scattered on the manure as it was piled ; this was more fatal than iron sulphate solution.

Dr L. O. Howard has kindly permitted me to secure the results of a series of experiments carried out by the Bureau of Entomology, assisted by the Bureau of Chemistry, of the U.S. Department of Agriculture, Washington, D.C., during the summer and fall of 1913 to discover an insecticide which was not only cheap enough for the farmer to use but which also did not injure the fertilising properties of the manure. It appears that borax (sodium borate in the crude form) used either in the powdered form (2 lbs. to 8 bushels of manure) or in solution (one-eighth of a pound to one gallon of water, using 40 quarts of the solution to 8 bushels of manure) gives the best results. The full report of this investigation will be published shortly by the U.S. Department of Agriculture, to which important report the reader should refer.

store large quantities of manure in the stable yards. The difficulties of the farmer's case have been somewhat unnecessarily magnified owing largely to a failure to appreciate the facts incident to the storage of manure. Experiments which my colleague, Dr F. T. Shutt, Dominion Chemist of the Canadian Department of Agriculture, has carried out have conclusively shown that from a fertilising standpoint it is a greater advantage to haul the manure directly on to the soil than to store it in the stable yard where a considerable proportion of its value is lost by leaching and other processes. In the control of the house-fly on the farm the most effective measure which can be adopted is the immediate hauling of the manure on to the land. If this is done flies will not breed in the manure as scattered manure, owing to its desiccated state, does not readily permit their breeding. The increase in the fertilising value of the manure thus treated serves as an additional reason for the more general adoption of the practice. If it cannot be scattered on the land at once it is an advantage to compost it in heaps and cover the same with a layer of soil. Flies are attracted to and deposit their eggs in fresh manure. Rotted and cold manure does not attract and only in exceptional cases have I ever found flies breeding in such manure either under natural or experimental conditions[1].

The insanitary privy is the greatest menace to the public health. Not only does this mediaeval institution, when not properly cared for, serve as a favourite breeding place for flies but it is the commonest source of infection. Fortunately, it cannot remain long; medical officers of health are unanimous in its condemnation and the more general institution of water-carriage systems is having a pronounced effect on its abolition. The majority of people have no conception of the state of affairs in regard to the occurrence and conditions of insanitary privies. Excellent testimony on the subject in relation to infantile mortality is given by Newsholme (1910), to whose valuable report the reader desiring further information is referred. In his general summary

---

[1] Washburn (1912) suggests the spraying of piles of horse manure with a solution of 8 ounces of sodium arsenite in about 20 gallons of water to which about half-a-pint of treacle has been added. A similar solution used to kill grasshoppers was found very attractive to flies.

Newsholme states "Infant mortality is highest in those counties where, under urban conditions of life, filthy privies are permitted, where streets and yards are to a large extent not 'made up' or paved." In his recommendations he says "Sanitary authorities in compactly populated districts should decide to remove all dry closets if a water-carriage system is practicable."

For the dry closet to be maintained in a sanitary condition great care and attention is required. It cannot be rendered fly-proof, however, owing to the ability of flies to emerge through the soil after having developed from eggs deposited on the faeces before they were covered.

DESTRUCTION OF REFUSE.

It is a remarkable fact that in many of our towns and cities, professedly progressive in sanitary measures, public "dumps," "tips" and garbage heaps are maintained. Frequently, they are located on a vacant piece of land surrounded by thickly populated districts in which there is almost invariably a plague of flies during the summer months. On many occasions during my investigations I have examined such accumulations of organic and other rubbish and have found flies breeding in vast numbers. While such heaps contain a large proportion of mineral matter, such as clinkers, cinders and ashes, etc., they invariably include cart-loads of organic rubbish of various kinds, especially domestic refuse, in which flies can breed (see p. 94). Not only do they breed but they also infect themselves with putrefactive and some-times pathogenic bacteria. The greatest danger of such heaps, however, lies in their fly-productive rather than infective character. Cities and towns can no longer afford to permit these methods of disposing of organic refuse. Its prompt destruction by means of an incinerator is the only measure which will prevent its being a public nuisance and a serious breeding ground for flies. Further, the storage of the domestic organic refuse by the householder in fly-proof receptacles frequently and regularly emptied or removed by the civic authorities must necessarily form an integral part of any sanitary system of refuse disposal.

The Protection of Infants and the Sick.

By their helpless nature and by the force of circumstances infants are peculiarly subject to the attentions of flies. One has only to visit the populous districts of our towns and cities, where the poorer classes are compelled to live, to observe the advantageous conditions under which flies are able to disseminate infection. All too insufficient attention is paid to the careful disposal of human excreta which occur in alleys and odd corners; privies are largely unprotected and the children lie and crawl about. One is impressed by the possibility of the fly-carriage hypothesis of summer diarrhoea and the frequent occurrence of intestinal myiasis becomes no longer surprising. The protection of such infants from the attentions of flies is absolutely essential under such conditions. Nor is it much less essential under most circumstances in the summer when infants are wheeled around in their carriages, subject to the attentions of flies whose previous visitations may have been of the foulest description.

The protection of sick persons, the exclusion of flies from the sickroom or the hospital, are precautions of so obvious a nature in the light of our knowledge of the dangers incurred by the presence of flies, that it should be unnecessary to do more than indicate the necessity for such protective measures.

Protection of Food.

In the home, and where it is exposed for or preparatory to sale, food which is liable to become infected by virtue of its being attractive to flies should be protected. Certain foods such as milk, cooked meats, etc., furnish excellent media for the transference of micro-organisms. Not infrequently one is filled with disgust at the sight of such eatables as cakes, confectionery and fruit liberally fouled with fly specks, the significance of which I have already discussed in a former chapter. The exposing of food on the street, in the dairy or cowshed, and in the shop should be prohibited, and it is an encouraging sign to find that sanitary authorities in many of our towns and cities both in Europe and America are not only passing bye-laws to this effect but are enforcing such requirements by fining the offenders.

THE DESTRUCTION OF ADULT FLIES.

Many means have been devised whereby the adult flies may be destroyed and, while the prevention of their breeding constitutes the fundamental principle of their control, the destruction of the flies must naturally form a part of any system of eradication.

The methods of destruction may be divided into two classes: trapping and poisoning.

Trapping.

There are on the market and in use a great variety of fly-traps the majority of which have proved successful. Such forms as the glass fly-traps baited with beer and the balloon wire trap baited with any attractive bait are well-known. The sticky fly papers, ribbons and wires need no description.

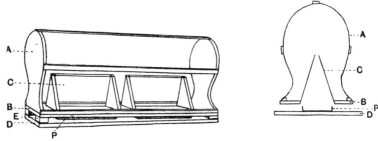

FIG. 101. The Minnesota Fly Trap. General view and cross section.
(After photograph by Washburn.)

Parrott has devised a trap which consists of a shallow tin box having sides about three-quarters of an inch deep; it is made long enough to fit in the bottom of the window pane. After these shallow boxes have been fitted they are filled about two-thirds full of some insecticidal substance such as paraffin or kerosene, or an emulsion of the same, or a sticky mixture of equal parts of castor oil and resin made by boiling the two constituents together.

The "Minnesota Fly Trap" which has been devised by Washburn (1912) has proved successful in the trapping of flies in large numbers. On the back porch of a dwelling, not far from a stable where a few horses are kept, 8700 flies were caught in two days, 12,000 in one day and 18,000 in one and a half days. The trap is illustrated in fig. 101, and the following is a description of its construction: It measures twenty-four inches long, twelve inches high and eight inches wide. The screen used for the large receptacle A and the roof-shaped entrances C is ordinary wire fly or mosquito screen. The upper portion A serves as a receptacle for the captured flies; it rests on a board B which carries two roof-shaped entrances C open at the top to permit the entry of the flies into A. The middle board B rests on a base board D from which it is separated by means of small pieces of wood E about half an inch in thickness fastened at the corners as shown. Upon the base board D rest two tin bait-pans P, separated by about one-quarter of an inch from the middle board B. The trap may be baited with any attractive substance such as meat scraps, raw fish scraps, bread and milk, etc. The captured flies may be killed by immersing the trap in hot water or by pouring boiling water over it.

Hodge (1911) has taken advantage of the attractive power which the garbage or refuse can has for flies and has devised a trap which can be attached to the lid of the garbage can; the lid is so arranged that it does not fit tightly on to the can but a quarter of an inch space permits the flies to enter and on leaving they naturally fly upwards and enter the trap which consists of an ordinary balloon wire fly-trap. In such a trap baited with attractive food such as fish heads, etc., he has caught as many as 2500 flies in fifty minutes. He has also made another device consisting of overlapping screens of wire-gauze (see Hodge, 1913) which can be fitted to the window of the stable or cowshed and catches the flies as they leave the same.

Poisoning.

The arsenical fly-papers or pads were among the earliest fly poisons.

Formalin or formaldehyde has been employed in recent years with marked success in most cases. Sometimes complete failure has been reported, but in such cases I have often found that it was not correctly used. R. J. Smith (1911) has demonstrated, and it has been subsequently confirmed, that if sweet milk is added to the formalin it proves very attractive to flies and the mixture makes an excellent and fatal bait. The solution is made as follows: one ounce or two tablespoonfuls of forty per cent. formalin is mixed with sixteen ounces, that is, one pint of equal parts of milk and water. If this mixture is exposed in shallow plates in the middle of each of which a piece of bread is placed for the flies to alight upon, the flies will be attracted to the solution and poisoned. The formalin has also the advantage of being a disinfectant. Houston (1913) has used the following method of application in the jail kitchen at Rajkot, India, with great success. Instead of exposing the mixture in shallow plates it is sprinkled about the room in tiny pools of one quarter to one inch in diameter from which pools the flies readily drink it. The substitution of buttermilk for the ordinary milk has also been suggested.

Berlese (1913) has for two years carried out observations and experiments on the control of flies at S. Vincenzo (Pisa) by means of poisoned baits, deriving the idea from his work on the control of the olive fruit-fly by means of sweetened arsenical solutions. He rightly emphasizes the importance of destroying the flies outside the houses and near the breeding places where they congregate. He sprayed plants in the gardens and orchards near dwellings, and also heaps of manure and other likely breeding places with the following mixture: 10 parts of treacle, 2 parts of arsenite of potash or soda and 100 parts of water. The operation was performed every ten days and repeated after rain, and the manure heaps were sprayed when a fresh surface was exposed. In 1912 he experimented with small bunches of straw suspended for protection from the weather under conical zinc covers. The straw was dipped in the following mixture: honey, 1 part; treacle, 1 part; sodium arsenite, $\frac{1}{2}$ part; water, 10 parts. These baits were hung round the houses in such places as the porches and verandas. Berlese states that by the use of these methods he succeeded in totally destroying the flies in each of the two years during the period of his residence in the village. By the perfection of this method, which enabled him to enjoy a flyless meal, he believes that it would prove to be a thoroughly practical method of dealing with the fly nuisance in larger places.

The burning of pure and fresh pyrethrum powder and also the dropping of twenty to thirty drops of carbolic acid on a hot iron plate or shovel have been recommended as means of ridding rooms of flies. As the fumes do not always kill all the flies but only stupefy a certain proportion, it is important that the flies should be swept up and burnt before they have an opportunity to recover[1].

[1] In a paper on "Experiments with House-fly Baits and Poisons," read by A. W. Morrill of Phoenix, Arizona, before the Amer. Ass. of Economic Entomologists at Atlanta, Ga., U.S.A., on 2nd Jan., 1914, the author described the results of tests of various chemical and fruit baits. He found that the attractiveness of formalin varied from day to day, that formalin with vinegar, and vinegar alone, were excellent baits, and that potassium bichromate, which is sometimes recommended, has little value as an insecticide. The discussion on this paper elicited various experiences among which were: that the formalin bait is more successful if water is absent from the room and that a mixture of sour milk and formalin sometimes gives good results.

CHAPTER XXVII

ORGANISED EFFORT IN CONTROL MEASURES

WHILE individual effort and example will materially alleviate the fly nuisance and minimize the danger, the problem can only be attacked by corporate and coordinated action. In the previous chapter I have endeavoured to show that the problem of fly control is fundamentally one of good sanitation. Where cities or towns have adopted the necessary standards of sanitation which the health of the community demands, the fly problem hardly exists. In support of this fact I will quote one of many examples which might be given, namely, the report of G. N. Ifft, United States Consul at Nuremburg, Germany. He states[1]: "There are so few flies in Bavaria that they can in no way be regarded as a pest. This is perhaps due to the extreme cleanliness of Bavarian cities. Courtyards, alleys, vacant lots, &c., are kept clean and the hallways and entrances to houses are as fresh as soap and water can make them. There are no quarters that could be justly designated as slums, not even in districts where buildings hundreds of years old are the rule. Garbage is collected in closed tin or zinc cans and regularly removed in closed waggons in such a manner as to be inoffensive to either sight or smell." Of how many cities in other parts of the world could the same be said ? Nevertheless, the last few years have witnessed an honest effort in the right direction on the part of local authorities, in spite of opposition from those who are interested in maintaining stables in thickly populated sections of cities and towns. It is a singular fact that prominent among such opponents to reforms relating to

---

[1] *Daily Consular and Trade Reports, Bur. of Manufactures, U.S. Dept. Commerce and Labour, Washington.* 12th March, 1912. 15th year, No. 60, p. 1031.

HOW TO DEAL WITH THE
FLY NUISANCE.

House flies are now recognized as **MOST SERIOUS CARRIERS OF THE GERMS OF CERTAIN DISEASES** such as typhoid fever, tuberculosis, infantile diarrhœa, etc.

They infect themselves in filth and decaying substances, and by carrying the germs on their legs and bodies they pollute food, especially milk, with the germs of these and other diseases and of decay.

NO FLY IS FREE FROM GERMS.

THE BEST METHOD IS TO PREVENT THEIR BREEDING.

House flies breed in decaying or decomposing vegetable and animal matter and excrement. **THEY BREED CHIEFLY IN STABLE REFUSE.** In cities this should be stored in dark fly-proof chambers or receptacles, and it should be **REGULARLY REMOVED WITHIN SIX DAYS** in the summer. Farm-yard manure should be regularly removed within the same time and either spread on the fields or stored at a distance of not less than a quarter of a mile, the further the better, from a house or dwelling.

House flies breed in such decaying and fermenting matter as kitchen refuse and garbage. Garbage receptacles should be kept tightly covered.

ALL SUCH REFUSE SHOULD BE BURNT OR BURIED within a few days **BUT AT ONCE IF POSSIBLE. NO REFUSE SHOULD BE LEFT EXPOSED** If it cannot be disposed of at once it should be sprinkled with chloride of lime.

FLIES IN HOUSES.

Windows and doors should be properly screened, especially those of the dining room and kitchen. Milk and other food should be screened in the summer by covering it with muslin ; fruit should be covered also.

Where they are used, especially in public places as hotels, etc., spittoons should be kept clean, as there is very great danger of flies carrying the germs of consumption from unclean spittoons.

Flies should not be allowed to have access to the sick room, especially in the case of infectious disease.

The faces of babies should be carefully screened with muslin.

TO KILL FLIES IN HOUSES.

Mix two tablespoonfuls (one ounce) of 40 per cent. Formalin, (a solution which may be obtained from any drug store at about 40 cents per pound bottle) (20 ounce), with one pint (sixteen ounces) of equal parts of milk and water. This mixture should be exposed in shallow plates, and a piece of bread placed in the middle of each plate will enable the flies to alight and feed. All dead flies should be swept up and burnt. The burning of pyrethrum in a room, preferably at night, is sometimes effective : the flies should be swept up and burnt, as many are only stupefied by this substance.

HOUSE FLIES INDICATE THE PRESENCE OF FILTH IN THE NEIGHBOURHOOD OR INSANITARY CONDITIONS.

Copies of this circular, printed on paper or card, may be had on application to the

DOMINION ENTOMOLOGIST, CENTRAL EXPERIMENTAL FARM, OTTAWA.

DEPARTMENT OF AGRICULTURE, CANADA.

(Published by direction of the Hon. MARTIN BURRELL, Minister of Agriculture.)
(Revised edition, 2nd April, 1913.)

FIG. 102. Reproduction of card prepared by the author and widely distributed by the Canadian Department of Agriculture. (Actual size 12 in. × 8½ in.)

the control of the fly pest should be the vendors of bread, milk and other articles of food. Yet such has been the case in my experience. Their voices, however, are becoming less powerful in the council chambers where the stigma of insanitary conditions has so long been disregarded, and the increasing number of convictions for maintaining nuisances and exposing milk and food supplies to the attentions of flies and the dust of the street are evidences of an awakening of the public conscience to its duty.

While individuals or small numbers of people acting in cooperation are seriously handicapped if they cannot look to the local health authorities for assistance, the converse is equally true, for without the sympathy and the support of the people whom they serve the health authorities cannot succeed in their endeavours to ameliorate the sanitary conditions of the people. The zealous efforts of both individuals and of those in authority are essential to success.

It cannot be denied, however, in the light of what we now know in regard to the habits of the house-fly and the effect of its presence in numbers on the health of the community, that health authorities should not only enact the necessary bye-laws that will enable them to deal satisfactorily with the breeding places of the house-fly such as stable refuse, insanitary privies, collections of organic refuse and the protection of food supplies, but that they should enforce the same. There are signs that the time is slowly approaching when one's senses will cease to be so distracted with the reiterated statement that the nation's greatest asset is the health of the people ringing in one's ears while filthy fly breeding spots meet one's gaze and foul odours greet one's sense of smell. The conscientious health officer is becoming more powerful than the vote-seeking "representative of the people."

Educational work is the most potent factor in dealing with this question. This should begin in the schools. It is unnecessary to change the curriculum but it *is* necessary that the teachers should know the facts and know them correctly. The lessons can be given as nature study or hygiene or both, and experience has shown that children not only quickly appreciate the significance of the fly but are singularly active in making practical use of their knowledge. Such an organisation as the Boy Scouts can be

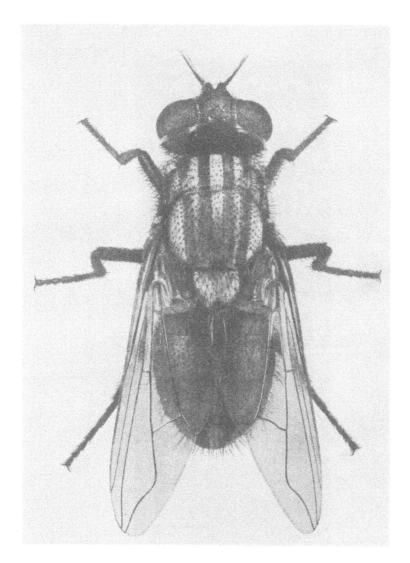

Fig. 103. Photograph of enlarged model of the House-fly, *Musca domestica*, in the American Museum of Natural History, New York.

(Reproduced by kind permission of the Museum authorities.)

made a powerful adjunct to an anti-fly campaign, as I have found in Canada and as others have found elsewhere. They can assist

Fig. 104. Photograph of enlarged model of the larva of the House-fly, *Musca domestica*, in the American Museum of Natural History, New York.

(Reproduced by kind permission of the Museum authorities.)

the sanitary authorities in locating the breeding grounds and in their own districts they can abolish the same.

Not less important is the work that can be accomplished by women's organisations. The women of the community suffer most from the fly pest. It is women who have to wage a constant war to protect the food in the home and the infants in the cradles. Women's organisations can do much to secure the adoption of anti-fly and sanitary measures and in many places they are succeeding in making the community more healthy. The boycotting of shopkeepers who do not protect the food they sell or of milk vendors whose premises are breeding grounds for flies is a cogent method of securing reform.

Citizens' organisations should begin campaigns for clean cities. Such organisations are able to render invaluable service to the health officers by cooperating with them in the locating and suppression of the breeding grounds. By lectures, by the distribution of circulars such as the one illustrated herewith (fig. 102) which I prepared four years ago, and of other literature, much good can be done. The press has shown its willingness to assist, and the aid of local newspapers should be enlisted. An excellent and educative cinematograph film illustrating the life-history of the house-fly and its method of spreading disease has been prepared and in my own experience has accomplished splendid work.

The statement cannot be too often repeated that flies and filth are synonymous terms. In a clean and sanitary community flies will be unable to exist in dangerous numbers and their absence may be taken as a measure of cleanliness. The time is coming when men will realise that it is easier to prevent disease than to cure it and less costly in terms of human lives.

BIBLIOGRAPHY

ABEL, R. "Einige Ergänzungen zu der in No. 5—12 dieser Zeitschrift erschienenen Abhandlung von Nuttall über die Rolle der Insekten bei der Verbreitung von Infektionskrankheiten des Menschen und der Thiere." Hygienische Rundschau, Berlin, vol. 9, pp. 1065—1070, 1899.

AINSWORTH, R. B. "The House-fly as a Disease Carrier." Journ. Roy. Army Med. Corps, vol. 12, pp. 485—498, 1909.

ALBERT, H. "Rôle of insects in transmission of disease." New York Med. Journ., vol. 81, pp. 220—225, 1905.

ALCOCK, A. Entomology for Medical Officers. (London : Gurney and Jackson), 347 pp., 136 figs., 1911.

ALDRICH, J. M. A Catalogue of North American Diptera. Smithsonian Misc. Coll., vol. 47, 680 pp., 1905.

ALDRIDGE, A. R. "The Spread of the Infection of Enteric Fever by Flies." Journ. Roy. Army Med. Corps, vol. 3, pp. 649—651, 1904.

—— "House-flies as Carriers of Enteric Fever Infection." Ibid., vol. 9, pp. 558—571, 1907.

ALTAMIRANO, F. "Investigaciones sobre la putrefacción de los Zurrones de los moscos." Ann. d. Inst. Med. Nat., Mexico, vol. 2, pp. 99—101, 1896.

ANDERSON, J. F. "The Differentiation of outbreaks of Typhoid Fever due to infection by water, milk and contact." Medical Record (Nov. 28, 1908), vol. 74, p. 909 ; also in American Journ. Pub. Health, vol. 19, pp. 251—259.

ANDERSON, J. F. and FROST, W. H. Transmission of Poliomyelitis by means of the Stable-fly, Stomoxys calcitrans. Public Health Reports, U.S. Public Health Service, vol. 27, pp. 1733—1735, 1912.

—— Poliomyelitis. Further attempts to transmit the disease through the agency of the Stable-fly (Stomoxys calcitrans). Public Health Reports, U.S. Public Health Service, vol. 28, No. 18, pp. 833—837, 1913.

ANDOUIN, V. "Recherches Anatomiques sur le thorax des Animaux Articulés et celui des Insectes Hexapodes en particulière." Ann. Sci. Nat. Zool., vol. 1, 1824.

ANDRÉ, C. Flies as agents in the dissemination of Koch's bacillus. (Paper before Anti-Tuberculosis Congress, Washington, D.C.) 1908.

ANTHONY. "The Suctorial Organs of the Blow-fly." Monthly Microsc. Journ., vol. 11, p. 242, 1874.

ARMSTRONG, H. E. House-flies and Disease ; and the duties of Sanitary Authorities in relation thereto. Special Rept. by Med. Officer of Health for Newcastle-upon-Tyne, 1908.

ARMSTRONG, D. B. Flies and Diarrhoeal Disease. Publication No. 79, New York Ass. for Improving the Conditions of the Poor, Bur. Public Health and Hygiene of Dept. of Local Welfare, New York, 29 pp., 3 pls., 1914.

—— "The House-fly and Diarrhoeal Disease among Children." Journ. Amer. Med. Ass., vol. 62, pp. 200—201, 1914.

AUSTEN, E. E. "The House-fly and Certain Allied Species as Disseminators of Enteric Fever among Troops in the Field." Journ. Roy. Army Med. Corps, vol. 2, pp. 651—668, 2 pls., 1904.

—— Illustrations of British Blood-sucking Flies, 74 pp., 34 col. plates. Brit. Mus. (Nat. Hist.), London, 1906.

—— Blood-sucking and other flies known or likely to be concerned in the spread of disease. In Allbutt's and Rolleston's System of Medicine, vol. 2, pp. 169—186, 1907.

—— "Notes on flies examined during 1908. How to distinguish the more important species of flies found in houses." Report Local Govnt. Bd. on Pub. Health and Med. Subjects, N.S. No. 5. Prelim. Repts. on Flies as Carriers of Infection, pp. 3, 4, 1909.

—— "Some Dipterous Insects which cause Myiasis in Man." Trans. Soc. Trop. Med. and Hygiene, vol. 3, pp. 215—242, 1910.

—— "Memorandum on the result of examinations of flies, etc., from Postwick Village and refuse deposit ; with a note on the occurrence of the Lesser House-fly at Leeds." Repts. Local Govnt. Bd. on Publ. Health and Med. Subjects, N.S. No. 53 ; Further Repts. (No. 4) on Flies as Carriers of Infection, pp. 11, 12, 1911.

—— "British Flies which cause Myiasis in man." Repts. Local Govnt. Bd. Publ. Health and Med. Subjects, N.S. No. 60 ; Further Reports (No. 5) on Flies as Carriers of Infection, pp. 5—15, 1912.

—— "The House-fly as a danger to Health. Its life-history and how to deal with it." Brit. Mus. (Nat. Hist.), Econom. Ser. No. 1, 11 pp., 7 figs., 1913.

AXENFELD, T. The Bacteriology of the Eye. (Translated by A. MacNab.) London, 402 pp., 87 figs., 3 pls, 1908.

AYLOTT, W. R. "Do Flies spread Tuberculosis?" Virginia Med. Semi-Monthly, June 26, 1896 ; also in Amer. Monthly Micros. Journ., Aug., 1896.

BACHMANN. "Ein Fall von lebenden Fliegenlarven im menschlichen Magen." Deutsche med. Wochenschr., vol. 24, pp. 193, 194.

BACHMETJEW, P. Experimentelle entomologische Studien. i. Temperatur-verhältnisse bei Insekten. Leipzig, 160 pp., 1901.

BACOT, A. W. " On the persistence of Bacilli in the gut of an insect during metamorphosis." Trans. Ent. Soc., Lond., 1911, pp. 497—500.

——— " The persistence of *Bacillus pyocyaneus* in pupae and imagines of *Musca domestica* raised from larvae experimentally infected with the bacillus." Parasitology, vol. 4, pp. 68—73. With addendum by J. C. G. Ledingham, pp. 73, 74, and an editorial note by G. H. F. Nuttall, p. 73, 1911.

BALFOUR, A. Third Report of the Wellcome Research Laboratories, Gordon College, Khartoum, pp. 218, 219.

BALZER, F., DANTIN and LANDESMANN. " Un cas de Myiase rampante due à l'*Hypoderma bovis*." Bull. Soc. Française de Dermatol. et de Syphiligraphe, vol. 24, pp. 219—226, 1913. Abstr. in Trop. Dis. Bull., vol. 2, (Nov. 15), pp. 526—527, 1913.

BANKS, N. " A Treatise on the Acarina or Mites." Proc. U.S. Nat. Mus., vol. 28, pp. 1—114, 201 figs., 1905.

——— " The Structure of Certain Dipterous larvae with particular reference to those in human Foods." Bull. No. 22, Techn. Series, Bur. Ent., U.S. Dept. Agric., 44 pp., 8 pls., 1912.

BARLOW, J. " The House-fly or Typhoid Fly." Bull. Extension Dept., Rhode Island State Coll. No. 3, 12 pp., 1912.

BATTERSBY, J. C. " Waterborne typhoid." Brit. Med. Journ., Part II, 10th Aug., 1895, p. 393.

BAUWERKER. (A campaign against flies.) Zeitschrift für Gestütkunde und Pferdezucht, vol. 8, pt. 6, pp. 121—129, June, 1913.

BECHER, E. F. " Insects as carriers of disease." Cheltenham Proc. Nat. Sci. Soc., N.S. No. 1, pp. 73—92, 1908.

BECKER, R. " Zur Kenntnis der Mundteile und des Kopfes des Dipterenlarven." Zool. Jahrb. (Anat.), vol. 29, pp. 281—314, 3 pls., 5 figs., 1910.

BÉCLARD, J. " Influence de la lumière sur les animaux." C. R. de l'Acad. d. Sc., vol. 56, pp. 441—453, 1858.

BERG, C. " Sobre los enemigos pequeños de la langosta peregrina *Schistocerca paranensis* (Burm.)." Com. Mus. Buenos Aires, vol. 1, pp. 25—30, 1898.

BERGEY, D. H. " The Relation of Insects to the dissemination of disease." New York Med. Journ., vol. 85, pp. 1120—1125, 1907.

BERLESE, A. " L' accoppiamento della Mosca domestica." Rev. Patolog. vegetale, vol. 9, pp. 345—357, 12 figs., 1902.

——— Gli Insetti. (Milan), vol. 2 (p. 18), 1912.

——— " La distruzione della Mosca domestica." Redia, vol. 8, pp. 462—470, 5 figs., 1912.

BERNSTEIN, J. Summary of literature relating to the Bionomics of the parasitic fungus of Flies; *Empusa muscae* Cohn, with special reference to the economic aspect. Repts. Local Govnt. Bd. on Public Health and Med. Subjects, N.S. No. 40, Further Repts. (No. 3) on Flies as Carriers of Infection, pp. 41—45, 1910.

BERRY. "Conjunctivitis set up by Flies." Brit. Med. Journ. 1892, p. 1114.

BERTARELLI, E. "Verbreitung des Typhus durch die Fliegen." Centralbl. Bakt. Parasitenk., vol. 53, pp. 486—495.

BEYER, H. G. "The Dissemination of Disease by the Fly." New York Med. Journ., vol. 91, pp. 677—685, 1910.

BEZZI, M. and STEIN, P. Katalog der paläarktischen Dipteren (Bearbeitet von Th. Becker, M. Bezzi, K. Kertesz und P. Stein), vol. 3, p. 828, Budapest, 1907.

BIGOT, J. M. F. "Diptères nouveaux ou peu connus." Bull. Soc. Zool. France, vol. 12, pp. 581—617, 1887.

BILLINGS. (Anthrax bacilli in flies collected on infected steer.) Twentieth Century Practice of Medicine, 1898.

BISHOPP, F. C. "The Stable-fly (Stomoxys calcitrans L.), an important live stock pest." Journ. Econ. Ent., vol. 6, pp. 112—126, 2 pls., 1913.

—— The Stable Fly. Farmers Bull. No. 540, U.S. Dept. Agriculture, 28 pp., 10 figs., 1913.

BLANCHARD, R. Traité Zoologie Médicale. Paris, 2 vols., 1890.

BLANKENMEYER, H. C. "Infection with fly larvae, Anthomyia canicularis." Journ. Amer. Med. Asstn., vol. 48, p. 1505, 1907.

BOGDANOW, E. A. "Zehn Generationen der Fliegen (Musca domestica) in veränderten Lebensbedingungen." Allg. Zeitschr. f. Entom., vol. 8, pp. 265—267, 1903.

BOLLINGER, O. "Experimentelle Untersuchungen über die Entstehung des Milzbrandes," 46 Versamml. d. D. Naturf. u. Aerzte zu Wiesbaden, September, 1873 ; and "Milzbrand," in von Ziemssen's Handb. d. spec. Pathol. u. Therapie, vol. 3, pp. 282 and 457, 1874.

BOOKER, W. H. "A working plan for an anti-fly and anti-mosquito campaign." Bull. North Carolina State Bd. Health, Raleigh, N. C., vol. 27, No. 4, pp. 126—133, July, 1912.

—— "Essential Facts about Flies," and "A Suggested anti-fly ordinance." Bull. North Carolina State Bd. Health, Raleigh, N. C., vol. 27, No. 4, pp. 133—141, 6 figs. (also issued as reprint), 1912.

BOUCHÉ, P. Fr. Naturgeschichte der Insekten besonders in Hinsicht ihrer ersten Zustande als Larven und Puppen. Berlin, 216 pp., 10 pls. (M. domestica, pp. 65—66, pl. 5, figs. 20—24), 1834.

BRAIN, C. K. Stomoxys calcitrans. Part I. Ann. Ent. Soc. Amer., vol. 5, pp. 421—430, 3 figs., 2 pls., 1912. Part II. Ibid., vol. 6, pp. 197—202, 8 figs., 1913.

BRAUER, F. "Die Zweiflügler des kaiserlichen Museums zu Wien : III. Systematische Studien auf Grundlage der Dipterenlarven nebst einer Zusammenstellung von Beispielen aus Literatur über dieselben und Beschreibung neuer Formen." Denkschr. der Kais. Akad. der Wiss., math.-naturwiss. Classe, Wien, vol. 47, pp. 1—100, 5 pls., 1883.

BRAUN. (Ophthalmia.) Centralbl. f. prakt. Augenheilk., p. 545, 1882. (A review ; cited by Abel, 1899.)

BRAUN, M. The Animal Parasites of Man. (London : Bale, Sons and Danielsson), 453 pp., 294 figs., 1906.

BREFELD, O. "Untersuchungen über die Entwickelung der *Empusa muscae* und *E. radicans.*" Abh. d. Naturf. Gesellsch. Halle, vol. 12, pp. 1—50, pls. 1—4, 1871.

BREWSTER, E. T. "The Fly, the Disease of the House." McClure's Magazine, vol. 33, pp. 564—568, Sept., 1909.

BRITTON, W. E. "The Common House-fly (*M. domestica*) in its relation to the public health." Yale Med. Journ., vol. 12, pp. 750—757, 1906.

—— "The House-fly and its Relation to Typhoid Fever." Proc. Sixth Conf. Health Officers of Connecticut, Hartford, Conn., April, 1912, 18 pp. (reprint).

—— "The rôle of the house-fly and certain other insects in the spread of human diseases." Pop. Sci. Monthly, July, 1912, pp. 36—49, 5 figs.

—— "The House-fly as a Disease-carrier and how controlled." 12 pp., 5 figs. (Publ. by Connecticut State Bd. of Health), 1912.

BRUES, C. T. "The Relation of the Stable-fly (*Stomoxys calcitrans*) to the transmission of Infantile Paralysis." Journ. Econ. Ent., vol. 6, pp. 101—109, 1913.

BRUES, C. T. and SHEPPARD, P. A. E. "The Possible Etiological Relation of Certain Biting Insects to the Spread of Infantile Paralysis." Journ. Econ. Ent., vol. 5, pp. 305—324, 1912. Abstracted previously in Monthly Bull. Mass. State Board of Health, Dec., 1911, pp. 337—340.

BUCHANAN, R. M. "The Carriage of Infection by Flies." Lancet, vol. 173, pp. 216—218, 5 figs., 1907.

—— "*Empusa muscae* as a carrier of bacterial infection from the House-fly." Brit. Med. Journ., Nov. 22, 1913; Reprint: 18 pp., 21 figs., 1913.

BUCHANAN, W. J. "Cholera Diffusion and Flies." Indian Med. Gaz., pp. 86, 87, 1897.

BUDD, W. "Observations on the occurrence (hitherto unnoticed) of malignant pustule in England." Lancet, pp. 164, 165, 1862.

BYER, H. G. "Dissemination of Disease by the Fly." New York Med. Journ., April 2, 1910.

CADET, G. The Pian. Thèse, Bordeaux, 1897.

CALANDRUCCIO, S. "Ulteriori ricerche sulla *Taenia nana.*" Boll. Accad. Gioenia, Catania, Fasc. 89, pp. 15—19, 1906.

CALDWELL. Manual of Military Hygiene. (Flies in trenches, p. 237), 1905.

CAMPBELL, C. "House-flies and Disease." Brit. Med. Journ., 1901 (2nd vol.), p. 980.

CANDIDO, G. "Anchilostomoanemia associata a Miasi intestinale." Ann. Med. Navale e Colon, vol. 1, pp. 394—407, 1 fig., 1913; Abstr. in Trop. Dis. Bull., vol. 2 (Nov. 15), p. 533, 1913.

CAO, G. "Sul passaggio dei micro-organismi attraverso l' intestino di alcuni insetti." Estratto dall' Ufficiale Sanitario, Rivista d' Igiene e di Medicina pratica. Anno 11, p. 98, 1898.

CAO, G. "Sul passaggio dei germi attraverso le larve di alcuni insetti."
Ann. d' Igiene sper., vol. 16, p. 645, 1906.

CARTER, H. J. "On a Bi-sexual Nematoid Worm which infests the Common
House-fly (*Musca domestica*) in Bombay." Ann. Mag. Nat. Hist., ser. (3),
vol. 7, pp. 29—33, 4 figs., 1861.

CARTER, HENRY F. and BLACKLOCK, B. "External Myiasis in a Monkey."
Brit. Med. Journ., 1913, Jan. 11, p. 72 ; also Abs. in Trop. Dis. Bull.,
vol. 1, no. 8, p. 471, 1913.

CASTELLANI, A. "Experimental Investigation on *Framboesia tropica* (Yaws)."
Journ. Hyg., vol. 7, pp. 558—599, 2 pls., 1 fig., 1907.

CATTLE, C. H. "A case in which large quantities of dipterous larvae were
passed per anum." Brit. Med. Journ., vol. 2, p. 1066, 1906.

CELLI, A. "Trasmissibilità dei germi patogeni mediante le dejecione delle
Mosche." Bull. Soc. Lancisiana d. ospedali di Roma, fasc. 1, pl. 1, 1888.

CHANTEMESSE. (Cholera and flies.) Report of Meeting of Acad. de Med.,
Paris, rev. in Med. Rec., vol. 68, p. 989, 1905.

CHAPMAN and JOHNSON. "House-flies and Disease." Brit. Med. Journ.,
1901 (2nd vol.), p. 126.

CHATTON, E., and LEGER, M. "L'Autonomie des Trypanosomes propre aux
Muscides démontrée par les élevages purs indéfinis." Comp. rend. Soc.
Biol., vol. 74, No. 10, pp. 549—551, 1913.

CHEVRIL, R. "Larves de mouches dans la vessie de l'Homme." Rev. Sci.,
Paris, vol. 11, pp. 621—624, 1909.

—— "Sur la myase des voies urinaires." Arch. de Parasitol., vol. 12,
pp. 369—450, 12 figs., 1909.

CHMELICEK, J. F. "My observations on the typhoid epidemic in southern
camps and its treatment." New York Med. Journ., vol. 70, pp. 193—198,
1899.

CLEAVER, EMMA O. "The Rôle of Insects in the Transmission of Disease :
a Résumé." Pennsylvania Med. Journ., vol. 4, p. 457, 1900.

CLELAND, J. B. The Relationship of Insects to Disease in Man in Australia.
Second Rept. Govt. Bur. Microbiology for 1910–11, pp. 141—158, 1912.

COATES, B. H. "Medical note on the more familiar flies." Trans. Coll.
Phys. Phila., N. S., vol. 3, p. 348, 1863.

COBB, J. O. "Is the Common House-fly a factor in the spread of Tuberculosis?" Amer. Med., No. 9, 1905, pp. 475—477.

COBB, N. A. "Fungus maladies of the sugar cane." Bull. No. 5. Expnt.
Sta., Hawaiian Sugar Planters' Asstn. Div. Pathology, Honolulu, 1906.

—— "The House-fly." National Geographic Mag., May, 1910, pp. 371—380,
4 figs.

COBBOLD, T. S. Parasites : A treatise on the entozoa of Man and Animals,
including some account of the ectozoa. 508 pp., 85 figs., London, 1879.

COCHRANE, E. W. W. "A small epidemic of Typhoid Fever in connection
with specifically infected Flies." Journ. Roy. Army Med. Corps, vol. 18,
pp. 271—276, March, 1912.

COCKERILL, J. W.   " Report on the Prevalence of Enteric Fever in Bermuda,"
with Tables and Diagrams.   Ibid., vol. 4, pp. 762—796, 1905.

COHN, F.   "*Empusa muscae* und die Krankheit der Stubenfliegen."   Nova
Acti Acad. Caes. Leop. Carol. Germ. Nat. Cur., vol. 25, p. 301, 1855.

COHN, M.   " Fliegeneier in den Entleerungen eines Säuglings."   Deutsche
med. Wochenschr., Jahrg. 24, pp. 191—193, 1898.

COKER, W. C.   " Necessity of water for flies."   Nature Study Review, vol. 7,
No. 9, Dec., 1911.

COLLINGE, W. E.   " House-flies and Public Health."   Abstract only, in Proc.
Ass. Econ. Biol., Journ. Econ. Biol., vol. 6, pp. 153—154, 1911.

COMPERE, G.   " A few facts concerning the Fruit-flies of the world, iii."
Monthly Bull. State Bd. Hort., Calif., U.S.A., vol. 1, pp. 907—911, 1912.

COMSTOCK, J. H., and NEEDHAM, J. G.   " The Wings of Insects."   Amer.
Nat., vol. 32, p. 43, etc., and continued through the vol. into vol. 33,
1898.

CONRADI, A. F.   Controlling Flies.   Circular No. 23, South Carolina Agric.
Export. Sta., Clemson Coll., S.C., 14 pp., 5 figs., 1913.

CONVERSE, G. M.   " Amoebiasis."   Bull. Cal. State Bd. of Health, October,
1910.

COPEMAN, S. M.   Report to the Local Government Board on the General
Sanitary Circumstances and Administration of the County Borough of
Wigan, with special reference to Infantile Mortality and to Endemic
Prevalence of Enteric Fever and Diarrhoea, 22 pp., London ; (Flies,
p. 18), 1906.

——— Memorandum on investigation into possible carriage of infection by
Flies.   Suggested " Plan of Campaign " in Urban Districts.   Repts.
Local Govnt. Bd. on Pub. Health and Med. Subjects, N. S. No. 16;
Further Prelim. Repts. on Flies as Carriers of Infection, pp. 1—4, 1909.

——— Note as to work in hand, but not yet published ; and as to proposed
further work in reference to Flies as Carriers of Infection.   Repts. Local
Govnt. Bd. Publ. Health and Med. Subjects, N. S. No. 40 ; Further
Repts. (No. 3) on Flies as Carriers of Infection, pp. 45—48, 1910.

——— Hibernation of House-flies (Preliminary Note).   Repts. Local Govnt.
Bd. Publ. Health and Med. Subjects, N. S. No. 85 ; Further Repts.
(No. 6) on Flies as Carriers of Infection, pp. 14—19, 1913.

COPEMAN, S. M., HOWLETT, F. M., and MERRIMAN, G.   An Experimental
Investigation on the range of Flight of Flies.   Repts. Local Govnt. Bd.
on Publ. Health and Med. Subjects, N. S. No. 53 ; Further Repts. (No. 4)
on Flies as Carriers of Infection, pp. 1—10, 1 map, 1 table, 1911.

COPLIN, W. M. S.   " The Propagation of Disease by means of Insects, with
special consideration of the common domestic types."   Pennsylvania
Med. Journ., vol. 3, p. 241, 1900.

COX, G. L., LEWIS, F. C., and GLYNN, E. E.   " The number and varieties of
Bacteria carried by the Common House-fly in sanitary and insanitary
city areas."   Journ. Hygiene, vol. 12, pp. 290—319, 2 pls., 2 figs., 1912.

CRAIG, C. F. "The rôle of insects in the propagation of disease." Philadelphia Med. Journ., vol. 3, p. 1381, 1899.

CRAWFORD, J. "Observations on the seats and causes of disease." Baltimore Med. and Physical Recorder, vol. 1, pp. 40, 81, 206 ; vol. 2, p. 31, 1808—1809.

CRICHTON-BROWNE, Sir J. "Malaria in relation to sanitation." Reprint of address before Sanit. Insp. Asstn., 6 Aug. 1902, 28 pp.

DANIELS, C. W. Studies in Laboratory work. London, 1907.

DARLING, S. T. "Experimental infection of the mule with *Trypanosoma hippicum* by means of *Musca domestica.*" Journ. Exp. Med., vol. 15, pp. 365—366, 1912.

—— The part played by flies and other insects in the spread of infectious diseases in the tropics, with special reference to ants and to the transmission of *Tr. hippicum* by *Musca domestica*. Trans. Fifteenth Internat. Congress Hyg. and Demog., Sect. v, Washington, Reprint 4 pp., 1913; Abstr. Rev. Appl. Ent., vol. 2, Ser. B, pp. 9—10, 1914.

DAVAINE, C. "Études sur la contagion du charbon chez les animaux domestiques." Bull. Acad. Méd. Paris, vol. 35, pp. 215—235, 1870.

DEADRICK, W. H. "Notes on intestinal myiasis." Arch. f. Schiffs- und Tropen-Hygiene, vol. 12, pp. 726—729, 1908.

DE GEER, CARL. "Mémoires pour servir à l'Histoire des Insectes." (*M. domestica*, vol. 6, pp. 71—78, pl. iv, figs. 1—11). Stockholm, 1776.

DELL, J. A. "On the Structure and Life-history of *Psychoda sexpunctata.*" Trans. Ent. Soc. London, pp. 293—311, 1905.

DEMETRIADES. (*Ophthalmia.*) Centralbl. f. prakt. Augenheilk., p. 412, 1894.

DICKENSON, G. K. "The House-fly and its connection with disease dissemination." Med. Rec., vol. 71, pp. 134—139, 1907.

DIESING, K. M. "Kleine helminthologische Mittheilungen." Sitz. Kais. Akad. d. Wiss. Wien, vol. 43, pp. 269—282, 1861.

DOANE, R. W. Insects and Disease. A popular account of the way in which insects may spread or cause some of our common diseases. Amer. Nature Series (New York : Holt and Co.), 227 pp., 112 figs., 1910.

—— "An annotated list of the literature on Insects and Diseases for the year 1910." Journ. Econ. Ent., vol. 4, No. 4, pp. 386—398, 1911.

—— "An annotated list of the literature on Insects and Diseases for the year 1911." Ibid., vol. 5, No. 5, pp. 268—285, 1912.

DONHOFF. "Beiträge zur Physiologie, I : Ueber das Verhalten Kaltblütiger Thiere gegen Frosttemperatur." Arch. f. Anat. und Phys. und Wiss. med. von Reichert und Du Bois-Reymond, p. 724, 1872.

DONOVAN, E. Natural History of British Insects. Vol. 6, p. 84, 1797.

DONOVAN, H. L. ("Can Flies carry Cholera ?") Indian Med. Journ., vol. 21, p. 318, 1886.

DREW, H. V. "A case of invasion by dipterous larvae." Brit. Med. Journ., vol. 2, p. 1066, 1906.

DUNKERLY, J. S. "On Some Stages in the Life-history of *Leptomonas muscae-domesticae*, with some remarks on the relationship of the Flagellate Parasites of Insects." Quart. Journ. Micros. Sci., vol. 56, pp. 645—655, 1911.

DUNNE, A. B. "Typhoid Fever in South Africa ; its Cause and Prevention." Brit. Med. Journ., March 8th, 1902, p. 622.

DUTTON, W. F. "Insect carriers of Typhoid Fever." Journ. Amer. Med. Asstn., vol. 53, pp. 1248—1252, 7 figs., 1909.

—— "Blue-bottle flies as carriers of infection." Journ. Amer. Med. Asstn., vol. 53, p. 1561, 1909.

EDGAR, C. L. "A case of Screw-worm in the Nose." Texas State Journ. of Med., vol. 9, p. 21, 1913.

ERCOLANI, G. B. "Sulla dimorfobiosi o diversi modi di vivere e riprodursi sotto duplice forma di una stessa specie animali. Osservazioni fatte sopra alcuni nematoelminti." Mem. Acad. d. Sc. Ist. di Bologna, Ser. 3, vol. 4, pp. 237—264, 2 pls.

ESTEN, W. M., and MASON, C. J. "Sources of Bacteria in Milk." Bull. No. 51, Storrs Agric. Exp. Sta., Storrs, Conn., U.S.A., 1908.

EWETZKIJ, T., and KENNEL, J. "Eine Fliegenlarve in der vorderen Augenkammer." Zeit. Augenheilk., Berlin, vol. 12, pp. 337—351, 1 pl., 1904.

EWING, H. E. "A new Parasite of the House-fly (Acarina, Gamasoidea)." Ent. News, vol. 24, pp. 452—456, 1 pl., 1913.

FAICHNE, N. "Fly-borne enteric fever ; the source of infection." Journ. Roy. Army Med. Corps, vol. 13, pp. 580—584, 1909.

—— "*Bacillus typhosus* in Flies." Ibid., pp. 672—675, Dec. 1909.

FARRAR, R. Reports of Medical Inspectors of the Local Govnt. Board, No. 216, p. 9, 1905. (Possible carriage of enteric fever by flies in Yorkshire.)

FELT, E. P. "The Economic Status of the House-fly." Journ. Econ. Ent., vol. 2, pp. 39—44, 1909.

—— Typhoid or House-fly. Twenty-fifth Report State Entomologist, Bull. No. 475, New York State Educ. Dept., pp. 12—17, 1910. Same observations also in Journ. Econ. Ent., vol. 3, pp. 24—26, 1910.

—— "Methods of controlling the House-fly and thus preventing the dissemination of Disease." New York Med. Journ., April 2, 1910, No. 91, pp. 685—687.

—— "Control of Flies and other household Insects." Bull. No. 465, New York State Educ. Dept., 53 pp. (*M. domestica*, pp. 6—16), 34 figs., 1910.

—— (Reports a case of Myiasis due to *M. domestica*). Notes for the year. Twenty-eighth Rept. State Entomologist on Injurious and other insects of the State of New York. Museum Bull. No. 165, N.Y. State Museum, p. 93, 1913.

FERMI, C. ("Lyssicide action and virus transmission by flies.") Centralbl. Bakt. 1 Abt. Orig., vol. 61, pp. 93—97, 1911.

FICKER, M. "Typhus und Fliegen." Arch. f. Hygiene, vol. 46, pp. 274—282, 1903.

FIELD, F. E. "Myiasis; with special reference to some varieties treated at the Georgetown Hospital." Brit. Guiana Med. Ann. for 1911, pp. 60—64, 1913.

FIRTH, R. H., and HORROCKS, W. H. "An Inquiry into the Influence of Soil, Fabrics, and Flies in the dissemination of enteric infection." Brit. Med. Journ., vol. 2, pp. 936—943, 1902.

FLETCHER, J. Report of the Entomologist and Botanist. Experimental Farms Report, Dept. Agric. Canada, Ottawa, pp. 225—226, 1900.

—— "Practical Entomology." (Rev. of Howard, 1900, Ins. Fauna of Human Excr.), Can. Ent., vol. 33, pp. 84—88, 1901.

FLEXNER, S. "Some problems in infection and control." Science, vol. 36, No. 934, pp. 685—702, 1912.

FLEXNER, S., and CLARKE, P. F. "Contamination of the house-fly with Poliomyelitis virus." Journ. Amer. Med. Asstn., vol. 56, pp. 1717—1718, 1911.

FLU, P. C. "Studien über die im Darm der Stubenfliege, *Musca domestica*, vorkommenden protozoären Gebilde." Centralbl. f. Bakt., vol. 57, pp. 522—534, 1911.

FLÜGGE, C. Grundriss der Hygiene, pp. 473 and 532, 1891.

—— "Die Verbreitungsweise und Verhütung der Cholera auf Grund der neueren epidemiologischen Erfahrungen und experimentellen Forschungen." Zeitschr. f. Hygiene, vol. 14, p. 165, 1893.

FONSSAGRIVES. "Les mouches au point de vue de l'hygiene." Gaz. hebdom. de méd., Paris, 2, S., 7, pp. 370—372.

FORMAN, R. H. "Indian Enteric and Latrines." Journ. Roy. Army Med. Corps, vol. 7, pp. 304—305, 1906.

FRANCAVIGLIA, M. C. "Altro caso di Myiasis nell' Uomo per larva cuticulare d' *Hypoderma bovis* (De Geer)." Policlinico, Sez. pratica, vol. 19, pp. 1593—1595, 1913; Abstr. in Trop. Dis. Bull., vol. 2, (Nov. 15), p. 526, 1913.

FRANCIS, C. F. "Cholera caused by a Fly (?)." Brit. Med. Journ., pt. 2, p. 65, 1893.

FRANKLIN, G. D. "Some Observations on the Breeding Ground of the Common House-fly." Indian Med. Gaz., vol. 41, pp. 349—350, 3 pls., 1906.

FRASER. (Epidemic diarrhoea in Portsmouth in relation to flies.) Hewlett (1905, p. 505) gives quotation from Fraser's Report for 1902, p. 47, cited by Nuttall and Jepson.

FROGGATT, W. W. "The House-fly and the diseases it spreads." Agric. Gazette, New South Wales, pp. 243—250, 1 fig., March, 1910; also issued separately as Miscellaneous Publication No. 1311.

FROST, W., and VORHEES, C. T. "The House-fly Nuisance." Country Life in America, May, 1908.

FUCHS. (Egyptian Ophthalmia.) Wien. klin. Wochenschr., No. 12, p. 211, 1894. (Cited by Abel, 1899.)

GADDIE, D. W. "What shall we do with the house-fly?" Ky. Med. Journ., May 1, 1911.

GALLI-VALERIO, B. "Les Insectes comme propagateurs des maladies." Ber. 14. Int. Congr. Hyg. u. Demogr., Berlin, 1907.

—— "L'état actuel de nos connaissances sur le rôle des mouches dans la dissemination des maladies parasitaires, et sur les moyens de lutte à employer contre elles." Centralbl. f. Bakt. Abt. 1, vol. 54, pp. 193—209, 1910.

GANN, T. W. "Beef-worm in the orbital cavity." Lancet, vol. 1, pp. 19—21, 1902.

GANON, J. "Cholera and Flies." Geneesk. Tijdschr. v. Nederl. Indië, vol. 48, No. 2, pp. 227—233, 1908 ; abs. in Journ. Trop. Med. and Hyg. London, vol. 12, No. 10, p. 158, 1909.

GARROOD, J. R. "Note on a case of intestinal myiasis." Parasitology, vol. 3, pp. 315—318, 1910.

GAYON, J. P. "A note concerning the transmission of Pathogenic fungus by flies and mosquitoes." Public Health, (U.S.A.), vol. 28, pp. 116—117, 1903.

GEDDINGS, H. D. Précis on the fly and mosquito as carriers of disease. Public Health Reports, U.S. Pub. Health and Mar. Hosp. Service, vol. 17, No. 35, 1903.

GENERALI, G. "Una larva di nematode della mosca commune." Atti Soc. d. Nat. di Modena, Rendic., Ser. (3), vol. 2, pp. 88—89, 1886.

GEOFFROY, E. L. Histoire abrégée des Insectes, vol. 2, p. 624, 1764.

GERHARD, W. P. "Bibliography on Flies and Mosquitoes as carriers of disease." Entom. News, Philadelphia, vol. 20, pp. 84—89, 1909.

—— "Additional Bibliography on Flies and Mosquitoes as carriers of disease." Ibid., pp. 207—211, 1909.

—— "Flies and Mosquitoes as carriers of disease." Author's Reprint from The Country Gentleman, Albany, N.Y. Publ. New York, 14 pp., 1911.

GERMAN. Ophthalmia.) Centralbl. f. prakt. Augenheilk. Suppl., p. 386, 1896. (Cited by Abel, 1899.)

GIARD, A. "Deux espèces d'Entomophthora nouveaux pour la flore française et la présence de la forme Tarichium sur une Muscide." Bull. Scient. du Department du Nord, ser. 2, second year, No. 11, pp. 353—363, 1879.

GILES, C. M. "The anatomy of biting flies of the genera Stomoxys and Glossina." Journ. Trop. Med., vol. 9, p. 99, 1906.

GIRAULT, A. A., and SANDERS, G. E. "The Chalcidoid Parasites of the Common House or Typhoid Fly (Musca domestica, L.) and its allies." Psyche, vol. 16, pp. 119—132, 5 figs., 1909.

—— "The Chalcidoid Parasites of the Common House or Typhoid Fly (Musca domestica, Linn.) and its allies." Ibid., vol. 17, pp. 9—28, Feb. 1910.

VON GLEICHEN, F. WILHELM. Geschichte der gemeinen Stubenfliege. 32 pp., 4 pls., Nurnberg, 1790.

GODFREY, R. "The False-scorpions of Scotland." Ann. Scot. Nat. Hist., No. 69, January, 1909, pp. 22—26.

GRAHAM-SMITH, G. S. Preliminary Note on Examinations of Flies for the presence of Colon Bacilli. Repts. Local Govnt. Bd. on Publ. Health and Med. Subjects, N. S. No. 16 ; Further Prelim. Repts. on Flies as Carriers of Infection, pp. 9—13, 1909.

—— Observations on the ways in which artificially infected flies (*Musca domestica*) carry and distribute pathogenic and other bacteria. Repts. Local Govnt. Bd. on Publ. Health and Med. Subjects, N. S. No. 40 ; Further Repts. (No. 3) on Flies as Carriers of Infection, pp. 1—41, 7 pls., 25 tables, 1910.

—— Further observations on the ways in which artificially infected flies (*Musca domestica* and *Calliphora erythrocephala*) carry and distribute pathogenic and other bacteria. Repts. Local Govnt. Bd. Publ. Health and Med. Subjects, N. S. No. 53 ; Further Reports (No. 4) on Flies as Carriers of Infection, pp. 31—48, 13 tables, 1911.

—— " Some observations on the anatomy and function of the oral sucker of the Blow-fly (*Calliphora erythrocephala*)." Journ. Hygiene, vol. 11, No. 3, pp. 390—408, 5 pls., Nov., 1911.

—— "House-flies." Bedrock, No. 2, pp. 205—223, 1912.

—— An Investigation of the incidence of the micro-organisms known as non-lactose fermenters in flies in normal surroundings and in surroundings associated with epidemic diarrhoea. In Ann. Rep. Local Govnt. Bd., Sup. Rept. Med. Officer, 1911—1912, pp. 304—320, 1912.

—— An Investigation into the possibility of pathogenic micro-organisms being taken up by the larva and subsequently distributed by the fly. In Ann. Rept. Local Govnt. Bd., Supp. Rept. Med. Officer, 1911—1912, pp. 330—335, 1912.

—— Flies and Disease : Non-blood sucking Flies. Camb. Pub. Health Ser., pp. xiv+292, 32 text figs., 24 pls., Cambridge, 1913.

—— Further observations on non-lactose fermenting Bacilli in flies, and the sources from which they are derived with special reference to Morgan's bacillus. Repts. to the Local Govnt. Board on Publ. Health and Med. Subjects, N. S. No. 85 ; Further Rept. (No. 6) on Flies as Carriers of Infection, pp. 43—46, 1913.

GRASSI, B. "Les méfaits des Mouches." Arch. ital. de Biologie, vol. 4, pp. 205—228, 1883.

GRIFFITH, A. "The Life-history of House-flies." Public Health, vol. 21, pp. 122—127, 1908.

GRIFFITH, F. "Description of a House-fly Parasite." Med. Brief, St Louis, vol. 35, pp. 59—63, 1907.

GRÜNBERG, K. Die Blutsaugenden Dipteren. Leitfaden zur allgemeinen Orientierung, mit besonderer Berücksichtigung der in den deutschen Kolonien lebenden Krankheitsüberträger, 188 pp., 127 figs. (Jena : Gustav Fischer), 1907.

GUDGER, E. W. "An early note on Flies as transmitters of disease." Science, vol. 31, pp. 31—32, 1910.

—— "A second early note on the transmission of yaws by Flies." Science, vol. 32, pp. 632—633, 4 Nov. 1910.

GÜSSOW, H. T. *Empusa muscae* and the extermination of the House-fly. Reports to the Local Govnt. Board on Public Health and Med. Subjects, N. S. No. 85; Further Rept. (No. 6) on Flies as Carriers of Infection, pp. 10—14, 1 pl., 1913.

GUYÉNOT, E. "L'appareil digestif et la digestion de quelques larves des mouches." Bull. scient. de la France et de la Belgique, vol. 41, pp. 353—369, 1907.

GUZMANN, E. "Ueber Ophthalmomyiasis." Klin. Monatsbl. Augenheilk., vol. 48, pp. 625—627. Stuttgart, 1910.

HAESER. (Plague and flies.) Geschichte der med. u. epidem. Krankh. 3. Aufl., vol. 3. (Cited by Nuttall, 1899.)

HAGEN, A. "On the larvae of insects discharged through the urethra." Proc. Boston Soc. Nat. Hist., vol. 20, pp. 101—118, 1879.

HALL, M. C. and MUIR, J. T. "A critical study of a case of Myiasis due to Eristalis." Arch. Internat. Med., vol. 12, pp. 193—202, 1913; Abstr. in Trop. Dis. Bull., vol. 2, (Nov. 15), p. 533, 1913.

HAMER, W. H. Nuisance from Flies. Report by the Medical Officer presenting a report by Dr Hamer, Medical Officer (General Purposes), on the extent to which the fly nuisance is produced in London by accumulations of offensive matter. 10 pp., 2 figs., 3 diagrams. Printed for the London County Council (Public Health Committee), London, 1908.

—— Nuisance from Flies. Report of the Medical Officer of Health presenting a further report by Dr Hamer, Medical Officer (General Purposes), on the extent to which the fly nuisance is produced in London by accumulations of offensive matter. 6 pp., 4 diagrams. Printed for the London County Council (Public Health Committee), London, 1908.

—— "The Breeding of Flies summarised." Am. Med. No. 3, 1908, p. 431.

—— Flies and vermin. Report by the Medical Officer of Health presenting reports by Dr Hamer, Medical Officer (General Purposes), on Nuisance from Flies and on the seasonal prevalence of vermin in common lodging-houses. Report of the Public Health Committee of the London County Council for 1909. Appendix No. 4, 9 pp., 5 charts, 1910.

HAMERTON, A. E. "Introduction to methods of study of the morbid histology of disease-carrying insects." Journ. Roy. Army Med. Corps, vol. 11, pp. 243—249, 1908.

HAMILTON, A. "The Fly as a Carrier of Typhoid; an inquiry into the part played by the common House-fly in the recent epidemic of typhoid fever in Chicago." Journ. Amer. Med. Ass., vol. 40, p. 567, 1903.

—— "The common House-fly as a carrier of typhoid fever." Journ. Amer. Med. Ass., vol. 42, p. 1034, 1904.

—— (Isolation of tubercle bacillus from flies caught in a privy.) Brit.

Med. Journ., p. 149, 1903 ; also see Journ. Amer. Med. Ass., 28th Feb., 1907.

HAMILTON, A. "The rôle of the House-fly and other insects in the spread of infectious diseases." Illinois Med. Journ., Springfield, vol. 9, pp. 583—587, 1906.

HAMMOND, A. "On the Thorax of the Blow-fly (*Musca vomitoria*)." Journ. Linn. Soc. (Zool.), vol. 15, pp. 9—31, 2 pls., 1881.

HARRINGTON, W. H. House-flies. Rep. Ent. Soc. Ont. 1882, pp. 38—44, 1 fig.

HARRISON, J. H. H. "A case of Myiasis." Journ. Trop. Med., vol. 11, p. 305, 1908.

HATCH, E. "The House-fly as a Carrier of Disease." Ann. Amer. Acad. Pol. Soc. Science, Mar. 1911, pp. 412—423.

—— Report of the Chairman of the Fly-fighting Committee of the American Civic Assn., Ann. Convention, Baltimore, Md., 20th Nov. 1912. Separately issued, 16 pp., 1913.

HAUTEFEUILLE. "Les Diptères parasites des cavités naturelles chez l'homme." Amiens Bull. soc. linn., vol. 18, pp. 81—90, 1907.

HAYWARD, E. H. "The Fly as a Carrier of Tuberculosis Infection." New York Med. Journ., vol. 80, pp. 643—644, 1904.

HECKENROTH, F. and BLANCHARD, M. "Note sur la présence et l'endémicité d'une Myiase furonculeuse au Congo français." Bull. Soc. Path. Exot., vol. 6, pp. 350—351, 1913.

HENNEGUY, L. F. Les Insectes. Paris, 804 pp., 1904.

HENSCHEN. (The larvae of flies as the cause of chronic pseudomembranous enteritis.) Wien. klin. Rundschau, No. 33, 1896. (Rev. in Amer. Journ. Med. Sci., vol. 113, p. 732, 1897.)

HEPWORTH, J. "On the Structure of the Foot of the Fly." Quart. Journ. Micr. Sci., vol. 2, pp. 158—160, 1854.

HERMS, W. B. "Essentials of House-fly Control." Bull. Berkeley (Cal.) Bd. of Health, 29 June, 9 pp., 1909.

—— "The House-fly Problem." Pacific Slope Asstn. Econ. Entomologists, Bull. Card No. 1, 1909.

—— "Medical Entomology, its scope and methods." Journ. Econ. Ent., vol. 2, pp. 265—268, 1909.

—— "The Berkeley House-fly Campaign." Calif. Journ. Technol., vol. 14, No. 2, 11 pp., 3 figs., 1909.

—— "Insects as they relate to rural hygiene." Calif. Cultivator, vol. 34, No. 2, pp. 35, 38, 39 and 43, 1910.

—— Insect Pests as they relate to rural hygiene, with special reference to control. Proc. Thirty-sixth Convention California State Fruit Growers, pp. 160—167, 1910.

—— "How to control the Common House-fly." Monthly Bull. Calif. State Bd. of Health, vol. 5, May, pp. 269—277, 5 figs., 1910.

—— "Fight the Fly,—Why,—When,—Why,—How ?" Bull. Berkeley (Cal.) Bd. of Health, 1910.

HERMS, W. B. "The House-fly in its relation to public health." Bull. 215, Calif. Agric. Exp. Sta., pp. 513—548, 15 figs., 1911.

HERVIEUX. (Report on carriage of smallpox by flies, read to Academy of Medicine, Paris, June 5th, 1904). Lancet, pt. 1, p. 1761, 16th June, 1904.

HESSE, E. "The parasitic fungus of the House-fly." Shrewsbury Chronicle, 29 Nov. (Reprint), 1912.

—— "Parasitic Mould of the House-fly." Brit. Med. Journ., Jan. 4th, 1913, p. 41.

HEWITT, C. G. "A Preliminary Account of the Life-history of the Common House-fly (*Musca domestica*, L.)." Manchester Mem., vol. 51, part i, 4 pp., 1906.

—— House-flies. Ann. Report and Trans. Manchester Micros. Soc., 1907, pp. 82—92, 1 pl.

—— "On the life-history of the Root-maggot, *Anthomyia radicum* Meigen." Journ. Econ. Biol., vol. 2, pp. 56—63, 1 pl., 1907.

—— "On the Bionomics of certain calyptrate Muscidae and their Economic significance with especial reference to Flies inhabiting houses." Journ. Econ. Biol., vol. 2, pp. 79—88, 1907.

—— "The Proboscis of the House-fly." Brit. Med. Journ., November 23rd, 1907, p. 1558.

—— "The Structure, Development, and Bionomics of the House-fly, *Musca domestica*, Linn. : Part I. The Anatomy of the Fly." Quart. Journ. Micr. Sci., vol. 51, pp. 395—448, pls. 22—26, 1907.

—— Idem, Part II. "The Breeding Habits, Development and the Anatomy of the Larva." Ibid., vol. 52, pp. 495—545, 4 pls., 1908.

—— "The Biology of House-flies in relation to Public Health." Journ. Roy. Inst. Public Health, vol. 16, pp. 596—608, 3 figs., 1908.

—— "The Structure, Development and Bionomics of the House-fly, *Musca domestica*, Linn. : Part III. The Bionomics, Allies, Parasites and the Relations of *M. domestica* to Disease." Quart. Journ. Micr. Sci., vol. 54, pp. 347—414, 1 pl., 1 fig., 1909.

—— "House-flies and Disease." Nature, vol. 84, pp. 73—75, 3 figs., 1910.

—— House-flies and their Allies. Fortieth Ann. Rep., Ent. Soc. Ontario, pp. 30—36, 4 figs., 1910.

—— "House-flies and the Public Health." Ottawa Naturalist, vol. 24, pp. 31—35, 1910.

—— The House-fly, *Musca domestica*. A Study of its Structure, Development, Bionomics and Economy. 195 pp., 10 pls., 1 fig. (Manchester, The University Press), 1910.

—— "The House-fly in Relation to Public Health." Public Health Journ. (Canada), vol. 2, pp. 259—261, 1911.

—— "The House-fly." (Review.) Canadian Ent., vol. 43, pp. 294—295, 1911.

—— Observations on the range of flight of Flies. Repts. Local Govnt. Bd. on

Publ. Health and Med. Subjects, N. S. No. 60 ; Further Reports (No. 5) on Flies as Carriers of Infection, pp. 1—5, with map, 1912.

HEWITT, C. G. An Account of the Bionomics and the Larvae of the Flies *Fannia (Homalomyia) canicularis* L. and *F. scalaris* Fab., and their relation to Myiasis of the intestinal and urinary tracts. Repts. Local Govnt. Bd. on Publ. Health and Med. Subjects, N. S. No. 60 ; Further Reports (No. 5) on Flies as Carriers of Infection, pp. 15—22, 3 figs., 1912.

—— "*Fannia (Homalomyia) canicularis* Linn. and *F. scalaris* Fab." Parasitology, vol. 5, pp. 161—174, 7 figs., 1 pl., 1912.

—— House-flies and how they spread disease. Cambridge University Press (in Cambridge Manuals of Science and Literature), 122 pp., 19 figs., 1912.

—— "On the predaceous habits of *Scatophaga*, a new enemy of *Musca domestica*." Can. Ent., vol. 46, pp. 2—3, 1914.

HEWLETT. "Insects as carriers of disease." Med. Press and Circ., N. S. vol. 76, pp. 439—442. London, 1903.

HEWLETT, H. T. "The etiology of epidemic diarrhoea." Journ. Prevent. Med., vol. 13, pp. 496—507, 1905.

HICKSON, S. J. "The Eye and Optic Tract of Insects." Quart. Journ. Micr. Sci., vol. 25, pp. 1—39, 3 pls., 1885.

—— "A Parasite of the House-fly." Nature, October 26th, 1905.

HINDLE, E. Note on the Colour Preference of Flies. Repts. Local Govnt. Bd. on Publ. Health and Med. Subjects, N. S. No. 85; Further Reports (No. 6) on Flies as Carriers of Infection, pp. 41—43, 1913.

—— "The Flight of the House-fly." Proc. Camb. Phil. Soc., vol. 17, pp. 310—313, 1914.

HIRSCH, C. T. W. "An account of two cases of Coko or Framboesia." Lancet, pt. 2, pp. 173—175, 1896.

HODGE, C. F. "A practical point in the study of typhoid or filth fly." Nature Study Review, vol. 6, pp. 195—199, 1910.

—— "A Plan to exterminate the Typhoid or Filth-Disease Fly." La Follette's Weekly Magazine, Madison, Wis., vol. 3, No. 15, pp. 7—8, (illus.), April 15, 1911.

—— "How you can make your home, town or city, flyless." Nature and Culture, Cincinnati, Ohio, vol. 3, Nos. 2 and 3, pp. 9—23, 7 figs., July—Aug., 1911.

—— "Exterminating the Fly." California Outlook, Sept. 30, 1911.

—— "A New Fly Trap." Journ. Econ. Ent., vol. 6, pp. 110—112, 1 pl., 1913.

—— "The Distance House-flies, Blue-bottles and Stable Flies may travel over water." Science, N. S., vol. 38, pp. 512—513, 1913.

HOFFMANN, E. "Ueber die Verbreitung der Tuberculose durch Stuben-fliegen." Correspondenzbl. d. ärztl. Kreis- und Bezirksvereine im Königr. Sachsen, vol. 44, pp. 130—135, 1888.

HOLMGREN, N. "Zur Morphologie des Insektenkopfes : II. Einiges über die Reduktion des Kopfes der Dipteren-larven." Zool. Anz., vol. 27, pp. 343—355, 12 figs., 1904.

Hope, F. "On the Insects and their larvae occasionally found in the human body." Trans. Ent. Soc. London, vol. 2, p. 256, pl. 22, 1840.

Houston, W. M. "Formalin against Flies." Indian Med. Gaz., Feb. 1913, p. 84.

Howard, C. W. and Clarke, P. F. "Experiments on insect transmission of the virus of Poliomyelitis." Journ. Exp. Med., vol. 16, pp. 850—859, 1912.

Howard, L. O. "House-flies" (in The Principal Household Insects of the United States, by L. O. Howard and C. L. Marlatt), United States of America Dept. of Agriculture, Washington, Division of Entomology. Bull. No. 4, N. S. (revised ed. 1902), pp. 43—47, and figs., 1896.

—— "Further notes on the House-fly," in "Some Miscellaneous Results of the work of the Division of Entomology." U.S. Dept. of Agriculture, Division of Entomology, Bull. No. 10, N. S., pp. 63—65, 1898.

—— "A Contribution to the Study of the Insect Fauna of Human Excrement (with especial reference to the spread of Typhoid Fever by Flies)." Proc. Wash. Acad. Sciences, vol. 2, pp. 541—604, figs. 17—38, pls. 30, 31, 1900.

—— "On some Diptera reared from cow manure." Can. Ent., vol. 33, pp. 42—44, 1901.

—— "Flies and Typhoid." Pop. Science Monthly, Jan. 1901, pp. 249—256.

—— "House-flies." Circular No. 35, Div. of Ent., U.S. Dept. Agric., 8 pp., 6 figs., 1898 ; also revised ed. Circular No. 71, 10 pp., 9 figs., 1906.

—— "How insects affect Health in Rural Districts." Farmers Bull. No. 155, U.S. Dept. Agric., pp. 1—19, 16 figs., 1908.

—— "Economic Loss to the People of the United States through Insects that carry Disease." Bull. No. 78, U.S. Dept. Agric., Bureau of Entomology, 40 pp., 3 tables, 1909.

—— "Flies as carriers of infection." Science, N. S., vol. 34, pp. 24—25, (Review), 1911.

—— "House-flies." Farmers Bull. No. 459, U.S. Dept. Agric., Washington, 16 pp., 9 figs., 1911.

—— The House-fly, disease-carrier: 312 pp., 39 figs. Stokes Co., New York ; John Murray, London, 1911.

Howe, L. (Egyptian *Ophthalmia.*) Seventh Internat. Congr. Ophthalmol., Wiesbaden, p. 323.

Huber, J. B. Insects and Disease. New York State Journ. Med., Nov., 1908.

Huber, J. C. Bibliographie der klinischen Entomologie. Jena, 1899.

Hutchison, R. H. "The Migratory Habit of House-Fly larvae as indicating a favourable remedial measure. An account of progress." Bull. No. 14, U.S. Dept. Agr., Washington, 11 pp., 1914.

Hutchinson, Woods. "How doth the little busy Fly ?" Country Life, (U.S.A.), vol. 20, pp. 31—33, Aug. 15, 1911.

—— "The story of the fly that does not wipe its feet." Sat. Even. Post, 7 March, 1908.

Hunter, W. (Occurrence of plague bacilli in alimentary tract of flies that had fed on infected material in Hong Kong.) Quoted in article on "The

Danger of the Common Fly." Nursing Times, 28th Sept., p. 842, 1907. (Cited by Nuttall and Jepson, 1909.)

HUNTER, W. D. "American Interest in Medical Entomology." Journ. Econ. Ent., vol. 6, pp. 27—39, 1913.

IMMS, A. D. " On the Larval and Pupal Stages of *Anopheles maculipennis*, Meigen." Journ. of Hygiene, vol. 7, pp. 291—318, 1 fig., pls. 4, 5, 1907.

JACKSON, D. D. Pollution of New York Harbour as a Menace to Health by the Dissemination of Intestinal Disease through the Agency of the Common House-fly. A Report to the Committee on Pollution of the Merchants' Association of New York, 22 pp., 2 maps, 3 charts, 3 figs., 1907.

—— "Conveyance of Disease by Flies summarised." Boston Med. and Surg. Journ. 1908, p. 451.

—— "The Disease-carrying House-fly." Review of Reviews, (U.S.A.), July, 1910.

JENNINGS, A. H. and KING, W. V. "An Intensive Study of Insects as a possible Etiologic factor in Pellagra." Amer. Journ. Med. Sci., vol. 46, (Sept.), p. 411, 1913.

JENYNS, L. Notice of a case in which the larvae of a dipterous insect supposed to be the *Anthomyia canicularis* Meig. were expelled in large quantities from the human intestines. Trans. Ent. Soc. London, vol. 2, p. 153, 1839.

JEPSON, J. P. Some observations on the breeding of *Musca domestica* during the winter months. Repts. Local Govnt. Bd. on Publ. Health and Med. Subjects, N. S. No. 3. Prelim. Repts. on Flies as Carriers of Infection, pp. 5—8, 1909.

—— "The Breeding of the Common House-fly (*Musca domestica*) during the winter months." Journ. Econ. Biol., vol. 4, pp. 78—82, 1909.

—— Notes on experiments in colouring Flies for purposes of identification. Repts. Local Govnt. Bd. on Publ. Health and Med. Subjects, N. S. No. 16, Further Preliminary Repts. on Flies as Carriers of Infection, pp. 4—9, 1909.

JOLY, P. R. Importance du rôle des insectes dans la transmission des maladies infectieuses et parasitaires. Du formol comme insecticide. Bordeaux, 90 pp., Thèse, 1898.

JONES. "The Common House-fly as the cause of disease." Maritime Med. News, Halifax (Canada), vol. 18, pp. 285—294, 1906.

JONES, F. W. C. "Notes on enteric fever prevention in India." Journ. Roy. Army Med. Corps, vol. 8, pp. 22—34, 1907.

JONES, G. I. 'Hepatic Abscess (Non-Amebic) and Gastro-intestinal myiasis." Journ. Amer. Med. Ass., vol. 61, p. 1457, 1913.

JOSEPH, G. "Ueber Fliegen als Schädlinge und Parasiten des Menschen." Theil 4, *Myiasis septica*. Deutsche Medicinal-Zeitung, No. 65, pp. 725—728, 1887.

JUDD, G. S. "Larvae discharged from the lower intestine of a boy." Amer. Nat., vol. 10, p. 374, 1876.

KAMMERER, P. "Regeneration des Dipterenflügels beim Imago." Arch. f. Entwick., vol. 25, pp. 349—360, 4 figs., 1908.

KEILIN, M. D. "Sur le parasitisme de la larve de *Pollenia rudis* Fab. dans *Allobophora chlorotica* Sav." Compt. rend. Soc. Biol., vol. 67, pp. 201—203.

—— "Structure du pharynx en fonction du régime chez les larves de Diptères cyclorhaphes." Comptes-rendus Acad. Sci., Paris, vol. 155, pp. 1548—1550, 1912.

KELLERS, H. C. "A sanitary garbage can holder." U. S. Naval Med. Bull., vol. 5, p. 45, 1911.

KELLY, H. A. "A historical note upon Diptera as carriers of diseases." Paré-Déclat. Johns Hopkins Hosp. Bull., vol. 12, 7 pp. Aug., 1901.

KENT, W. S. A Manual of the Infusoria, vol. 1, p. 245, pl. 13, figs. 29—34, 1880—1881.

KEW, H. W. "Lincolnshire Pseudo-scorpions: with an Account of the Association of such Animals with other Arthropods." Naturalist, No. 534, July, 1901, pp. 193—215.

KEYT, F. T. "A case of 'Beef-worm' (*Dermatobia noxialis*) in the orbit." Brit. Med. Journ., vol. 2, p. 316, 1900.

KIRBY and SPENCE. Introduction to Entomology. Vol. 4, pp. 228—229, 1826.

KLEIN, E. "Flies as carriers of the *Bacillus typhosus*." Brit. Med. Journ. (Oct. 17, 1908), pp. 1150—1151.

KOBAYASHI, S. "Nipponsan kajo no hassei oyobi shusei ni tsuite." (The metamorphosis and habits of the Japanese House-fly). Dobuts Z. Tokyo, vol. 21, pp. 335—341, 1909.

KOBER, G. M. "The etiology and prevention of infectious diseases." Virginia Med. Monthly (Reprint), April, 1892.

—— Report on the prevalence of typhoid in the district of Columbia. Report, Health Officer, D. C. for year ending June 30, 1905, pp. 253—292.

KOCH, R. "Bericht über die Thätigkeit der deutschen Cholera-Kommission in Aegypten und Ostindien." Wiener med. Wochenschr., No. 52, pp. 1548—1551, 1883.

KOWALEVSKI, A. "Beiträge zur Kenntniss der nachembryonalen Entwicklung der Musciden." Zeit. f. wiss. Zool., vol. 45, pp. 542—594, pls. 26—30, 1887.

KRAEPELIN, K. "Zur Anatomie und Physiologie des Rüssels von Musca." Zeit. f. wiss. Zool., vol. 39, pp. 683—719, 2 pls., 1883.

KRONTOWSKI, A. "Zur Frage über die Typhus- und Dysenterieverbreitung durch Fliegen." Centralbl. f. Bakt. Parasitenk. und Infektionskrankheiten, Jena, vol. 68, pp. 586—590, 1913; Abs. in Rev. Appl. Ent., vol. 1, Ser. B., pp. 117—118, 1913.

KUNCKEL D'HERCULAIS, J. Récherches sur l'organisation et le développement des Volucelles, insectes diptères de la famille des Syrphides. Paris, part i, 1875—81.

LABOULBÈNE. (A case of intestinal myiasis caused by *Fannia* larvae.) Comptes rendus Soc. Biol., p. 8, 1856.

LALLIER, P. Tableau des larves de Diptères évacuées par l'urèthra. Thésis, Paris, 1897.

LAMBOTTE, U. "Insectes et maladies infectieuses." Ann. Soc. Méd.-chir. de Liège, vol. 44, pp. 371—389, 1905.

LANGE, M. Rudimenta doctrinae de peste. 2nd edit., Offenbach, pp. 27—28, 1791. (Quoted by Abel.)

LANGFIELD, M. "The rôle of insects in the transmission of disease." Trained Nurse, etc., New York, vol. 35, pp. 195, 263 and 336, 1905.

LAVERAN, A. "Contribution à l'étude du bouton de Biskra." Ann. d. Dermatologie, 2nd ser., vol. 1, pp. 173—197, 1880.

LAWRENCE, S. M. "Dangerous dipterous larvae." Brit. Med. Journ., vol. 1, pp. 88, 1909.

LEBŒUF, A. "Dissemination du bacille de Hansen par la mouche domestique." Bull. Soc. Path. Exot., vol. 5, No. 10, pp. 860—868, 1912.

LEDINGHAM, J. G. "On the survival of specific microorganisms in pupae and imagines of *Musca domestica* raised from experimentally infected larvae. Experiments with *Bacillus typhosus*." Journ. Hygiene, (Camb.), vol. 11, No. 3, pp. 333—340, 1911.

LÉGER, L. "Sur la structure et le mode de multiplication des Flagelles du genre Herpetomonas, Kent." C. R. Ac. Sci., vol. 134, p. 781, 7 figs., 1902.

—— "Sur quelques Cercomonadines nouvelles ou peu connues parasites de l'intestin des insectes." Arch. f. Protistenk., vol. 2, pp. 180—189, 4 figs., 1903.

LEIDY, J. "Flies as a means of communicating contagious diseases." Proc. Acad. Nat. Sc. of Philadelphia (Meeting of 21st Nov., 1871), p. 297, 1872.

—— "On a parasitic worm of the House-fly." Proc. Acad. Nat. Sci. Philadelphia, vol. 26, p. 139, 1874.

—— "Researches in Helminthology and Parasitology." Smithsonian Miscell. Collections, Washington, in vol. 46, 281 pp., 28 figs. (Reprint), 1904.

LELEAN, P. S. "Notes on Myiasis." Brit. Med. Journ., vol. 1, pp. 245—246, 1904.

LEUCKART, R. "Die Fortpflanzung und Entwicklung der Pupiparen. Nach Beobachtungen an *Melophagus ovinus*." Abhandl. Naturf.-Gesell., Halle, vol. 4, pp. 147—226, 3 pls., 1858.

LEVY, E. C. and FREEMAN, A. E. "Certain conclusions concerning typhoid fever in the south, as deduced from a study of typhoid fever in Richmond, Va." Old Dominion Journ. Med. and Surg., vol. 8, (Reprint), 39 pp., 3 maps, 3 charts, 1908.

LEVY, E. C. and TUCK, W. T. "The maggot-trap—A new weapon in our warfare against the typhoid fly." Amer. Journ. Pub. Health, vol. 3, No. 7, pp. 657—660, 1 fig., 1913.

LINGARD, A. and JENNINGS, E. Some Flagellate Forms found in the Intestinal Tracts of Diptera and other Genera. London (Adlard and Son), 25 pp., 5 pls., 1906.

LINNEUS, C. DE. Systema naturae (10th ed.), vol. 1, p. 596, 1758, and Fauna suecica, (ed. ii), Holmiae, 1761.

VON LINSTOW. "Beobachtungen an neuen und bekannten Helminthen." Arch. f. Naturgesch., pp. 183—207, 1875.

LOCHMANN, R. "Eine epidemische auftretende Krankheit der Stubenfliege verursacht durch *Empusa muscae*." Pharm. Reformer, Wien, vol. 4, p. 127, 1899.

LOEB, J. Der Heliotropismus der Thiere und seine Uebereinstimmung mit dem Heliotropismus der Pflanzen. Wurzburg, 118 pp., 6 figs., 1890.

LORD, F. T. "Flies and tuberculosis." Boston Med. and Surg. Journ., vol. 101, pp. 651—655, 15th Dec., 1904.

LOWNE, B. T. The Anatomy and Physiology of the Blowfly (*Musca vomitoria*). 121 pp., 10 pls., London, 1870.

—— "On the Compound Vision and the Morphology of the Eye in Insects." Trans. Linn. Soc. (Zool.), vol. 2, pt. 11, 1884.

—— The Anatomy, Physiology, Morphology and Development of the Blowfly (*Calliphora erythrocephala*). 2 vols., London, 1895.

LUBBOCK, J. (LORD AVEBURY). "The fly in its sanitary aspect." Lancet, pt. 2, 1871, p. 270.

LUMSDEN, L. L. and ANDERSON, J. F. "The origin and prevalence of Typhoid Fever in the District of Columbia (1909—1910)." Pub. Health and Marine Hosp. Service, Hyg. Lab. Bull. No. 78, Oct., 1911.

LUMSDEN, L. L. Sanitation of Flood-stricken towns and cities, with special reference to conditions observed in river towns and cities of Kentucky. Public Health Repts., U.S. Public Health Service, vol. 28, No. 24, pp. 1195—1220, 1913.

LYONET, P. Traité anatomique de la chenille qui ronge le bois de saule. 2nd éd., La Haye, 18 pls., 1762.

McCAMPBELL, E. F. and COOPER, H. J. "*Myiasis intestinalis* due to infection with three species of dipterous larvae." Journ. Amer. Med. Assn., vol. 53, Oct. 9, pp. 1160—62, 1909.

MacDOUGALL, R. S. "Insects and Arachnids in relation to the spread of disease." Pharm. Journ., London, Ser. 4, vol. 22, p. 60, 1906.

—— "Sheep Maggot and related flies. Their classification, life-history and habits." Trans. Highland Soc. Scot. (Reprint), 42 pp., 1909.

McFARLAND, J. "Relation of insects to the spread of disease." Medicine, vol. 8, 15 pp., 12 figs., 1902.

MACKINNON, D. L. "Herpetomonads from the alimentary tract of certain Dung-flies." Parasitology, vol. 3, pp. 255—274, 1 pl., 1910.

MACLOSKIE, G. "The Proboscis of the House-fly." Amer. Nat., vol. 5, pp. 153—161, 1880.

MACRÆ, R. "Flies and Cholera Diffusion." Indian Med. Gaz., 1894, pp. 407—412.

MADDOX, R. L. "Experiments in Feeding some Insects with the Curved or *Comma* Bacillus, and also with another Bacillus (*B. subtilis* ?)." Journ. Roy. Micros. Soc., ser. 2, vol. 5, pp. 602—607, 941—952, 1885.

MANEWARING, W. H. "Flies as carriers of Bacteria." Journ. Applied Micr. (Rochester, N. Y.), vol. 6, p. 2402, 1903.

MANSA, F. V. Bidrag til Folkesygdommenes og Sundhedspleiens Historie i Danmark fra de aeldste Tider till Begyndelsen af det attende Aarhundrede. Copenhagen, 1872. (Cited by Nuttall, 1899, *re* plague and insects; see pp. 126, 212, 312.)

MANSON, P. Tropical Diseases. (4th ed.), 876 pp., 241 figs., 7 col. pls., London, 1907.

MARPMANN, G. "Die Verbreitung von Spaltpilzen durch Fliegen." Arch. f. Hygiene, vol. 2, pp. 560—563, 1884.

MARTIN, A. W. "Flies in relation to typhoid fever and summer diarrhœa." Public Health (London), vol. 15, p. 652, 1903.

MARTIN, C. J. Horace Dobell Lectures on Insect Porters of Bacterial Infections, delivered before the Roy. Coll. of Physicians. Brit. Med. Journ., Jan. 4th, pp. 1—8, and Jan. 11th, pp. 59—68, 1913. (*M. domestica* discussed in the first lecture.)

MARTINI, E. "Insekten als Krankheitsüberträger." Moderne ärztliche Bibliothek, 39 pp., 27 figs. (Reprint), 1904.

MASON, C. F. "The spread of disease by insects." Internat. Clin. Phila., Ser. 14, vol. 2, pp. 1—24, 1904.

MAYS, T. J. "The Fly and Tuberculosis." New York Med. Journ. and Phila. Med. Journ., vol. 82, pp. 437—438, 1905.

MÉGNIN, J. P. "Du transport et de l'inoculation du virus charbonneux et autres par les mouches." Compt. rend. de l'Acad. des Sci., Paris, vol. 69 pp. 1338—1340, 1874.

—— "Memoire sur la question du transport et de l'inoculation du virus par les mouches." Journ. de l'anat. et de physiol. etc., Paris, vol. 11, pp. 121—133, 1 pl. Also in Journ. de méd. vétér. mil., Paris, vol. 12, pp. 461—675, 1875.

—— Les parasites articulés (Paris : Masson), 2 vols., 210 pp., 91 figs., 26 pls., 1895.

—— Les insectes buveurs de sang et colporteurs de virus. Paris, 150 pp., 1906.

DE MEIJERE, J. C. H. "Ueber die Prothorakalstigmen der Dipterenpuppen." Zoöl. Jahrb. (Anat.), vol. 15, pp. 623—692, pls. 32—35, 1902.

MERCURIALIS. De pestilentia. Venice, 1577. (Cited by Abel, 1899.)

MERK, A. "On the relation of *M. domestica* to vaccine." Hyg. Rundschau, vol. 20, pp. 233—235, 1910.

MERLIN, A. A. C. E. "The Foot of the House-fly." Journ. Quekett Club (2), vol. 6, p. 348, 1897.

MERLIN, A. A. C. E. "Supplementary Note on the Foot of the House-fly." Journ. Quekett Club (2), vol. 9, pp. 167—168, 1905.

MILLER, R. T. "*Myiasis dermatosa* due to ox-warble flies." Journ. Amer. Med. Ass., vol. 55, pp. 1978—1979, 1910.

MILLIKEN, F. B. "Another Breeding-Place for the House-fly." Journ. Econ. Ent., vol. 4, p. 275, 1911.

MINCHIN, E. A. "Report on the Anatomy of the Tsetse-fly (*Glossina morsitans*)." Proc. Roy. Soc. (Ser. B.), vol. 76, pp. 531—547 and figs., 1905.

MINETT, E. P. "The Question of Flies as Leprosy Carriers." Journ. Lond. School Trop. Med., vol. 1, pp. 31—35, 1911.

MITZMAIN, M. B. "*Stomoxys calcitrans* Linn. A Note giving a summary of its Life-history." Public Health Reports, U.S. Pub. Health Service, vol. 28, No. 28, pp. 345—346, 1913.

—— "The Rôle of *Stomoxys calcitrans* in the transmission of *Trypanosoma evansi*." Philippine Journ. Science, Sect. B (Philippine Journ. Trop. Med.), vol. 7, pp. 475—518, with 5 pls., Dec. 1912. Abstr. in Trop. Dis. Bull., vol. 2, pp. 130—133, 1913.

—— "Experimental Insect Transmission of Anthrax." Public Health Repts., U.S. Pub. Health Service, vol. 29, (Jan. 9, 1914), pp. 75—77, 1914.

MONIEZ, R. "Apropos des publications récentes sur le faux parasitisme des Chernétides sur différents Arthropodes." Rev. Biol. du Nord de la France, vol. 6, pp. 47—54, 1874.

MOORE, W. "Diseases probably caused by flies." Brit. Med. Journ., pt. 1, p. 1154, 3rd June, 1893; also in Med. Magaz., July, 1893.

MORGAN, H. DE R. "Upon the Bacteriology of the Summer Diarrhoea of Infants." Brit. Med. Journ., April 21, 1906, 12 pp., and July 6, 1906, 11 pp.

MORGAN, H. DE R. and LEDINGHAM, K. C. G. "The Bacteriology of Summer Diarrhoea." Proc. Roy. Soc. Med., Mar., 1909, pp. 1—17 (separate pagination in Reprint).

MOUCHET, R. "Myase intestinale chez l'homme." Bul. Soc. Path. Exot., vol. 5, pp. 508—511, 1912.

DE MOURA, C. "Myiase do Seio." Revista Med. de S. Paulo, vol. 16, p. 1, 1913 ; Abstr. in Trop. Dis. Bull., vol. 2, (Nov. 15), p. 530, 1913.

MUNSON, E. L. The theory and practice of Military Hygiene. (London : Ballière, Tindall and Cox), 1901.

MURRAY, A. Economic Entomology. London, p. 129, 1877.

NASH, J. C. T. "The Etiology of Summer Diarrhoea." Lancet, p. 330, 1903.

—— "The seasonal incidence of typhoid fever and of diarrhoea," etc. Trans. Epidemiol. Soc., London, N. S. vol. 22, pp. 110—138, 1903.

—— "Some points in the prevention of epidemic diarrhoea." Lancet, p. 892, 1904.

—— "The waste of infant life." Journ. Roy. Sanit. Inst., vol. 26, pp. 494—498, 1905.

NASH, J. C. T. Annual Report of the Medical Officer of Health, Borough of Southend-on-Sea, 1906.

—— Special Report on Epidemic Diarrhoea, Borough of Southend-on-Sea. 16 pp., 1 chart, 1906.

—— Second Report on same. 28 pp., 1906.

—— " The Prevention of Summer or Epidemic Diarrhoea." The Practitioner, May, 1906, 12 pp.

—— " House-flies as carriers of Disease." Journ. of Hygiene, vol. 9, pp. 141—169, 1909.

—— " Flies as a Nuisance and Flies as a ' Dangerous Nuisance.' " Lancet, 1908, pp. 131, 132.

—— " A note on the Bacterial contamination of Milk as illustrating the connection between Flies and epidemic diarrhoea." Lancet, part 2, pp. 1668—1669, 1908.

—— " House-flies as carriers of disease." Norwich Rep. Mus. Assn., vol. 2, pp. 14—19, 1908—09.

NEIVA, A. and GOMES DE FABRIA. "Myiasis humana, verursacht durch Larven von *Sarcophaga pyophila* n. sp." Mem. Inst. Oswaldo Cruz, vol. 5, pp. 16—22, 1913 ; Abstr. in Trop. Dis. Bull., vol. 2, (No. 15), pp. 529—530, 1913.

NEWPORT, G. " Insecta," in Todd's Cyclopaedia of Anatomy and Physiology, vol. 2, pp. 853—994, 1839.

NEWSHOLME, A. Annual Report on Health of Brighton (p. 21, infection of milk by flies), 1903.

—— " Domestic infection in relation to epidemic diarrhoea." Journ. of Hygiene, vol. 6, pp. 139—148, 1906.

—— Enteric fever in Durham. Quotation of Newsholme's report in Times, Weekly Edition, August 5th, 1910.

—— A Report on infant and child mortality, being a Supplement to the Report of the Medical Health Officer in the 39th Ann. Rept. of the Local Government Board 1909—10, (separate, 110 pp.).

NEWSTEAD, R. " On the Life-history of *Stomoxys calcitrans* Linn." Journ. Econ. Biol., vol. 1, pp. 157—166, 1 pl., 1906.

—— Preliminary Report on the Habits, Life-Cycle and Breeding Places of the Common House-fly (*Musca domestica* Linn.) as observed in Liverpool, with suggestions as to the best means of checking its increase. Liverpool, 23 pp., 14 figs., 1907.

—— Second Interim Report on the House-fly as observed in the city of Liverpool, 4 pp., 1909.

NICHOLAS, G. E. "The Fly in its Sanitary Aspect." Lancet, 1873, vol. 2, p. 724.

NICHOLLS, L. St Lucia Laboratory report for the half year ending Sept. 30th, 1911. Report to the Advisory Committee of the Trop. Disease Research Fund, App. VI, No. 14, p. 199, 1911.

—— " The transmission of pathogenic micro-organisms by flies in St Lucia." Bull. Ent. Research, vol. 3, p. 81, 1912.

NICHOLSON, J. L. "Myiasis ; a report of three cases of primal rectal infection." Journ. Amer. Med. Ass., May 21, 1910.

NICOLL, W. "On the varieties of *Bacillus coli* associated with the House-fly (*Musca domestica*)." Journ. Hygiene, vol. 11, No. 3, pp. 381—389, 1911.

—— On the part played by Flies in the dispersal of the eggs of parasitic worms. Repts. Local Govnt. Bd. on Publ. Health and Med. Subjects, N.S. No. 53 ; Further Repts. (No. 4) on Flies as Carriers of Infection, pp. 13—30, 1911.

NIEWENGLOWSKI, G. H. "La Transmission des Maladies par les Mouches." Naturaliste Canadien, vol. 40, pp. 33—38, 1913.

NIVEN, J. Annual Reports on the Health of the City of Manchester, 1904—1912.

—— "The House-fly in relation to Summer Diarrhoea and Enteric Fever." Proc. Roy. Soc. of Med., April, 1910. (Reprint, 83 pp.)

NOEL, P. "La Guerre aux Mouches." Bull. d. lab. régional d'Ent. agric., Rouen, 1913, pp. 4—5.

NOVY, F. G. "Disease carriers." Science, vol. 36, No. 914, pp. 1—9, 1912.

NUTTALL, G. H. F. "Zur Aufklärung der Rolle, welche Insekten bei der Verbreitung der Pest spielen—Ueber die Empfindlichkeit verschiedener Thiere für dieselbe." Centralbl. f. Bakteriol., vol. 22, pp. 87—97, 1897.

—— On the Rôle of Insects, Arachnids, and Myriapods, as Carriers in the Spread of Bacterial and Parasitic Diseases in Man and Animals : a Critical and Historical Study. Johns Hopkins Hospital Reports, vol. 8, 154 pp., 3 pls. (A very full bibliography is given.) 1899.

—— "The part played by Insects, Arachnids and Myriapods in the propagation of infective diseases of Man and Animals." Brit. Med. Journ., 9 Sept. (4 pp. reprint), 1899.

—— "The rôle of Insects, Arachnids and Myriapods in the propagation of infective diseases of Man and Animals." Journ. Trop. Med., pp. 107—110, Nov. 1899.

—— "Die Rolle der Insekten, Arachniden (Ixodes) und Myriapoden als Träger bei der Verbreitung von durch Bakterien und thierische Parasiten verursachten Krankheiten des Menschen und der Thiere." Hyg. Rundschau, vol. 9, (72 pp. reprint), 1899.

—— "Insects as carriers of disease." Reprint from Bericht über den xiv. Intern. Kongress f. Hyg. u. Demographie, Berlin, 1907.

NUTTALL, G. H. F., and JEPSON, F. P. The part played by *Musca domestica* and allied (non-biting) Flies in the spread of infective diseases. A summary of our present knowledge. Repts. Local Govnt. Bd. on Publ. Health and Med. Subjects, N.S. No. 16 ; Further Prelim. Repts. on Flies as Carriers of Infection, pp. 13—41, 1909.

NUTTALL, G. H. F., HINDLE, E. and MERRIMAN, G. The Range of Flight of *Musca domestica*. Repts. Local Govnt. Bd. on Publ. Health and Med. Subjects, N.S. No. 85 ; Further Repts. (No. 6) on Flies as Carriers of Infection, pp. 20—41, 11 charts, 1913.

ODLUM, W. H. "Are flies the cause of enteric fever?" Journ. Roy. Army Med. Corps, vol. 10, pp. 528—530, 1908.

OLIVE, E. W. "Cytological Studies on the Entomophthoreæ: i. The Morphology and Development of *Empusa*." Bot. Gaz., vol. 41, p. 192, 2 pls., 1906.

OLSEN, A. B. "Only a Flyspeck." Good Health, vol. 10, No. 7, pp. 195—200, 3 figs., July, 1912.

OSMUND, A. E. The Fly: an etiological factor in intestinal diseases. Cincinnati, 1909.

OSTEN-SACKEN, C. R. "On Mr Portchinski's publications on the larvae of Muscidae, including a detailed abstract of his last paper : 'Comparative Biology of Necrophagous and Coprophagous Larvae.'" Berl. Ent. Zeit., vol. 31, pp. 17—28, 1887.

PACKARD, A. S. "On the Transformations of the Common House-fly, with notes on allied forms." Proc. Boston Soc. Nat. Hist., vol. 16, pp. 136—150, 1 pl., 1874.

PAINE, J. H. "The House-fly in relation to city garbage." Psyche, vol. 19, pp. 156—159, 1912.

PALMER, J. W. "The relation the House-fly bears to typhoid and other infectious diseases." Atlantic Journ. Rec. of Med., Aug., 1910.

PARANT, G. "Un procédé de destruction des mouches." Bull. Soc. Autun, vol. 17, pp. 118—124, 1905.

PARIS, P. "Un cas de myase intestinale." C. R. 41ᵐᵉ Session Ass. Franç. avancement d. Sciences (Nîmes, 1912). Paris, p. 447, 1913.

PARKES, L. C. "The Common House-fly." Journ. Roy. Sanit. Inst., May, 1911.

PARROTT, P. J. "To rid the house of flies." Bull. No. 99, Kansas State Expnt. Sta., 1900.

PATTON, W. S. "The Life-Cycle of a species of *Crithidia* parasitic in the intestinal tract of *Gerris fossarum* Fabr." Arch. f. Protistenk., vol. 12, pp. 131—146, 1908.

—— "*Herpetomonas lygæi*." Ibid., vol. 13, pp. 1—18, 1 pl., 1908.

—— "The Parasite of Kala-Azar and Allied Organisms." Lancet, January 30th, 1909, pp. 306—309, 2 figs., 1909.

—— "A critical review of our present knowledge of the Haemoflagellates and allied forms." Parasitology, vol. 2, pp. 91—139, 1909.

—— Preliminary Report on an Investigation into the Oriental Sore in Cambay. Scientific Mem. by Officers of the Med. and Sanit. Depts. of the Govnt. of India, N. S. No. 50, 21 pp., 1912. Abs. in Kala-Azar Bull. No. 3, pp. 163—167, Trop. Dis. Bur., London, 1912.

—— Studies on the Flagellates of the Genera *Herpetomonas*, *Crithidia* and *Rynchoidomonas*. No. 1. The Morphology and Life-history of *Herpetomonas culicis*, Novy, Macheal and Torrey. Scient. Mem. by Officers of the Med. and Sanit. Depts., Govnt. of India, N. S. No. 57, pp. 1—21, 1 pl., 2 figs. (Calcutta), 1912.

PATTON, W. S. and STRICKLAND, C. "A critical review of the relation of blood-sucking invertebrates to the life-cycles of the Trypanosomes of the vertebrates, with a note on the recurrence of a species of *Crithidia*, *C. ctenophthalmi* in the alimentary tract of *Ctenophthalmus agyrtes*, Heller." Parasitology, vol. 1, pp. 322—346, 12 figs., 1908.

PATTON, W. S. and CRAGG, F. W. A Text-book of Medical Entomology. xxxiv + 764 pp., 89 pls. (Madras and Calcutta), 1913.

PEASE, H. D. "Relation of flies to the transmission of infectious disease." Long Island Med. Journ., Dec., 1910.

PÉREZ, C. (Metamorphosis of Blowfly.) Arch. Zool. Expér., vol. 4, pp. 1—274, 1910.

PERRIN, W. S. "Note on the possible transmission of *Sarcocystis* by the Blowfly." Spol. Zeyl., Columbo, vol. 5, pp. 58—61, 1 pl., 1907.

PETERS. "Observations upon the natural history of Epidemic Diarrhoea." Journ. Hygiene, vol. 10, p. 602, 1910.

PIANA, G. P. "Osservazioni sul *Dispharagus nasutus* Rud. dei polli e sulle Nematoelmintiche delle mosche e dei Porcellioni." Atti della Soc. Ital. d. Sci. Nat., vol. 36, pp. 239—262, 21 figs., 1896.

PICKARD-CAMBRIDGE, O. "On the British Species of False-Scorpions." Proc. Dorset Nat. Hist. and Antiq. Field Club, vol. 13, pp. 199—231, 3 pls., 1892.

PIETER, H. "Un Cas de Myase Vulvo-Vaginale." Rev. de Méd. et d'hyg. Trop., vol. 9, pp. 176—177, 1912.

PINKUS, H. "The Life-History and habits of *Spalangia muscidarum*. A parasite of the Stable-fly." Psyche, vol. 20, pp. 148—158, 1913.

POORE, G. V. "Flies and the science of scavenging." Lancet, pt. 1, pp. 1389—1391, 18th May, 1901.

PORTCHINSKY, J. A. "Recherches biologiques sur le *Stomoxys calcitrans* L. et biologie comparée des mouches coprophagues." (In Russian.) Publications of the Entomological Bureau of the Russian Dept. of Land Administr. and Agriculture, vol. 8, No. 8, 91 pp., 97 figs., 1910.

—— "*Hydrotaea dentipes* F. Sa biologie et la destruction par ses larves de celles de *Musca domestica*, L." Publications Entomological Bureau, Russian Dept. of Land Administr. and Agriculture, vol. 9, No. 5, 30 pp., 25 figs., 1911.

—— "*Muscina stabulans* Fall., Mouche nuisible à l'homme et à son ménage, etc., état larvaire destructeuse des larves de *Musca domestica*." Publications Entomological Bureau, Russian Dept. of Land Administr. and Agriculture, vol. 10, No. 1, 39 pp., 32 figs. Abstr. in Rev. Appl. Ent., Imp. Bur. Ent., vol. 1, Ser. B, pp. 108—110, 1913.

—— "*Oestrus ovis*, sa biologie et son rapport à l'homme." Publications of the Entomological Bureau of the Russian Department of Land Administr. and Agriculture, vol. 10, No. 3, 63 pp., 28 figs., 1913.

PORTER, A. "The life-cycle of *Herpetomonas jaculum* (Léger), parasitic in the alimentary tract of *Nepa cinerea*." Parasitology, vol. 2, pp. 367—391, 1 pl., 1909.

PORTER, A. "The Structure and Life-history of *Crithidia pulicis* n. sp., parasitic in the alimentary tract of the human flea *Pulex irritans*." Parasitology, vol. 4, pp. 237—254, 1 pl., 1911.

PRATT, F. C. "Insects bred from Cow-manure." Canadian Ent., vol. 44, pp. 180—184, 1912.

PROTO GOMEZ Y. DURÁN BORDA, G. "Sobre la causa de la meurre de los moscas en Bogotá." Rev. Med., Bogotá, vol. 12, pp. 65—74, 1888—9.

PROWAZEK, S. "Die Entwicklung von *Herpetomonas*, einem mit den Trypanosomen verwandten Flagellaten." Arb. aus dem. Kaiserl. Gesundheitsamte, vol. 20, pp. 440—452, 7 figs., 1904.

—— "Notiz zur *Herpetomonas*-Morphologie sowie Bemerkung zu der Arbeit von Wenyon." Arch. f. Protistenkunde, vol. 31, pp. 37—38 ; Abstr. in Trop. Diseases Bull., vol. 2, No. 9, p. 465, 1913.

PRUVOT, G. Contribution à l'étude des larves de diptères trouvées dans le corps humain. Thèse No. 267, Faculté de médecine de Paris, 84 pp., 2 pls., 1882.

PURDY, J. S. "Flies and fleas as factors in disease." Journ. Roy. Sanit. Inst., Trans. vol. 30, pp. 496—503, 1910.

—— "Flies and fleas as factors in dissemination of disease ; effect of petroleum as an insecticide." Med. Press and Circular, Jan. 12, 1910.

QUILL, R. H. Report on an Outbreak of Enteric Fever at Diyatalawa Camp, Ceylon, among the 2nd King's Royal Rifles. Army Med. Dept. Report, Appendix 4, p. 425, 1900.

RAIMBERT, A. "Recherches expérimentales sur la transmission du charbon par les mouches." C. R. Ac. Sci. Paris, vol. 69, pp. 805—812, 1869.

RAMIREZ, R. "The Diptera from a hygienic point of view." Public Health (U.S.A.), vol. 24, pp. 257—259, 1898.

RANSOM, B. A. "The life-history of a parasitic nematode (*Habronema muscae*)." Science, N. S., vol. 34, pp. 690—692, 1911.

—— "The life-history of *Habronema muscae* (Carter). A parasite of the horse transmitted by the House-fly." Bull. No. 163, Bur. Animal Indust., U.S. Dept. Agric., 36 pp., 41 figs., 1913.

RÉAUMUR, R. A. F. DE. Mémoires pour servir à l'Histoire des Insectes. Paris, vol. 4 ; *M. domestica*, p. 384, 1738.

REED, W. (Flies the cause of typhoid outbreak in army in 1899.) War Dept. Ann. Rept. (Washington), pp. 627—633, 1899.

REED, W., VAUGHAN, V. C. and SHAKESPEARE, E. O. Report on the origin and spread of typhoid fever in U.S. military camp during the Spanish war of 1898. Vol. 1, text, 720 pp., vol. 2, maps and charts, 1904.

REUM, W. "Der weisse Tod der *Musca domestica*." Societas Entomologica, Zurich, vol. 29, pp. 13—14, 1914.

RICHARDSON, C. H. "An undescribed Hymenopterous parasite of the House-fly." Psyche, vol. 20, pp. 38—39, 1 pl., 1913.

RICHARDSON, C. H. "Studies on the habits and development of a hymenopterous parasite, *Spalangia muscidarum* Richardson." Journ. Morphology, vol. 24, pp. 513—557, 16 figs., 1913.

RIDLON, J. R. An investigation of the prevalence of typhoid fever at Charlestown, W. Va. Public Health Reports, Public Health and Marine Hospital Service, Washington, vol. 26, pp. 1789—1799, 1911.

RILEY, W. A. The Relation of Insects to Disease. 4 pp. Author's Reprint (source not given), Ent. Labty., Cornell Univ., Ithaca, N. Y.

—— "Earlier references to the relation of flies to disease." Science, N. S. vol. 31, pp. 263—264, 1910.

ROBERTSON, J. Report of the Medical Officer of Health of the City of Birmingham for the year 1909, 144 pp.

ROBERTSON, A. "Flies as carriers of contagion in Yaws (*Framboesia tropica*)." Journ. Trop. Med. and Hyg., vol. 11, No. 14, p. 213, 1908.

ROGERS, L. "The Conditions affecting the Development of Flagellated organisms from Leishman bodies and their bearing on the probable mode of infection." Lancet, June 3rd, 1905, pp. 1484—1487, 1905.

ROSENAU, M. J. and BRUES, C. T. "Some experimental observations upon Monkeys concerning the transmission of Poliomyelitis through the agency of *Stomoxys calcitrans*." Bull. Mass. State Board of Health, Sept., 1912, pp. 314—317; also see Public Health Repts., U. S. Public Health Service, vol. 27, p. 1593, 1912.

ROSENAU, M. J., LUMSDEN and KASTLE. Report No. 3 on Origin and Prevalence of typhoid fever in the District of Columbia. Bull. No. 52, Hyg. Lab. U.S. Pub. Health and Mar. Hosp. Service, Washington, p. 30, 1909.

ROSENBUSCH, F. "Ueber eine neue Encystierung bei *Crithidia muscaedomesticae*." Centr. f. Bakt. 1. Abt. Orig. vol. 53, pp. 387—393, 1 pl., 1910.

ROSS, E. H. The Reduction of Domestic Flies. London, John Murray, 103 pp., illustr., 1913.

RUDOLPHI, R. Entozoorum sive vermium intestinarum historia naturalis, p. 524, 1808—1810.

RYDER, J. A. "Cholera and Flies." Entom. News, vol. 3, p. 210, 1896.

SAMUELSON, J. and HICKS, J. B. "The Earthworm and the Common House-fly." Humble Creatures, pt. 1, 79 pp., 8 pls., London, The House-fly, pp. 26—79, pls. 3—8, 1860.

SANDILANDS, J. E. "Epidemic Diarrhoea and the bacterial content of Food." Journ. Hyg., vol. 6, pp. 77—92, 1906.

SANDWITH, F. M. "The Danger of the House-fly." Clinical Journ., vol. 39, No. 4, Nov. 1, 1911.

SANGREE, E. B. "Flies and Typhoid Fever." New York Med. Record, vol. 55, pp. 88—89, 4 figs., 1899.

SANTORI, S. "La mosche domestiche come causa di diffusione delle malattie infettive intestinali." Il Policlinico (Sez. pratica).

SAWTCHENKO, J. G. "Le rôle des mouches dans la propagation de l'épidémie cholérique." Vratch, St Petersburg. (Reviewed in Ann. de l'Institut Pasteur, vol. 7, p. 222, 1892.)

SAVAGE, W. G. "Recent work on typhoid fever bacteriology." Public Health (U.S.A.), Oct., 1907.

SCHILLING, C. "Die Uebertragung von Krankheiten durch Insekten und ihre Bekämpfung." Gesundh. Ingenieur, vol. 30, pp. 300—303, 1907.

SCHINER, J. R. Fauna Austriaca: Die Fliegen. 2 vols., Wien, 1862.

SCHLEULEN. "Geschichtliche und experimentelle Studien über den Prodigiosus." Arch. f. Hyg., vol. 26, pp. 16—17, 1896.

SCHUBERG and BÖING. (Further investigations of the transmission of pathogenic micro-organisms by native blood-sucking flies.) Centralbl. Bakt. 1 Abt., Ref., 57, pp. 301—303, 1913.

SCHUBERG, A. and KUHN, P. (Dissemination of disease by Stomoxys calcitrans.) Arbeit. k. Gesundheitsamte, vol. 31, pp. 377—393, 1911.

SCOTT, J: "The dangerous house-fly." Indian Public Health, vol. 5, pp. 292—298, 1909.

SEDGWICK, W. T. and WINSLOW, C. E. A. "Statistical studies on the seasonal prevalence of Typhoid Fever in various countries and its relation to seasonal temperature." Mem. Am. Acad. Sci., vol. 12, Nov., pp. 521—577, 8 charts, 1902.

SENATOR. "Ueber lebende Fliegenlarven im Magen und in der Mundhöhle." Berlin. klin. Wochenschr., No. 7, 1890.

SERGENT, EDM. and ET. "La 'Tamné,' Myiase humaine des Montagnes Sahariennes Tonareg, identique à la 'Thimni' des Kabyles, due à Oestrus ovis." Bull. Soc. Path. Exot., vol. 6, pp. 487—484, 1913.

SERIZIAT. Études sur l'oasis de Biskra. Paris, 1875. (Cited by Laveran, 1880.)

SHARP, D. "Insects," part 2, Cambridge Nat. History, London, 1895.

SHARPE, W. S. "Influence of dust and flies in the contamination of food and the dissemination of disease." Lancet, pt. 1, 2nd June, 1900.

SHIPLEY, A. E. Infinite torment of Flies. Camb. Univ. Press. (Printed privately.) 23 pp., 1905.

—— "Insects as carriers of Disease." Nature, vol. 73, pp. 235—238, 4th Jan. (Abs. of address before Brit. Assn., Pretoria, S. A.), 1906.

—— Infinite Torment of Flies, and The Danger of Flies. In Pearls and Parasites. (London, John Murray), pp. 155—173 and 174—182, 1908.

—— "The Danger of Flies." Science Progress, vol. 1, pp. 723—729, April, 1907.

SHOEMAKER, E. M. and WAGGONER, |A. "Flies as carriers of Bacteria." School Science, April, 1903.

SIBTHORPE, E. H. "Cholera and Flies." Brit. Med. Journ., Sept., 1896, p. 700.

SIMMONDS, M. "Fliegen und Choleraübertragen." Deutsch. med. Wochenschr., No. 41, p. 931, 1892.

SIMPSON, R. J. S. "Medical history of the South African War." Journ. Roy. Army Med. Corps, vol. 15, pp. 257 and 260—1, 1910.

SIMPSON, W. J. R. The Principles of Hygiene as applied to Tropical and Subtropical Climates. London, 1908.

SKINNER, H. "The Relation of House-flies to the spread of Disease." New Orleans Med. and Surgical Journ., vol. 61, pp. 950—959, 1909.

—— "How does the House-fly pass the Winter ?" Entom. News, vol. 24, pp. 303—304, 1913.

SLATER, J. W. "On Diptera as spreaders of disease." Journ. of Science, London, ser. 3, pp. 533—539, 1881.

SMIT, R. "Die Fliegenkrankheit und ihre Behandlung." D. med. Wochenschr., Leipzig, vol. 32, pp. 763—764, 1906.

SMITH, A. J. "Notes upon several larval insects occurring as parasites in Man." Med. News, vol. 81, p. 1060, 2 figs., 1902.

SMITH, F. "House-flies and their ways at Benares." Journ. Roy. Army Med. Corps, vol. 9, pp. 150—155 and p. 447, 1907.

—— "Municipal Sewage." Journ. Trop. Med., vol. 6, pp. 285—291, 304—308, 330—334, 353—355, 381—383, 1903.

SMITH, J. L. "An investigation into the conditions affecting the occurrence of typhoid fever in Belfast." Journ. of Hyg., vol. 4, pp. 407—433, 1904.

SMITH, R. I. "Formalin for poisoning house-flies proves very attractive when used with sweet milk." Journ. Econ. Ent., vol. 4, pp. 417—419, 1911.

SMITH, T. "The House-fly as an agent in Dissemination of infectious diseases." Journ. Pub. Hyg., Aug., 1908, pp. 312—317.

SNELL. (Infantile diarrhoea and flies, 1906.) Cited by Ainsworth, 1909, p. 487, no reference given.

SOLTAU, A. B. "Note on a case of intestinal infection of man, with the larva of *Homalomyia canicularis*." Parasitology, vol. 3, p. 314, 1910.

SOLMS-LAUBACH, GRAF DU. "Über die liebstliche Pilzkrankheit der Stubenfliege." Abh. d. Naturf.-Ges., Halle, 37, 1870.

SPEISER, P. "Insekten als Krankheitsüberträger," in Krancher's Entomol. Jahrg., 7 pp., 1904.

SPILLMANN and HAUSHALTER. "Dissemination du bacille de la tuberculose par les mouches." C. R. Ac. Sci., vol. 105, pp. 352—353, 1887.

STALLMAN, G. P. "Ants destroying the larvae of Flies." Military Surgeon, vol. 31, No. 3, pp. 325—326, 1912.

STEIN, F. R. Der Organismus des Infusionsthiere, III. Abtheilung—Die Naturgeschichte des Flagellaten oder Geisselinfusorien. 154 pp., 24 pls., Leipzig, 1878.

STEPHENS, J. W. W. Two Cases of Intestinal Myiasis. Thompson Yates and Johnstone Laboratories Report, vol. 6, part 1, pp. 119—121, 1905.

—— "Transmission of disease by insects." Bart's. Hosp. Journ., vol. 12, pp. 131—134, 1905.

STEPHENS, J. W. W. and NEWSTEAD, R. "The Anatomy of the Proboscis of

Biting Flies. Part 2. *Stomoxys*." Ann. Trop. Med. and Parasit., vol. 1, pp. 171—182, 8 pls., 1907.

STEPHENSON, S. Report on the prevalence of Ophthalmia in the Metropolitan Poor Law Schools. Blue Book, October 2nd, 1897. (Reviewed in Lancet, October 16th, pp. 990—991, 1897.)

STERNBERG, G. M. "Sanitary Lessons of the War." Philad. Med. Journ., June 10th and 17th, 1899.

STILES, C. W. "Insects as disseminators of disease." Virginia Med. Semimonthly, vol. 6, pp. 53—58, 10th May, 1901.

—— "The Sanitary Privy : its purpose and construction." Public Health Bull. No. 37, U.S. Pub. Health and Mar. Hosp. Service, 24 pp., 12 figs., 1910.

—— Contamination of Food Supplies. Public Health Reports. U.S. Public Health Service, vol. 28, pp. 290—291, 1913.

STILES, C. W. and KEISTER, WM S. "Flies as Carriers of *Lamblia* spores. The contamination of food with human excreta." U.S. Public Health Rep., vol. 28, No. 48, pp. 2530—2534, 1913.

STILES, C. W. and LUMSDEN, L. L. "The Sanitary Privy." Farmers Bull., No. 463, U.S. Dept. Agric., Washington, 9 figs., 1911.

STILES, C. W. and MCMILLER, H. The ability of fly larvae to crawl through sand. Public Health Reports, Public Health and Marine Hosp. Service, vol. 26, p. 1277, 1911.

STRATTON, C. H. "The prevention of enteric fever in India." Journ. Roy. Army Med. Corps, vol. 8, p. 224, 1907.

SURCOUF, J. "La Transmission du ver macaque par un Moustique." Compte Rend. Ac. Sci., vol. 156, pp. 1406—1408, 2 figs., 1913.

SYDENHAM, T. Sydenham's Works. Syd. Soc. Ed., vol. 1, p. 271, 1666.

SYKES, G. F. The distribution of flies in Providence, R.I. Ann. Rept. Supt. Health, Providence, for 1909, pp. 7—15.

TASCHENBERG, E. L. Praktische Insektenkunde, part IV ; (*M. domestica*, pp. 102—107, fig. 27), 1880.

TAYLOR, T. "*Musca domestica* as a carrier of contagion." Proc. Am. Assoc. Adv. Sci., vol. 31, p. 528, 1883.

TEBBUTT, H. "On the influence of the metamorphosis of *Musca domestica* upon which bacteria administered in the larval stage." Journ. of Hygiene, vol. 12, pp. 516—526, 1913.

TERRY, C. E. "Extermination of the House-fly in cities ; its necessity and possibility." Amer. Journ. Pub. Health, vol. 2, pp. 14—22, 1912.

TERRY, C. E. Fly-borne typhoid fever and its control in Jacksonville (Florida, U.S.A.). Public Health Reports, U.S. Public Health Service, vol. 28, pp. 68—73, 1913.

THAXTER, R. "The Entomophthoreae of the United States." Mem. Boston Soc. Nat. Hist., vol. 4, pp. 133—201, pls. 14—21, 1888.

THÉBAULT. "Hémorrhagie intestinale et infection typhoïde causée par des larves de Diptères." Arch. de Parasitol., vol. 4, p. 353, 1901.

THEOBALD, F. V. Swarms of Flies bred in house-refuse. Second Rept. on Economic Zool. Brit. Mus. Nat. Hist., pp. 125—126, 1904.

—— "Flies and ticks as agents in the distribution of disease." Proc. Ass. Econ. Biol., vol. 1, pt. I, pp. 17—26, 1905.

—— "Flies in distribution of disease." Nursing Times, London, vol. 1, p. 461, 1905.

—— The House-fly. Rept. on Economic Zoology, S.E. Agric. Coll., Wye, pp. 109—111, 2 figs., 1905.

—— The House-fly Annoyance. Rept. on Economic Zoology, S.E. Agric. Coll., Wye, pp. 141—143, 1 fig., 1907.

—— House-flies, their destruction and prevention. Rept. on Economic Zoology, S.E. Agric. Coll., Wye, pp. 133—137, 1 fig., 1911.

THOMAS, F. G. and PARSONS, J. H. "Dipterous Larva in the anterior chamber of the eye." Lancet, 1908, Oct. 24, pp. 1217—1218.

THOMSON, F. W. "The House-fly as a carrier of Typhoid infection." Journ. Trop. Med. and Hyg., vol. 15, pp. 273—277, 1912.

THOMSON, J. A. Darwinism and Human Life. (London: Andrew Melrose), 245 pp., 1909.

TIZZONI, G. and CATTANI, J. "Untersuchungen über Cholera." Centralbl. f. d. med. Wissensch., Berlin, vol. 24, pp. 769—771, 1886.

TOOTH, H. H. "Enteric Fever in the Army in South Africa." Brit. Med. Journ., November 10th, 1900, (Reprint 5 pp.).

—— "Some personal Experiences of the Epidemic of Enteric Fever among the Troops in South Africa, in the Orange River Colony." Trans. Clin. Soc., vol. 34, 64 pp. (Reprint), 1901.

TOOTH, H. H. and CALVERLEY, J. E. G. A Civilian War Hospital. (London: John Murray), 1901.

TORREY, J. C. "Numbers and types of bacteria carried by city flies." Journ. Infect. Diseases, vol. 10, No. 2, pp. 166—177, 1912.

TSUZUKI, J. "Bericht über meine epidemiologischen Beobachtungen und Forschungen wahrend der Choleraepidemie in Nordchina im Jahre 1902, etc." Arch. f. Schiffs- u. Tropen-Hyg., vol. 8, pp. 71—81, 1904.

TULLOCH, F. "The Internal Anatomy of *Stomoxys*." Proc. Roy. Soc. (Ser. B.), vol. 77, pp. 523—531, 5 figs., 1906.

TULPINS, N. Observationes medicae, vol. 2, pp. 173—174, 1672.

UFFELMANN, J. "Beiträge zur Biologie der Cholerabacillus." Berl. klin. Wochenschr., 1892, pp. 1213—1214.

UNDERWOOD, W. L. House-fly as carrier of disease. Boston, 1903.

VACHER, F. Report of County Medical Officer of Health on "Some recent Investigations regarding the propagation of disease by flies." Cheshire County Council, 1909.

VAILLARD. "Au sujet des mesures à prendre contre les mouches." Bull. Mens. Office Internat. d'hyg. Publique, Paris, pp. 1313—1336, 1913.

VANEY, C. "Contributions à l'étude des Larves et des metamorphoses des

Diptères." Ann. de l'Univ. de Lyon, N. S., 1. Sciences méd., fasc. 9, 178 pp., 4 pls., 1902.

VARLEY, C. "Microscopical observations on a malady affecting the common house-fly." Trans. Micros. Soc., London, vol. 3, pp. 55—57, 1 pl., 1852.

VEAU DE LAUNY. "Observations sur les vers rendus avec l'urine." Observ. de phys. de Rozier, vol. 50, p. 158, 1792.

VEEDER, M. A. "Flies as Spreaders of Disease in Camp." New York Med. Record, vol. 54, September 17th, p. 429, 1898.

—— "The Spread of Typhoid and Dysenteric Diseases by Flies." Public Health (U.S.A.), vol. 24, pp. 260—262, 1898.

—— "The relative importance of Flies and water supply in spreading disease." Med. Record, vol. 55, pp. 10—12, 1899.

—— "Typhoid Fever from sources other than water supply." Med. Record, vol. 62, pp. 121—124, July 26, 1902.

VOGLER, C. H. "Weitere Beiträge zur Kenntnis von Dipteren-larven." Illust. Zeitschr. f. Entom., vol. 5, pp. 273—276, 8 figs., 1900.

VIGNON, P. "Recherches de Cytologie générale sur les Epithéliums, l'appareil pariétal protecteur ou moteur; le rôle de la co-ordination biologique." Arch. Zool. Exp. et Gén., vol. 9, pp. 371—720, pls. 15—18, 1901.

WAHL, B. "Ueber die Kopfbildung Cyclorapher Dipterenlarven und die postembryonale Entwicklung des Fliegenkopfes." Arb. zool. Inst. Wien, vol. 20, 114 pp., 20 figs., 3 pls., 1914.

WAHLGREN, E. "Insekten som sjukdomsspridare." Fauna och Flora Uppsala, vol. 3, pp. 12—22, 1908.

WALLMAN, E. "Contribution à la connaissance du rôle des microbes dans les voies digestives." Ann. Inst. Pasteur, vol. 24, pp. 1—96, 1911.

WALSH, B. D. "Larvae in the human bowels." Amer. Entom., vol. 2, pp. 137—141, 1 fig., 1870.

WALKER, F. Insecta Saundersiana, i, Diptera, p. 345, 1856.

WANHILL, C. F. "An investigation into the causes of the prevalence of enteric fever among the troops stationed in Bermuda, giving details of the measures adopted to combat the disease, and showing the results of these measures, during the years 1904—1906." Journ. Roy. Army Med. Corps, vol. 12, pp. 28—45, 1909.

WARD, H. B. "The Relation of Animals to Disease." Science, N. S. vol. 22, pp. 193—203, 1905; also Trans. Amer. Micro. Soc., vol. 27, pp. 5—20, 1907.

WASHBURN, F. L. The Typhoid Fly on the Minnesota Iron Range. Thirteenth Report of the State Entomologist of Minnesota, 1909—1910, pp. 135—141, 6 figs., 1910.

—— "Grasshopper work in Minnesota during 1911." Journ. Econ. Ent., vol. 5, pp. 111—118, 1912. (*M. domestica*, suggested poison, p. 114.)

—— "The Minnesota Fly Trap." Journ. Econ. Ent., vol. 5, pp. 400—402, 3 figs., 1912.

WEISMANN, A. "Die Entwickelung der Dipteren im Ei, nach Beobachtungen

370 BIBLIOGRAPHY

an *Chironomus* spec., *Musca vomitoria* und *Pulex canis*." Zeit. f. wiss. Zool., vol. 13, pp. 107—220, pls. 7—13, 1863.

WEISMANN, A. "Die nachembryonalen Entwickelung der Musciden nach Beobachtungen an *Musca vomitoria* und *Sarcophaga carnaria*." Zeit. f. wiss. Zool., vol. 14, pp. 185—336, pls. 21—27, 1864.

WELANDER. (Gonorrhoeal ophthalmia conveyed by flies.) Wien. klin. Wochenschr., No. 52, 1896. (Cited by Abel, 1899.)

WELCH, P. S. "Observations on the life-history of a new species of *Psychoda*." Annals Ent. Soc. Amer., vol. 5, pp. 411—418, 2 pls., 1912.

WENYON, C. M. "Oriental sore in Bagdad, together with observations on a gregarine in *Stegomyia fasciata*, the Haemogregarine of dogs and the Flagellates of House-flies." Parasitology, vol. 4, pp. 273—344, 5 pls., 1911.

—— "Observations on *Herpetomonas muscae-domesticae* and some allied Flagellates with special reference to the structure of their nuclei." Arch. f. Protistenkunde, vol. 31, pp. 1—36, 3 pls., 6 figs., 1 diagram ; Abst. in Trop. Diseases Bull., vol. 2, No. 9, pp. 463—465, 1913.

WERNER, H. "Ueber eine eigeisselige Flagellatenform im Darm der Stubenfliege." Arch. f. Protistenk., vol. 13, pp. 19—22, 2 pls., 1908.

WESCHE, W. "The Genitalia of both Sexes in the Diptera, and their relation to the Armature of the Mouth." Trans. Linn. Soc., vol. 9, pp. 339—386, 8 pls., 1906.

WESTCOTT, S. "Flies and Disease in the British Army." Journ. State Med., vol. 21, pp. 480—488, 1913.

WHEELER, W. M. Ants. Their structure, development and behaviour. Columbia Univ. Press, Biol. Ser. No. 9 (New York), 663 pp., 286 figs., 1910.

WHERRY, W. B. "Insects and infection." Calif. State Journ. Med., Nov., 1907, pp. 281—285.

—— Notes on rat leprosy and on the fate of human and rat lepra bacilli in flies. Public Health Reports, Pub. Health and Marine Hosp. Service, Washington, U.S.A., vol. 23, No. 42, 8 pp. (Reprint), 1908.

—— "Further notes on the rat leprosy and on the fate of the human and the rat leper bacillus in flies." Journ. Infect. Diseases, vol. 5, No. 5, 1908.

WHIPPLE, G. C. Typhoid Fever, its Causation, Transmission and Prevention. (New York: Wiley and Sons, London: Chapman and Hall). 1908.

WHITE, T. C. "On *Empusa muscae*." Journ. Quekett Micros. Club, London, vol. 4, pp. 211—213, 1874—77.

WILCOX, E. V. "Fighting the House-fly." Country Life in America, May, 1908.

WILKERSON. "Flies as carriers of disease." Mobile Med. and Surg. Journ., vol. 4, pp. 125—141, 1904.

WINTER, G. "Zwei neue Entomophthoren." Bot. Centralbl., vol. 5, p. 62, 1881.

WIRSING, E. "Ueber *Myiasis intestinalis*." Zeits. klin. Med., Berlin, vol. 60, pp. 122—133, 1906.

WISE, K. S. and MINNETT, E. P. "Experiments with crude carbolic acid as a larvicide in British Guiana." Ann. Trop. Med. and Parasitol., vol. 6, pp. 327—330, 1912.

WOODHOUSE, T. P. "Notes on the causation and prevention of enteric fever in India." Journ. Roy. Army Med. Corps, vol. 10, p. 616, 1910.

YERSIN. "La peste bubonique à Hongkong." Ann. Inst. Pasteur, vol. 8, pp. 662—667, 1894.

ZEPEDA, P. "Nouvelle note concernant les Moustiques qui propagent les larves de Dermatobia cyaniventris et de Chrysomia macellaria et peut-être celle de Lund, et de la Cordilobia anthropophaga." Rev. de Med. et d'Hyg. Trop., vol. 10, pp. 93—95, 1913.

ZETEK, J. "Dispersal of Musca domestica Linne." Ann. Ent. Soc. Amer., vol. 7, pp. 70—72, 2 figs., 1914.

ZWICK, K. G. "Massnahmen gegen die Uebertragung von Infektionskrankheiten durch die Hausfliegen." Schweizerischen Rundschau für Med., vol. 14; (Reprint, pp. 1—16), No. 13, 1914.

ANONYMOUS. "Gangrenous Fly." Lancet, pt. 2, p. 114, 1863.

—— "The pestiferous fly." Lancet, pt. 1, p. 156, 25th Jan., 1873.

—— "Les mouches considérées comme agents de propagation des maladies contagieuses, des épidémies et des parasites." Bull. Soc. Linn. Nord France, No. 152, pp. 215—217, 1884—5.

—— "The Fly Plague." Lancet, pt. 2, p. 418, Aug., 1903.

—— Bulletin Chicago School of Sanitary Instruction. (Dept. of Health, Chicago), 27th June, 1908.

—— "Flies and typhoid fever." Editorial in Journ. Amer. Med. Assn., Nov. 19, 1910.

—— "Abating the Fly Nuisance." Comment in Journ. Amer. Med. Assn., Aug. 20, 1910.

—— "The Domestic Flies." Ed. in Brit. Med. Journ., Aug. 26, p. 449, 1911.

—— "Literature on Flies." Journ. Amer. Med. Assn., June 24, 1911, p. 1900.

—— "Breeding Places for Flies as nuisances; disposal of wastes in a non-sewered town." Journ. Amer. Med. Assn., Sept. 23, 1911, p. 1076.

—— "Dangerous Flies." E. P. W. Nature, vol. 29, pp. 482—483.

—— Transmission of Trypanosoma hippicum by Musca domestica. Report Dept. Sanit., Smithsonian Canal Com., Dec., 1911, pp. 42—43.

—— "A Campaign against Flies." Nature Study Review, Jan., 1911.

—— "The Flyless City Campaign." Ed. in The Outlook, Aug. 19, 1912, pp. 857—858.

—— "The Etiology of Kala-Azar." Nature, vol. 89, pp. 386—388, 1912.

—— "The House-fly." Monthly Bull. Indiana State Bd. Health, May, 1908.

—— The House-fly at the Bar. Indictment Guilty or Not Guilty? 48 pp., illustr. (Merchants' Asstn. of New York, Chairman Ed. Hatch, Jr.) April, 1909.

—— "Flies as Carriers of Disease." Good Health, vol. 5, No. 14, pp. 424—426, 2 figs., July, 1907.

ANONYMOUS. "A parasitic mould of the House-fly." Brit. Med. Journ., No. 2714, Jan. 4, 1913, pp. 41—42.

—— "Deadly Typhoid Fly an old acquaintance in Disguise." Weekly News Letter, U.S. Dept. Agriculture, Washington, vol. 1, No. 30, Mar. 4, p. 4, 1914.

—— "House-flies and Disease : Flies in Delhi ; Flies in Poona." Trop. Dis. Bull., vol. 3, No. 7 (Sanitation No.), pp. 377—379, 1914.

AUTHORS' INDEX

SUBJECT INDEX